13th edition

Study Guide for **MEMMLER'S**

The HUMAN BODY
in Health
and Disease

13th edition

Study Guide for **MEMMLER'S**

The HUMAN BODY
in Health and Disease

Kerry L. Hull
Professor, Department of Biology
Bishop's University, Sherbrooke, Quebec, Canada

Barbara Janson Cohen

Philadelphia • Baltimore • New York • London
Buenos Aires • Hong Kong • Sydney • Tokyo

Senior Acquisitions Editor: Michael Nobel
Product Development Editor: Staci Wolfson
Editorial Assistant: Tish Rogers
Production Project Manager: Marian Bellus
Design Coordinator: Terry Mallon
Illustration Coordinator: Jennifer Clements
Artist: Dragonfly Media Group
Manufacturing Coordinator: Margie Orzech
Prepress Vendor: SPi Global

13th edition

ISBN 978-1-4511-9348-0

Preface

The *Study Guide for Memmler's The Human Body in Health and Disease*, 13th edition, helps students learn foundational concepts in anatomy and physiology required for success in allied health occupations. Although it will be more effective when used in conjunction with the 13th edition of *Memmler's The Human Body in Health and Disease*, the *Study Guide* may also be used to supplement other textbooks on basic anatomy and physiology. The questions in this edition reflect revisions and updating of the text. The labeling and coloring exercises are taken from the illustrations designed for the book.

The exercises are designed to facilitate student learning, not merely to test knowledge. Each chapter contains three main components. The first section, "Addressing the Learning Objectives," can be completed as students read through the chapter. It contains exercises in many formats, including labeling, coloring, matching, and short answer, all designed to foster active learning. The second section, "Making the Connections," asks students to complete a concept map integrating information from multiple learning objectives. Finally, "Testing Your Knowledge" includes multiple choice, true/false, completion, short answer, and essay questions to identify areas requiring further study. Within this section, "Practical Applications" questions use clinical situations to test students' mastery of a subject.

All answers to the Study Guide questions are available on the instructor's website on thePoint.

Learning about the Human Body

You already have some ideas about the human body that will influence how you learn the information in this textbook. Many of your theories are correct, and this Study Guide, created to accompany the 13th edition of *Memmler's The Human Body in Health and Disease*, will simply add detail and complexity to these ideas. Other theories, however, may be too simplistic. It can be difficult to replace these ingrained beliefs with more accurate information. For instance, many students think that the lungs actively inflate and deflate as we breathe, but it is the diaphragm and the rib cage muscles that accomplish all of the work. Learning physiology or any other subject therefore involves:

1. **Construction**: Adding to and enhancing your previous store of ideas.
2. **Reconstruction**: Replacing misconceptions (prior views and ideas) with scientifically sound principles.
3. **Self-monitoring**: Construction and reconstruction require that you also monitor your personal understanding of a particular topic and consciously formulate links between what you are learning and what you have previously learned. Rote learning is not an effective way to learn anatomy and physiology (or almost anything else, for that matter). **Metacognition** is monitoring your own understanding. Metacognition is very effective if it takes the form of self-questioning during the lectures. Try to ask yourself questions during lectures, such as "What is the prof trying to show here?" "What do these numbers really mean?" or "How does this stuff relate to the stuff we covered yesterday?" Self-questioning will help you create links between concepts. In other words, try to be an active learner during the lectures. Familiarity with the material is not enough. You have to internalize it and apply it to succeed. You can greatly enhance your ability to be an active learner by reading the appropriate sections of the textbook before the lecture.

Each field in biology has its own language. This language is not designed to make your life difficult; the terms often represent complex concepts. Rote memorization of definitions will not help you learn. Indeed, because biological terms often have different meanings in everyday conversation, you probably hold some definitions that are misleading and must be revised. For example, you may say that someone has a "good metabolism" if they can eat enormous meals and stay slender. However, the term "metabolism" actually refers to all of the chemical reactions that occur in the body, including those that build muscle and fat. We learn a new language not by reading about it but by using it. The Study Guide you hold in your hands employs a number of learning techniques in every chapter to help you become comfortable with the language of anatomy, physiology, and disease.

Addressing the Learning Objectives

The exercises in this section will help you master the material both verbally and visually. Work through the section as you read the textbook chapter; completing the exercises will help you actively learn the material, improving your chances of remembering it at exam time.

The labeling and coloring exercises will be especially useful for mastering anatomy. You can use these exercises in two ways. First, follow the instructions to label and color (when appropriate) the diagram, using your textbook if necessary. Second, use the diagrams for exam preparation by covering up the label names and practicing naming each structure. Coloring exercises are fun and have been shown to enhance learning.

"Making the Connections:" Learning through Concept Maps

This learning activity uses concept mapping to master definitions and concepts. You can think of concept mapping as creating a web of information. Individual terms have a tendency to get lost, but a web of terms is more easily maintained in memory. You can make a concept map by following these steps:

1. Select the concepts to map (6–10 is a good number). Try to use a mixture of nouns, verbs, and processes.
2. If one exists, place the most general, important, or overriding concept at the top or in the center, and arrange the other terms around it. Organize the terms so that closely related terms are close together.
3. Draw arrows between concepts that are related. Write a short phrase to connect the two concepts.

For instance, consider a simple concept map composed of three terms: student learning, professors, and textbooks. Write the three terms at the three corners of a triangle, separated from each other by 3 to 4 inches. Next is the difficult part: devising connecting phrases that explain the relationship between any two terms. What is the essence of the relationship between student learning and professors? An arrow could be drawn from professors to student learning, with the connecting phrase "*can explain difficult concepts to facilitate.*" The relationship would be "*Professors* **can explain difficult concepts to facilitate** *student learning.*" Draw arrows between all other term pairs (*student learning* and *textbooks*, *textbooks* and *professors*) and try to come up with connecting phrases. Make sure that the phrase is read in the direction of the arrow.

There are two concept mapping exercises for most chapters. The first exercise consists of filling in boxes and, in the later maps, connecting phrases. The guided concept maps for Chapters 1 through 7 ask you to think of the appropriate term for each box. The guided concept maps for Chapters 8 through 25 are more traditional concept maps. Each pair of terms is linked together by a connecting phrase. The phrase is read in the direction of the arrow. For instance, an arrow leading from "*genes*" to "*chromosomes*" could result in the phrase "*Genes are found on pieces of DNA called chromosomes.*" The second optional exercise provides a suggested list of terms to use to construct your own map. This second exercise is a powerful learning tool, because you will identify your own links between concepts. The act of creating a concept map is an effective way to understand terms and concepts.

Testing Your Knowledge

These questions should be completed after you have read the textbook and completed the other learning activities in the Study Guide. Try to answer as many questions as possible without referring to your notes or the text. As in the end-of-chapter questions, there are three different levels of questions. Type I questions (Building Understanding) test simple recall: How well have you learned the material? Type II questions (Understanding Concepts) examine your ability to integrate and apply the information in simple practical situations. Type III questions (Conceptual Thinking) are the most challenging. They ask you to apply your knowledge to new situations and concepts. There is often more than one right answer to Conceptual Thinking questions. The answers to all questions are available from your instructor.

Learning from the World around You

The best way to learn anatomy and physiology is to immerse yourself in the subject. Tell your friends and family what you are learning. Discover more about recent health advances from television, newspapers, magazines, and the Internet. Our knowledge about the human body is constantly changing. The work you will do using the Study Guide can serve as a basis for lifelong learning about the human body in health and disease.

Contents

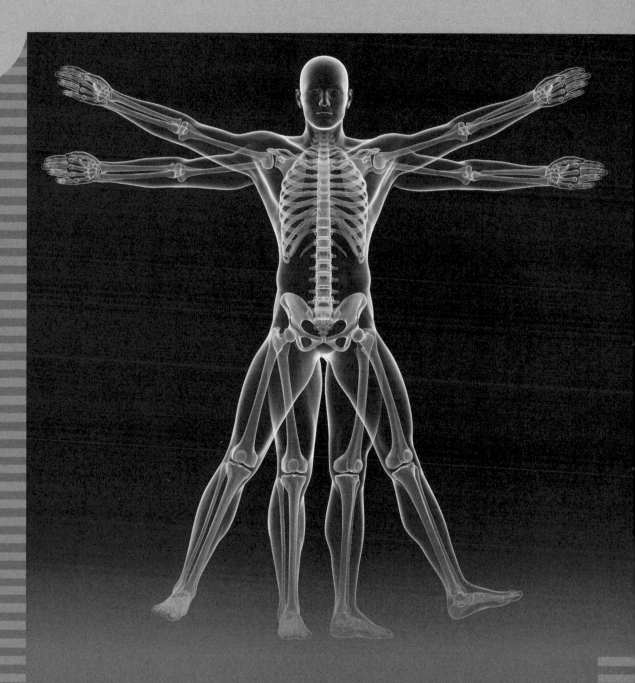

UNIT I

The Body as a Whole

Organization of the Human Body

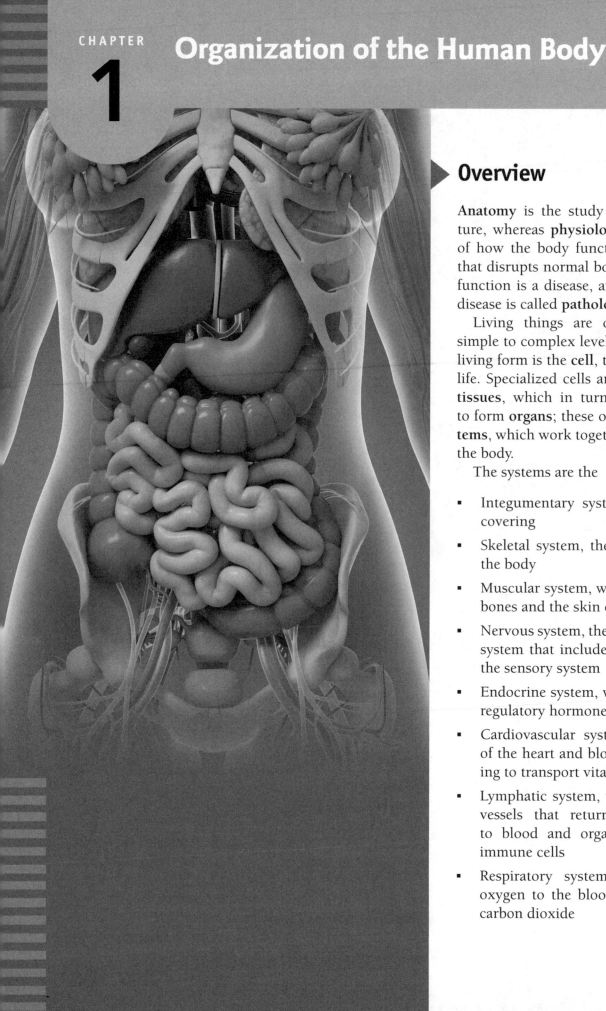

Overview

Anatomy is the study of body structure, whereas **physiology** is the study of how the body functions. Anything that disrupts normal body structure or function is a disease, and the study of disease is called **pathology**.

Living things are organized from simple to complex levels. The simplest living form is the **cell**, the basic unit of life. Specialized cells are grouped into **tissues**, which in turn are combined to form **organs**; these organs form **systems**, which work together to maintain the body.

The systems are the

- Integumentary system, the body's covering

- Skeletal system, the framework of the body

- Muscular system, which moves the bones and the skin of the face

- Nervous system, the central control system that includes the organs of the sensory system

- Endocrine system, which produces regulatory hormones

- Cardiovascular system, consisting of the heart and blood vessels, acting to transport vital substances

- Lymphatic system, which includes vessels that return tissue fluids to blood and organs that house immune cells

- Respiratory system, which adds oxygen to the blood and removes carbon dioxide

- Digestive system, which converts raw food materials into products usable by cells
- Urinary system, which removes wastes and excess water
- Reproductive system, by which new individuals of the species are produced

All the cellular reactions that sustain life together make up **metabolism**, which can be divided into **catabolism** and **anabolism**. In catabolism, complex substances are broken down into simpler molecules. When the nutrients from food are broken down by catabolism, energy is released. This energy is stored in the compound **ATP** (adenosine triphosphate) for use by the cells. In anabolism, simple compounds are built into substances needed for cell activities.

All the systems work together to maintain a state of balance or **homeostasis**. The main mechanism for maintaining homeostasis is **negative feedback**, by which the state of the body is the signal to keep conditions within set limits. In this process, a **sensor** gathers information about a regulated parameter. A **control center** compares the sensor input with its set point and, if necessary, alters the activity of an **effector** to return the parameter closer to the set point. Hormones and electrical impulses act as signals between the different components.

The human body is composed of large amounts of fluid, the amount and composition of which must be constantly regulated. The **extracellular fluid** consists of the fluid that surrounds the cells as well as the fluid circulating in blood and lymph. The fluid within cells is the **intracellular fluid**.

Study of the body requires knowledge of directional terms to locate parts and to relate various parts to each other. Planes of division represent different directions in which cuts can be made through the body. Separation of the body into areas and regions, together with the use of the special terminology for directions and locations, makes it possible to describe an area within the human body with great accuracy.

The large internal spaces of the body are cavities in which various organs are located. The **dorsal cavity** is subdivided into the **cranial cavity** and the **spinal cavity (canal)**. The **ventral cavity** is subdivided into the **thoracic** and **abdominopelvic cavities**. Imaginary lines are used to divide the abdomen into regions for study and diagnosis.

Addressing the Learning Objectives

1. DEFINE THE TERMS *ANATOMY, PHYSIOLOGY,* AND *PATHOLOGY.*

EXERCISE 1-1

Write a definition of each term in the spaces below.

1. Anatomy _____

_____

2. Physiology _____

_____

3. Pathology _____

_____

2. DESCRIBE THE ORGANIZATION OF THE BODY FROM CHEMICALS TO THE WHOLE ORGANISM.

EXERCISE 1-2: Levels of Organization (Text Fig. 1-1)

1. Write the name or names of each labeled level of organization on the numbered lines in different colors.
2. Color the different structures on the diagram with the corresponding colors. For instance, if you wrote "cell" in blue, color the cell blue.

1. _____

2. _____

3. _____

4. _____

5. _____

6. _____

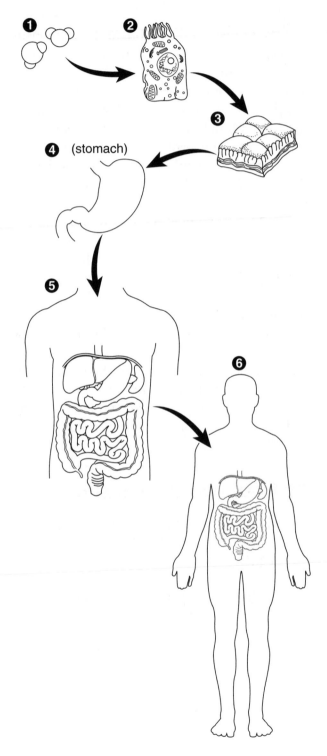

3. LIST 11 BODY SYSTEMS, AND GIVE THE GENERAL FUNCTION OF EACH.

EXERCISE 1-3

Write the appropriate term in each blank from the list below. Not all terms will be used.

nervous system   integumentary system   cardiovascular system

respiratory system   skeletal system   urinary system

endocrine system   lymphatic system   digestive system

1. The system that processes sensory information   _____

2. The system that delivers nutrients to body tissues   _____

3. The system that breaks down and absorbs food   _____

4. The system that includes the fingernails   _____

5. The system that includes the bladder   _____

6. The system that includes the joints   _____

7. The system that delivers oxygen to the blood   _____

8. The system that includes the tonsils   _____

4. DEFINE AND GIVE EXAMPLES OF HOMEOSTASIS.

See Exercises 1-4 and 1-5.

5. USING EXAMPLES, DISCUSS THE COMPONENTS OF A NEGATIVE FEEDBACK LOOP.

EXERCISE 1-4

Use the terms below to complete the paragraph.

control center   sensor   effector   negative feedback

homeostasis   body cells   insulin   pancreas.

The maintenance of a consistent internal body state, known as (1) _____, is critical for health. Different body parameters, such as body temperature and blood glucose concentration, are kept constant using (2) _____. This process includes several components. A(n) (3) _____ measures the level of the regulated parameter. A(n) (4) _____ compares the input from component (3) with the set point and sends signals to a(n) (5) _____ in order to reverse any changes in the regulated parameter. In the control of blood glucose, the (6) _____ fulfills the roles of sensor and control center. The effector is (7) _____. The signal that passes from the control center to the effector is (8) _____.

EXERCISE 1-5

The body tightly regulates body fluid volume and composition. Fill in the blank after each statement—does it apply to extracellular fluid (EC) or intracellular fluid (IC)?

1. Includes lymph and blood   _____

2. Refers to fluids inside cells   _____

3. Includes fluid between cells   _____

6. DEFINE *METABOLISM*, AND NAME THE TWO TYPES OF METABOLIC REACTIONS.

EXERCISE 1-6

Use the terms below to complete the paragraph.

ATP   metabolism   catabolism   anabolism

The term (1) _____ refers to all life-sustaining reactions that occur within the body. The reactions involved in (2) _____ assemble simple components into more complex ones. The reactions of (3) _____ break down substances into simpler components, generating energy in the form of (4) _____. This energy can be used to fuel cell activities.

7. LIST AND DEFINE THE MAIN DIRECTIONAL TERMS FOR THE BODY.

EXERCISE 1-7: Directional Terms (Text Fig. 1-6)

1. Write the name of each directional term on the numbered lines in different colors.
2. Color the arrow corresponding to each directional term with the appropriate color.

1. _____
2. _____
3. _____
4. _____
5. _____
6. _____
7. _____
8. _____

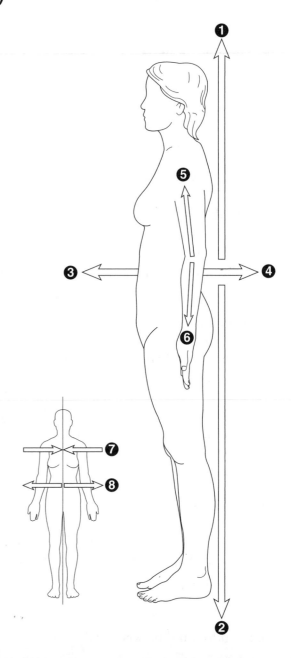

EXERCISE 1-8

Write the appropriate term in each blank from the list below. Not all terms will be used.

posterior   anterior   medial   distal

proximal   lateral   horizontal

1. A term that indicates a location toward the front   _____

2. A term that means farther from the origin of a part   _____

3. A directional term that means away from the midline
   (toward the side)   _____

4. A term that describes the position of the ankle in relation
   to the toes   _____

5. A term that describes the position of the shoulder blades in
   relation to the collar bones   _____

8. LIST AND DEFINE THE THREE PLANES OF DIVISION OF THE BODY.

EXERCISE 1-9: Planes of Division (Text Fig. 1-7)

1. Write the names of the three planes of division on the correct numbered lines in different colors.
2. Color each plane in the illustration with its corresponding color.

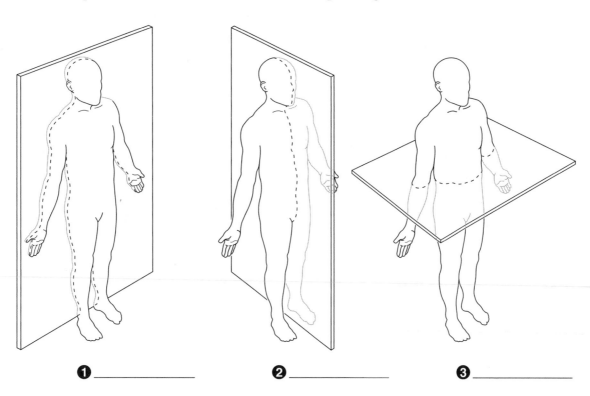

❶ _____   ❷ _____   ❸ _____

9. NAME THE SUBDIVISIONS OF THE DORSAL AND VENTRAL CAVITIES.

EXERCISE 1-10: Lateral View of Body Cavities (Text Fig. 1-10)

1. Write the names of the different body cavities and other structures in the appropriate spaces in different colors. Try to choose related colors for the dorsal cavity subdivisions and for the ventral cavity subdivisions.
2. Color parts 2, 3, and 6 to 9 with the corresponding colors.

1. _____

2. _____

3. _____

4. _____

5. _____

6. _____

7. _____

8. _____

9. _____

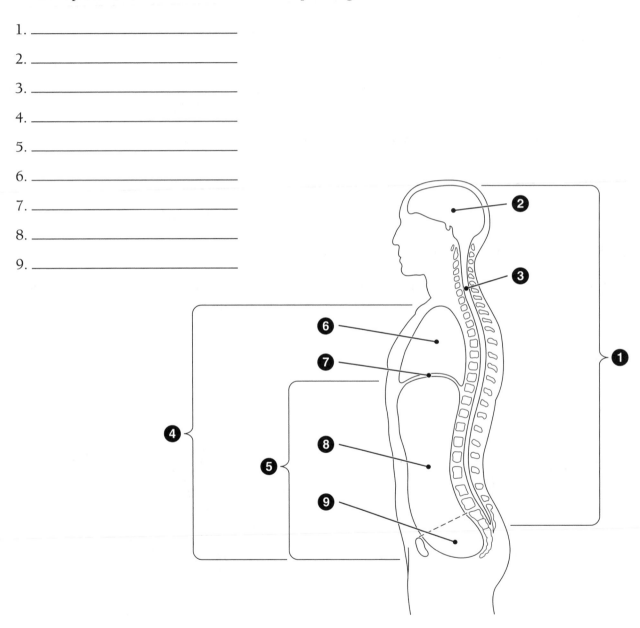

10. NAME AND LOCATE THE SUBDIVISIONS OF THE ABDOMEN.

EXERCISE 1-11: Regions of the Abdomen (Text Fig. 1-12)

1. Write the names of the nine regions of the abdomen on the appropriate numbered lines in different colors.
2. Color the corresponding regions with the appropriate colors.

1. _____

2. _____

3. _____

4. _____

5. _____

6. _____

7. _____

8. _____

9. _____

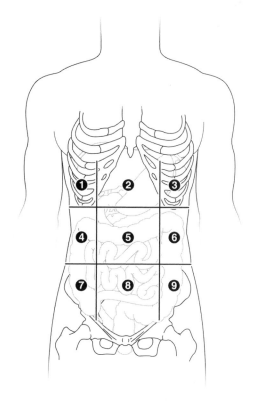

EXERCISE 1-12: Quadrants of the Abdomen (Text Fig. 1-13)

1. Write the names of the four quadrants of the abdomen on the appropriate numbered lines in different colors.
2. Color the corresponding quadrants in the appropriate colors.

1. _____

2. _____

3. _____

4. _____

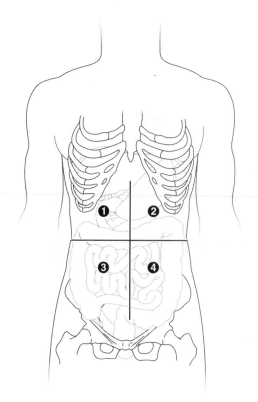

11. CITE SOME ANTERIOR AND POSTERIOR BODY REGIONS ALONG WITH THEIR COMMON NAMES.

EXERCISE 1-13

Complete the following table by writing in the missing terms.

Common Name	Anatomic Adjective
Thigh	
	Antecubital
	Inguinal
Arm	
Forearm	
	Axillary
	Tarsal
Shoulder blade	
	Acromial

12. FIND EXAMPLES OF ANATOMIC AND PHYSIOLOGIC TERMS IN THE CASE STUDY.

EXERCISE 1-14

Read through the case study at the beginning of the chapter and the case study discussion at the end of the chapter. Find an example of each type of medical term listed below and write it in the blank.

a. A term describing one of four abdominal regions   _____

b. A term describing a particular region of the upper limb   _____

c. A term describing a body cavity   _____

d. A term describing one of nine abdominal regions   _____

e. A directional term   _____

13. SHOW HOW WORD PARTS ARE USED TO BUILD WORDS RELATED TO THE BODY'S ORGANIZATION.

EXERCISE 1-15

Complete the following table by writing the correct word part or meaning in the space provided. Write a word that contains each word part in the "Example" column.

Word Part	Meaning	Example
1. -tomy	_____	_____
2. -stasis	_____	_____
3. _____	nature, physical	_____
4. home/o	_____	_____
5. _____	apart, away from	_____
6. _____	down	_____
7. _____	upward	_____
8. path/o	_____	_____
9. -logy	_____	_____

Making the Connections

The following concept map deals with the body's cavities and their divisions. Complete the concept map by filling in the blanks with the appropriate word or term for the cavity, division, subdivision, or region.

Internally the body is divided into two main cavities.

They are the **❶** [   ] and the **❷** [   ]

The two subdivisions of this cavity are the

The two subdivisions of this cavity are the

❸ [   ] and the [spinal cavity]   [thoracic cavity] and the **❹** [   ]

They are separated by
a muscle called the
❺ [   ]

This cavity can be
divided into the
[abdominal cavity]
and the
❻ [   ]

For the purpose of examination the abdominal cavity can be divided into nine regions. From superior to inferior they are:

Right Lateral	Central	Left Lateral
❼ [ ]	Epigastric region	Hypochondriac region
Lumbar region	**❾** [ ]	**⓫** [ ]
❽ [ ]	**❿** [ ]	Iliac region

Testing Your Knowledge

BUILDING UNDERSTANDING

I. MULTIPLE CHOICE

Select the best answer.

1. Which of these phrases describes a wart on the fingertip?   1. _____
   a. phalangeal wart
   b. pedal wart
   c. tarsal wart
   d. axillary wart
2. Which two body cavities are separated by the diaphragm?   2. _____
   a. cranial and the spinal cavities
   b. dorsal and ventral cavities
   c. thoracic and abdominal cavities
   d. abdominal and pelvic cavities
3. What term describes the breakdown of complex molecules into
   simpler ones?   3. _____
   a. anabolism
   b. synthesis
   c. negative feedback
   d. catabolism
4. Blood plasma is an example of which type of fluid?   4. _____
   a. extracellular
   b. intracellular
   c. superior
   d. extraneous
5. Which body system consists of the skin and accessory organs?   5. _____
   a. circulatory system
   b. nervous system
   c. integumentary system
   d. digestive system
6. Which of these terms describes the right superior region of the abdomen?   6. _____
   a. right lumbar region
   b. right hypochondriac region
   c. right iliac region
   d. right inguinal region
7. Which of these terms describes the study of normal body structure?   7. _____
   a. physiology
   b. pathology
   c. anatomy
   d. chemistry

8. Which of these sections is created when you cut a banana right down the
   middle, making two identical halves?   8. _____
       a. longitudinal section
       b. horizontal section
       c. cross-section
       d. coronal section

9. In a negative feedback loop, which component receives signals from
   the sensor?   9. _____
       a. effector
       b. set point
       c. control center
       d. receptor

II. COMPLETION EXERCISE

➤ Group A: General Terminology

Write the word or phrase that correctly completes each sentence.

1. In the anatomic position, the body is upright and palms are facing _____.

2. Fluid inside cells is called _____.

3. Catabolism releases energy in the form of _____.

4. Homeostasis is maintained by a form of feedback known as _____.

5. The sum of all catabolic and anabolic reactions in the body is called _____.

6. The abbreviation ATP stands for _____.

➤ Group B: Body Cavities, Directional Terms, and Planes of Division

1. The term that means nearer to the point of origin is _____.

2. The term that means farther from the body's midline is _____.

3. The abdomen may be divided into four regions, each of which is called
   a(n) _____.

4. The cavity that houses the brain is the _____.

5. The plane that divides the body into left and right parts is the _____.

6. The ventral body cavity that contains the stomach, most of the intestine, the liver, and the
   spleen is the _____.

7. The abdomen may be subdivided into nine regions, including three along the midline. The
   region closest to the sternum (breastbone) is the _____.

8. The space between the lungs is called the _____.

9. The diaphragm separates the thoracic cavity from the _____.

➤ **Group C: Body Regions**

Write the anatomic adjective that corresponds to each body region.

1. Buttock _____

2. Wrist _____

3. Back of knee _____

4. Hip _____

5. Forearm _____

UNDERSTANDING CONCEPTS

I. TRUE/FALSE

For each statement, write T for true or F for false in the blank to the left of each number. If a statement is false, correct it by replacing the underlined term, and write the correct statement in the blank below the statement.

_____ 1. Proteins are broken down into their component parts by the process of catabolism.

_____

_____ 2. The calcaneal tendon is found in the heel.

_____

_____ 3. The spinal cavity is superior to the cranial cavity.

_____

_____ 4. Your umbilicus is lateral to the left lumbar region.

_____

_____ 5. The right hypochondriac region is found in the right lower quadrant.

_____

II. PRACTICAL APPLICATIONS

➤ **Group A: Directional Terms**

Study each discussion. Then write the appropriate word or phrase in the space provided.

1. The gallbladder is located just above the colon. The directional term that describes the position of the gallbladder with regard to the colon is _____.

2. The stomach is closer to the midline than the kidneys. The directional term that describes the kidneys with regard to the stomach is _____.

3. The entrance to the stomach is nearest the point of origin or beginning of the stomach, so this part is said to be _____.

4. The knee is located closer to the hip (the point of origin for the lower limb) than is the ankle. The term that describes the position of the ankle with regard to the knee is _____.

5. A child suffered from a burn extending from the umbilicus to the sternum. The term that describes the umbilical region in relation to the hypogastric region is _____.

6. The sternum is closer to the front of the body than the vertebrae are. The term that describes the vertebrae with regard to the sternum is _____.

7. The head of the pancreas is nearer to the midsagittal plane than is its tail portion, so the head part is more _____.

➤ Group B: Body Cavities and Body Regions

Study the following cases and answer the questions based on the nine divisions of the abdomen and the anatomic terms for body regions.

1. Mr. A bruised his ribs in a dirt buggy accident. He experienced tenderness in the upper left side of his abdomen. In which of the nine abdominal regions are the injured ribs located? _____

2. Ms. D had a history of gallstones. One of her symptoms was referred pain between the shoulder blades. The adjective describing the shoulder blade is _____.

3. The operation to remove these stones involved the upper right part of the abdominal cavity. Which abdominal division is this? _____

4. Ms. C is eight weeks' pregnant. Her uterus is still confined to the most inferior division of the abdomen. This cavity is called the _____.

5. Ms. C is experiencing heartburn as a result of her pregnancy. The discomfort is found just below the breastbone in the _____.

6. When she went into labor, she experienced significant pain in the small of her back. What is the anatomical adjective describing this region? _____

7. Following the birth of her child, Ms. C opted for a tubal ligation. The doctor threaded a fiber-optic device through a small incision in her navel as part of the surgery. Ms. C will now have a very small incision in which of the nine abdominal regions? _____

➤ Group C: Body Systems

The triage nurse in the emergency room was showing a group of students how she assessed patients with disorders in different body systems. Study each situation, and answer the following questions based on your knowledge of the 11 body systems.

1. One person was complaining of dizziness and blurred vision. Vision is controlled by the _____.

2. One person had been injured in a snowboarding accident, spraining his wrist joint. The wrist joint is part of the _____.

3. A woman had attempted a particularly onerous yoga pose and felt a sharp pain in her left thigh. Now she is limping. The nurse suspected a tear to structures belonging to the _____.

4. An extremely tall individual entered the clinic, complaining of a headache. The nurse suspected that he had excess production of a particular hormone. The specialized glands that synthesize hormones make up the _____.

5. A middle-aged woman was brought in unable to move the right side of her body. The nurse felt that a blood clot in a blood vessel of the brain was producing the symptoms. Blood vessels are part of the _____.

6. A man complaining of pain in the abdomen and vomiting blood was brought in by his family. A problem was suspected in the system responsible for taking in food and converting it to usable products. This system is the _____.

7. Each client was assessed for changes in the color of the outer covering of the body. The outer covering is called the skin, which is part of the _____.

8. A young woman was experiencing pain in her pelvic region. The doctor suspected a problem with her ovaries. The ovaries are part of the _____.

9. An older man was experiencing difficulty with urination. The production of urine is a function of the _____.

III. SHORT ESSAYS

1. Compare and contrast the terms *anabolism* and *catabolism*. List one similarity and one difference.

_____

_____

_____

_____

_____

2. What is homeostasis, and how does negative feedback help maintain it?

_____

_____

_____

_____

3. Explain why specialized terms are needed to indicate different positions, regions, and directions within the body. Provide a concrete example.

_____

_____

_____

_____

_____

CONCEPTUAL THINKING

1. In the lines below, rewrite this description replacing all underlined words with precise anatomic terms.

   "A 40-year-old man was brought into the ER suffering from multiple wounds. One laceration was in the <u>shoulder blade</u> region, just <u>above</u> a mole. The second was in his <u>calf</u> region, <u>farther from the midline than</u> a scar from a previous injury. Finally, he had multiple small cuts extending from the <u>abdominal region just below his sternum</u> to the region <u>overlying his left hip bone</u>."

   _____

   _____

   _____

   _____

   _____

2. Look back at the case study at the beginning of the chapter, and answer the following questions.
   a. What was the main challenge to Mike's homeostasis (hint: the numbers involved are 80 and 40)?
   b. How did his heart try to compensate for this challenge?
   c. Which body system contains the heart?
   d. Find two medical interventions *performed by the paramedics* that helped Mike deal with his homeostatic challenge.

   a. _____

   _____

   b. _____

   _____

   c. _____

   _____

   d. _____

   _____

3. A disease at the chemical level can have an effect on the whole body. That is, a change in a chemical affects a cell, which alters the functioning of a tissue, which disrupts an organ, which disrupts a system, which results in body dysfunction. Illustrate this concept by rewriting the following description in your own words using the different levels of organization in order of complexity: chemical, cell, tissue, organ, system, and body (hint: blood is a tissue). Your answer should address the following issues: (a) Which of the bold terms applies to each level of organization? (b) Which level of organization is not explicitly stated in the question?

Mr. S. experiences pain throughout his **body**. The movement of **blood** through his **blood vessels** is impaired. His **blood cells** are misshapen. A chemical found in red blood cells called **hemoglobin** is abnormal.

_____

_____

_____

_____

_____

4. Read the following description carefully. Then, identify the components of the negative feedback loop by writing the correct bolded term in each blank. NOTE: you will not use all of the bolded terms.

   Victoria was at the doctor's office for her annual checkup. As she was lying on the bed, her **systolic blood pressure** was measured at **110 mm Hg**. Her physician was happy with the reading, mentioning that the average systolic blood pressure for a woman her age was **between 100 mm Hg and 120 mm Hg**. When Victoria stood up, she felt dizzy for a second but then recovered. The physician reassured her, mentioning that **baroreceptors** found in some of her arteries were able to detect the decrease in blood pressure that happens when we stand up. The **vasomotor center** in her brain increases **heart rate** in response to low blood pressure.

   a. regulated parameter(s) _____

   b. sensor(s) _____

   c. effector(s) _____

   d. control center(s) _____

   e. set point _____

Expanding Your Horizons

As a student of anatomy and physiology, you have joined a community of scholars stretching back into prehistory. The history of biological thought is a fascinating one, full of murder and intrigue. We think that scientific knowledge is entirely objective. However, as the books below will show, theories of anatomy, physiology, and disease depend upon societal factors such as economic class, religion, and gender issues.

- Endersby J. A Guinea Pig's History of Biology. Cambridge, MA: Harvard University Press; 2009.
- History of Anatomy. Available at: http://www.historyworld.net/wrldhis/PlainTextHistories.asp?groupid=44&HistoryID=aa05&gtrack=pthc
- Luft EVD. History of Anatomy and Physiology: The Classical and Medieval Periods. Available at: http://www.bookrags.com/research/history-of-anatomy-and-physiology-t-wap/
- Magner LN. A History of the Life Sciences. Boca Raton, FL: CRC Press; 2003.

CHAPTER 2

Chemistry, Matter, and Life

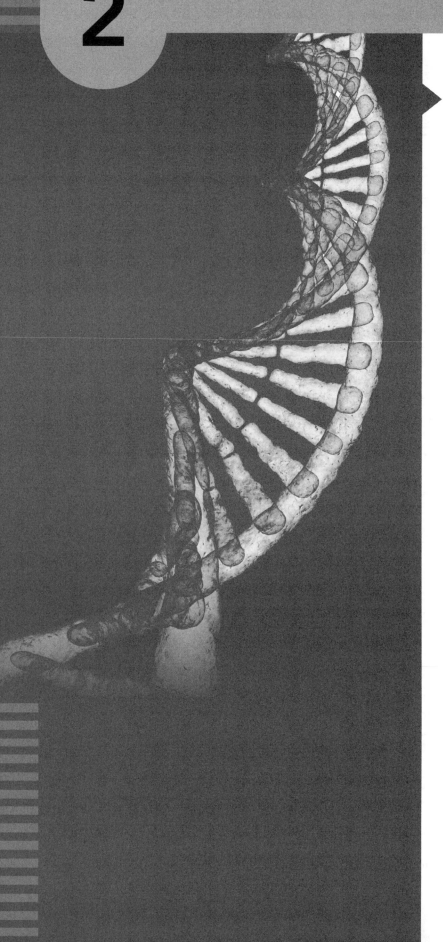

▶ Overview

Chemistry is the physical science that deals with the composition of matter. To appreciate the importance of chemistry in the field of health, it is necessary to know about elements, atoms, molecules, compounds, and mixtures. An **element** is a substance consisting of just one type of atom. Although exceedingly small particles, atoms possess a definite structure. The **nucleus** contains **protons** and **neutrons**, and the element's **atomic number** indicates the number of protons in its nucleus. The **electrons** surround the nucleus where they are arranged in specific orbits called **energy levels**.

If an atom does not have enough electrons to fill its outermost energy level, it will interact with other atoms, forming a **chemical bond**. The transfer of electrons from one atom to another results in an **ionic bond**. The participating atoms now have an uneven number of protons and electrons, so they have an electric charge and are called **ions**. Ions serve many important roles in the body. Ionically bonded substances tend to separate into their component ions in solution. Bonds that form when two atoms share electrons between them are called **covalent bonds**. Covalent bonds are usually very strong; they result in the formation of **molecules**. The atoms in the molecule may be alike (as in the oxygen molecule, O_2) or different (as in water, H_2O). A **compound** is any substance composed of more than one type of atom. So, oxygen (O_2) is a molecule but not a compound, but NaCl (which is formed by ionic bonds) is a compound but not a molecule. A combination of substances, each of which retains its separate properties, is a **mixture**. Mixtures include solutions, such as salt water, and suspensions.

Water is a vital substance composed of hydrogen and oxygen. It makes up more than half of the body and is needed as a solvent and a transport medium. Hydrogen, oxygen, carbon, and nitrogen are the elements that constitute about 96% of living matter, whereas calcium, sodium, potassium, phosphorus, sulfur, chlorine, and magnesium account for most of the remaining 4%.

Inorganic compounds include acids, bases, and salts. **Acids** are compounds that can release hydrogen ions (H^+) when dissolved in water. **Bases** can accept (bind with) hydrogen ions and usually contain the hydroxide ion (OH^-). The **pH scale** is used to indicate the strength of an acid or base. **Salts** are formed when an acid reacts with a base. The components of a salt are always joined by ionic bonds.

Isotopes are forms of elements that vary in neutron number. Isotopes that give off radiation are said to be **radioactive**. Because they can penetrate tissues and can be traced in the body, they are useful in diagnosis. Radioactive substances also have the ability to destroy tissues and can be used in the treatment of many types of cancer.

Proteins, carbohydrates, and lipids are the organic compounds characteristic of living organisms. Each type of organic compound is built from characteristic building blocks. **Enzymes**, an important group of proteins, function as catalysts in metabolism. **Nucleotides** are a fourth type of organic compound. Important molecules built from nucleotides include DNA, RNA, and ATP.

Addressing the Learning Objectives

1. DEFINE A CHEMICAL ELEMENT.

See Exercise 2-1.

2. DESCRIBE THE STRUCTURE OF AN ATOM.

EXERCISE 2-1

Write the appropriate term in each blank from the list below. Not all terms will be used.

nucleus	proton	element	electron
neutron	atom	energy level	

1. A positively charged particle inside the atomic nucleus _____

2. The smallest complete unit of matter _____

3. An uncharged particle inside the atomic nucleus _____

4. A substance composed of one type of atom _____

5. The part of the atom containing protons and neutrons _____

6. A negatively charged particle outside the atomic nucleus _____

EXERCISE 2-2: Parts of the Atom, Molecule of Water (Text Figs. 2-2 and 2-6)

1. This figure illustrates two hydrogen atoms and one oxygen atom. Write the names of the parts of the atom (electron, proton, neutron) on the appropriate lines in different, darker colors.
2. Color the electrons, protons, and neutrons on the figure in the appropriate colors. You should find 10 electrons, 10 protons, and 8 neutrons in total.
3. Use two contrasting, light colors to shade the first energy level (4) and the second energy level (5).

1. _____

2. _____

3. _____

4. _____

5. _____

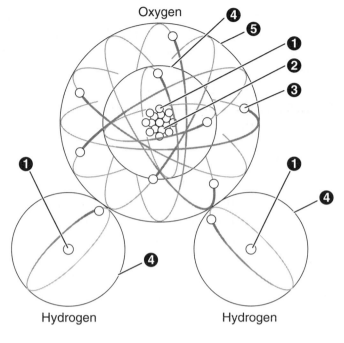

3. DIFFERENTIATE BETWEEN IONIC AND COVALENT BONDS.

See Exercise 2-3.

4. DEFINE AN ELECTROLYTE.

EXERCISE 2-3

Write the appropriate term in each blank from the list below.

cations   ionic   nonpolar covalent

electrolytes   anions   polar covalent

1. Negatively charged ions   _____

2. A bond formed by the equal sharing of electrons between two atoms   _____

3. Compounds that form ions when in solution   _____

4. Positively charged ions   _____

5. A bond formed by the transfer of electron(s) from one atom to another   _____

6. A bond formed by the unequal sharing of electrons between two atoms   _____

5. DIFFERENTIATE BETWEEN MOLECULES AND COMPOUNDS.

EXERCISE 2-4

Fill in the blank after each statement—does it apply to molecules (M), compounds (C), or both (B)?

1. Contain two or more atoms   _____

2. Contain two identical atoms   _____

3. Always contain two different atoms   _____

4. Can consist of two identical covalently bonded atoms   _____

5. Can consist of two different covalently bonded atoms   _____

6. Can consist of two different ionically bonded atoms   _____

6. DEFINE *MIXTURE*; LIST THE THREE TYPES OF MIXTURES, AND GIVE TWO EXAMPLES OF EACH.

EXERCISE 2-5

Write the appropriate term in each blank from the list below. Not all terms will be used.

solution   suspension   solute   solvent

aqueous   mixture   colloid

1. The substance in which another substance is dissolved   _____

2. A substance that is dissolved in another substance   _____

3. A mixture in which substances will settle out unless the
   mixture is shaken   _____

4. Term used to describe a solution mostly formed of water   _____

5. Cytosol and blood plasma are examples of this type of
   suspension   _____

6. Any combination of two or more substances in which each
   constituent maintains its identity   _____

7. EXPLAIN WHY WATER IS SO IMPORTANT IN METABOLISM.

EXERCISE 2-6

Which of the following properties are NOT true of water? There is one correct answer.

  a. All substances can dissolve in water.
  b. Water participates in chemical reactions.
  c. Water is a stable liquid at ordinary temperatures.
  d. Water carries substances to and from cells.

8. COMPARE ACIDS, BASES, AND SALTS.

See Exercise 2-7.

9. EXPLAIN HOW THE NUMBERS ON THE pH SCALE RELATE TO ACIDITY AND ALKALINITY.

See Exercise 2-7.

10. EXPLAIN WHY BUFFERS ARE IMPORTANT IN THE BODY.

EXERCISE 2-7

Use the terms below to complete the paragraph. Not all terms will be used.

pH scale   salt   acid   base   buffer   hydroxide   alkali

hydrogen   high   low

Any substance that can release a hydrogen ion is called a(n) (1)_____. Any substance that can accept a hydrogen ion is called a(n) (2) _____ or a(n) (3)_____. Many of these contain a(n) (4) _____ ion. A reaction between a hydrogen-accepting substance and a hydrogen-releasing substance produces a(n) (5) _____. The (6)_____ measures the concentration of hydrogen ions in a solution. A solution with a large concentration of hydrogen ions will have a(n) (7) _____ pH; a solution with a large concentration of hydroxide ions will have a(n) (8) _____ pH. A substance that helps maintain a stable hydrogen ion concentration in a solution is called a(n) (9) _____; these substances are critical for health.

11. DEFINE *RADIOACTIVITY*, AND CITE SEVERAL EXAMPLES OF HOW RADIOACTIVE SUBSTANCES ARE USED IN MEDICINE.

EXERCISE 2-8

Which of the following statements about radioactivity are TRUE? There are four right answers.

a. All isotopes are radioactive.
b. Isotopes have the same atomic weight.
c. Isotopes have the same number of protons.
d. Isotopes have the same number of electrons.
e. Isotopes have the same number of neutrons.
f. Radioactive isotopes disintegrate easily.
g. Radioactive isotopes can be used for diagnosis and treatment.

12. NAME THE THREE MAIN TYPES OF ORGANIC COMPOUNDS AND THE BUILDING BLOCKS OF EACH.

EXERCISE 2-9

Write the appropriate term in each blank from the list below. Not all terms will be used.

carbon   protein   amino acid   phospholipid   carbohydrate

monosaccharide   steroid   disaccharide   nitrogen

1. A building block always containing nitrogen _____

2. The nutrient formed by amino acids _____

3. A lipid containing a ring of carbon atoms _____

4. A lipid that contains phosphorus in addition to carbon, hydrogen, and oxygen _____

5. A category of organic compounds that includes simple sugars and starches _____

6. The element found in all organic compounds _____

7. A building block for complex carbohydrates _____

EXERCISE 2-10: Carbohydrates (Text Fig. 2-8)

1. Write the terms *disaccharide, monosaccharide,* and *polysaccharide* in the appropriate numbered boxes 1 to 3.
2. Write the terms *sucrose, glycogen,* and *glucose* in the appropriate numbered boxes 4 to 6 in different colors.
3. Color the glucose, sucrose, and glycogen molecules with the corresponding colors. To simplify your diagram, only use the glucose color to shade the glucose molecule in the monosaccharide.

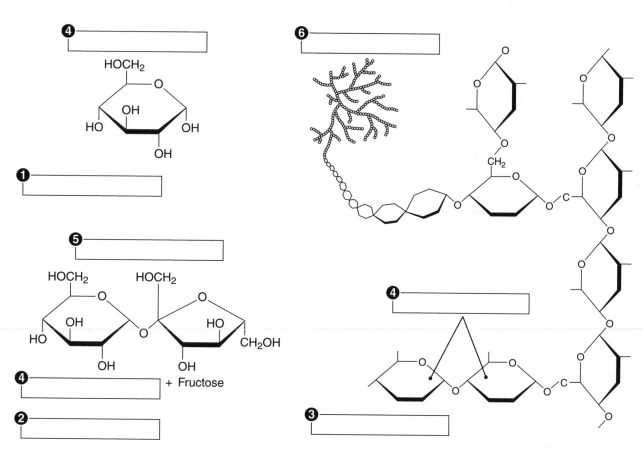

EXERCISE 2-11: Lipids (Text Fig. 2-9)

1. Write the terms *fatty acids* and *glycerol* on the appropriate lines in two different colors.
2. Find the boxes surrounding these two components on the diagram, and color them lightly in the appropriate colors.
3. Write the terms *cholesterol* and *triglyceride* in the boxes under the appropriate diagrams.

1. _____

2. _____

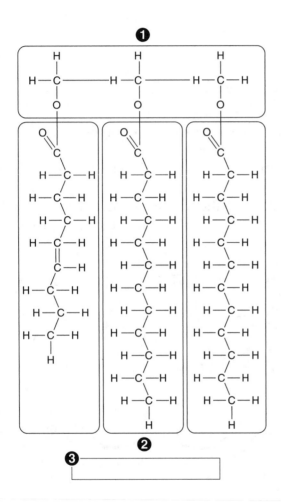

EXERCISE 2-12: Proteins (Text Fig. 2-10)

1. Write the terms *acid group* and *amino group* on the appropriate numbered lines in different colors.
2. Find the shapes surrounding these two components on the diagram, and color them lightly in the corresponding colors.
3. Place the following terms in the appropriate numbered boxes: helix, amino acid, globular protein, fibrous protein.

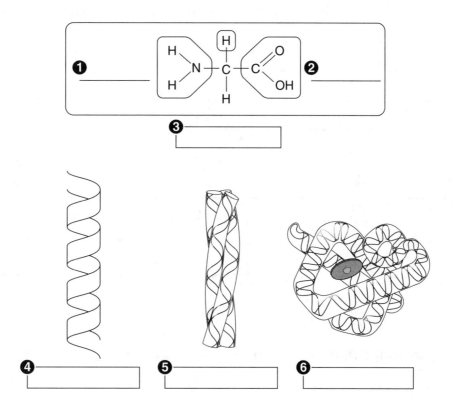

13. DEFINE *ENZYME*; DESCRIBE HOW ENZYMES WORK.

EXERCISE 2-13: Enzyme Action (Text Fig. 2-11)

1. Write the terms *enzyme, substrate 1,* and *substrate 2* on the appropriate numbered lines in different colors (red and blue are recommended for the substrates).
2. Color the structures on the diagram with the corresponding colors.
3. What color will result from the combination of your two substrate colors? Write "product" in this color on the appropriate line, and then color the product.

1. _____

2. _____

3. _____

4. _____

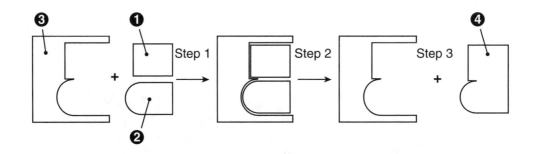

14. LIST THE COMPONENTS OF NUCLEOTIDES, AND GIVE SOME EXAMPLES OF NUCLEOTIDES.

EXERCISE 2-14

Complete the following table by providing the missing information.

Building Blocks	Finished Product	Purpose of Finished Product or Example
	Triglyceride (simple fat)	Purpose:
Two monosaccharides		Example:
One monosaccharide	Monosaccharide	Example:
Many monosaccharides		Examples: starch, glycogen
	Protein	Purpose:
One nucleotide containing three phosphate groups		Purpose:
	DNA or RNA	Purpose:

EXERCISE 2-15

Fill in the blank after each component—does it apply to an amino acid (A) or a nucleotide (N)?

1. Phosphate group   _____

2. Amino group   _____

3. Nitrogenous base   _____

4. Sugar   _____

5. Acid group   _____

6. Side chain (R group)   _____

15. USE THE CASE STUDY TO DISCUSS THE IMPORTANCE OF REGULATING BODY FLUID QUANTITY AND COMPOSITION.

EXERCISE 2-16

Use the terms below to complete the paragraph. Not all terms will be used.

potassium   increased   decreased   sodium   high blood pressure   low blood pressure

Margaret Ringland presented many signs of dehydration. Without adequate water, her blood volume (1) _____, resulting in hypotension, or (2) _____. Her body tried to compensate, so it (3) _____ her heart rate. Without enough water, the concentration of (4) _____, abbreviated as Na^+, was elevated.

16. SHOW HOW WORD PARTS ARE USED TO BUILD WORDS RELATED TO CHEMISTRY, MATTER, AND LIFE.

EXERCISE 2-17

Complete the following table by writing the correct word part or meaning in the space provided. Write a word that contains each word part in the "Example" column.

Word Part	Meaning	Example
1. _____	fear	_____
2. _____	to like	_____
3. glyc/o	_____	_____
4. _____	different	_____
5. hydr/o	_____	_____
6. hom/o-	_____	_____
7. _____	many	_____
8. -ase	_____	_____
9. sacchar/o	_____	_____
10. _____	together	_____

Making the Connections

The following concept map deals with the three major types of nutrients. Complete the concept map by filling in the blanks with the appropriate word or term.

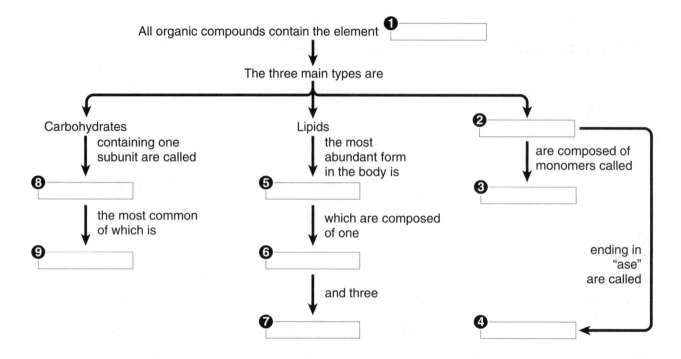

All organic compounds contain the element **❶** [_____]

The three main types are

Carbohydrates
containing one
subunit are called

❽ [_____]

the most common
of which is

❾ [_____]

Lipids
the most
abundant form
in the body is

❺ [_____]

which are composed
of one

❻ [_____]

and three

❼ [_____]

❷ [_____]

are composed of
monomers called

❸ [_____]

ending in
"ase"
are called

❹ [_____]

Optional Exercise: Make your own concept map based on the chemistry of the body. Use the following terms and any others you would like to include: atom, nucleus, electron, proton, neutron, energy level, covalent bond, and ionic bond.

Testing your Knowledge

BUILDING UNDERSTANDING

I. MULTIPLE CHOICE

Select the best answer.

1. Which of these terms best describes a solution with a pH of 3?
   a. basic
   b. acidic
   c. neutral
   d. hydrophobic

2. Which of these terms describes the smallest particle of an element that has all the properties of that element?
   a. proton
   b. neutron
   c. electron
   d. atom

1. _____

2. _____

3. Which of these terms describes the particle that results when
an electron is added to an atom?   3. _____
   a. proton
   b. anion
   c. cation
   d. electrolyte

4. Which of these phrases best describes the different orbits
that electrons can occupy?   4. _____
   a. energy levels
   b. ellipses
   c. pathways
   d. isotopes

5. What sort of macromolecule is ATP?   5. _____
   a. protein
   b. amino acid
   c. nucleotide
   d. lipid

6. Which of these is a building block for lipids?   6. _____
   a. amino acids
   b. free fatty acids
   c. nucleotides
   d. monosaccharides

7. Which of these phrases best describes covalent bonds?   7. _____
   a. usually formed between two ions
   b. can be classified as polar or nonpolar
   c. never used in organic molecules
   d. usually very unstable

8. What type of solution results when salt completely dissolves
in water?   8. _____
   a. colloid
   b. suspension
   c. aqueous solution
   d. isotope

9. What is an alternate term for a complex carbohydrate?   9. _____
   a. polysaccharide
   b. lipid
   c. inorganic molecule
   d. isotope

10. Chlorine is an atom with seven atoms in its outermost
energy level. What is its valence?   10. _____
   a. zero
   b. one
   c. seven
   d. eight

II. COMPLETION EXERCISE

Write the word or phrase that correctly completes each sentence.

1. All organic compounds contain the element _____.

2. The number of electrons lost or gained by an atom in a chemical reaction is called its _____.

3. Water can dissolve many different things. For this reason, it is called the _____.

4. The atoms in hydrogen gas (H_2) and in water (H_2O) are joined together by covalent bonds. Both of these substances are thus composed of _____.

5. An isotope that disintegrates, giving off rays of atomic particles, is said to be _____.

6. A mixture that is not a solution but does not separate because the particles in the mixture are so small is a(n) _____.

7. Many essential body activities depend on certain compounds that form ions when in solution. Such compounds are called _____.

8. The name given to a chemical system that prevents changes in hydrogen ion concentration is _____.

9. The study of the composition and properties of matter is called _____.

10. Metabolic reactions require organic catalysts called _____.

UNDERSTANDING CONCEPTS

I. TRUE/FALSE

For each statement, write T for true or F for false in the blank to the left of each number. If a statement is false, correct it by replacing the underlined term, and write the correct statement in the blank below the statement.

_____ 1. Sodium chloride (NaCl) is an example of an underlined element.

_____

_____ 2. If a neutral atom has 12 protons, it will have underlined 11 electrons.

_____

_____ 3. An atom with an atomic number of 22 will have underlined 22 protons.

_____

_____ 4. Sugar dissolves easily in water. Sugar is thus an example of a <u>hydrophobic</u> substance.

_____

_____ 5. When table salt is dissolved in water, the sodium ion donates one electron to the chloride ion. The chloride ion has 17 protons, so it will have <u>16</u> electrons.

_____

_____ 6. You put some soil in water and shake well. After 10 minutes, you note that some of the dirt has settled to the bottom of the jar. With respect to the dirt at the bottom of the jar, your mixture is a <u>colloid</u>.

_____

_____ 7. A pH of 7.0 is <u>basic</u>.

_____

_____ 8. Oxygen gas is composed of two oxygen atoms bonded by the sharing of electrons. Oxygen gas is formed by <u>polar covalent bonds</u>.

_____

_____ 9. Glycogen, a <u>polysaccharide,</u> is composed of many glucose molecules.

_____

_____ 10. Nucleotides contain three building blocks: a nitrogenous base, a sugar, and an <u>amino acid</u>.

_____

II. PRACTICAL APPLICATIONS

Study each discussion. Then write the appropriate word or phrase in the space provided.

1. Young Ms. M was experiencing intense thirst and was urinating more than usual. Her doctor suspected that she might have diabetes mellitus. This diagnosis was confirmed by finding of ketones in her urine. Ketones result from the partial breakdown of fatty acids. Fatty acids, along with glycerol, form simple fats, which are also known as _____.

2. The pH of Ms. M's urine was lower than normal, reflecting the presence of the ketones. Ketones release hydrogen ions when dissolved in water, so they can be described as _____.

3. The presence of ketones changed the pH of Ms. M's urine. The pH of her urine was _____ than normal.

4. Ms. M's urine also contained large amounts of glucose. Each glucose molecule consists of a single sugar unit, so glucose is an example of a(n) _____.

5. Because of Ms. M's excessive urination, she was suffering from dehydration. Her intense thirst reflects a shortage of the most abundant compound in the body, which is _____.

6. Mr. Q has been experiencing diarrhea, intestinal gas, and bloating whenever he drinks milk. He was diagnosed with a deficiency in an enzyme called lactase, which digests the sugar in milk. Enzymes, like all proteins, are composed of building blocks called _____.

7. Daredevil Mr. L was riding his skateboard on a high wall when he fell. He had pain and swelling in his right wrist. His examination included a procedure in which rays penetrate body tissues to produce an image on a photographic plate. The rays used for this purpose are called _____.

8. Ms. F. was given an intravenous solution, containing sodium, potassium, and chloride ions. These elements come from salts that separate into ions in solution and are referred to as _____.

III. SHORT ESSAYS

1. Describe the structure of a protein. Make sure you include the following terms in your answer: globular protein, fibrous protein, helix, protein, peptide bond, and amino acid.

_____

_____

_____

_____

_____

2. What is the difference between a solvent and a solute? Name the solvent and the solute in salt water.

_____

_____

_____

_____

3. Why is the shape of an enzyme important in its function?

_____

_____

_____

_____

_____

4. Compare and contrast colloids and suspensions. Name at least one similarity and one difference.

_____

_____

_____

_____

_____

CONCEPTUAL THINKING

1. Using the periodic table of the elements in Appendix 1 of the textbook, answer the following questions:
   a. How many protons does calcium (Ca) have? _____
   b. How many electrons does nitrogen (N) have? _____
   c. Phosphorus (P) exists as many isotopes. One isotope is called P^{32}, based on its atomic weight. The atomic weight can be calculated by adding up the number of protons and the number of neutrons. How many neutrons does P^{32} have? _____
   d. How many electrons does the magnesium ion Mg^{2+} have? (The 2+ indicates that the magnesium atom has lost two electrons). _____

2. There is much more variety in proteins than in complex carbohydrates. Explain why.

_____

_____

_____

3. Read each of the following descriptions. State whether each description pertains to molecules (M), compounds (C), and/or electrolytes (E) by writing the correct letter(s) in the blanks. More than one term may apply to each description.
   a. Hydrochloric acid (HCl) dissociates completely into an H^+ ion and a Cl^- ion when dissolved in water. _____
   b. Methane (CH_4) contains one atom of carbon that forms nonpolar covalent bonds with four different hydrogen atoms. _____
   c. Nitrous oxide (N_2O) is also known as laughing gas. It consists of one nitrogen atom that forms polar covalent bonds with two different oxygen atoms. _____
   d. If nitrogen gas (N_2) is cooled to a very low temperature, it forms a liquid that can freeze a grape solid in two seconds. _____

Expanding Your Horizons

1. You may have read the term *antioxidant* in newspaper articles or on the Web. But what is an antioxidant, and why do we want them? An understanding of atomic structure is required to answer this question. Radiation or chemical reactions can cause molecules to pick up or lose an electron, resulting in an unpaired electron. Molecules with unpaired electrons are called **free radicals**. For instance, an oxygen molecule can gain an electron, resulting in superoxide (O_2^-). Most free radicals produced in humans contain oxygen, so they are called **reactive oxygen species (ROS)**. Unpaired electrons are very unstable; thus, free radicals steal electrons from other substances, converting them into free radicals. This chain reaction disrupts normal cell metabolism and can result in cancer and other diseases. Antioxidants, including vitamin C, give up electrons without converting into free radicals, thereby stopping the chain reaction. The references listed below will tell you more about free radicals and antioxidants.

- Brown K. A radical proposal. Sci Am Presents 2000;11:38–43.
- National Cancer Institute. Antioxidants and cancer prevention. Available at: http://www.cancer.gov/cancertopics/factsheet/prevention/antioxidants
- Waris G, Ahsan H. Reactive oxygen species: role in the development of cancer and various chronic conditions. J Carcinog 2006;5:14. Available at: http://www.ncbi.nlm.nih.gov/pmc/articles/PMC1479806/

2. Proteins are the "nanomachines of life," accomplishing all of the functions needed to keep us alive. For decades, scientists have tried to create synthetic proteins of a particular shape to serve a particular function, such as blocking the sites the cholera toxin uses to create diarrhea. But despite extensive efforts, scientists still have no way of predicting the final shape of a protein based on its amino acid sequence. Without this information, it is difficult to develop proteins to serve a particular purpose. Scientists are turning to synthetic building blocks called bis-amino acids that assume more predictable conformations when they are joined together. You can read more about the prospects of these "molecular lego blocks" in the reference below.

- Schafmeister CE. Molecular Lego. Sci Am 2007;296:76–82B.

Cells and Their Functions

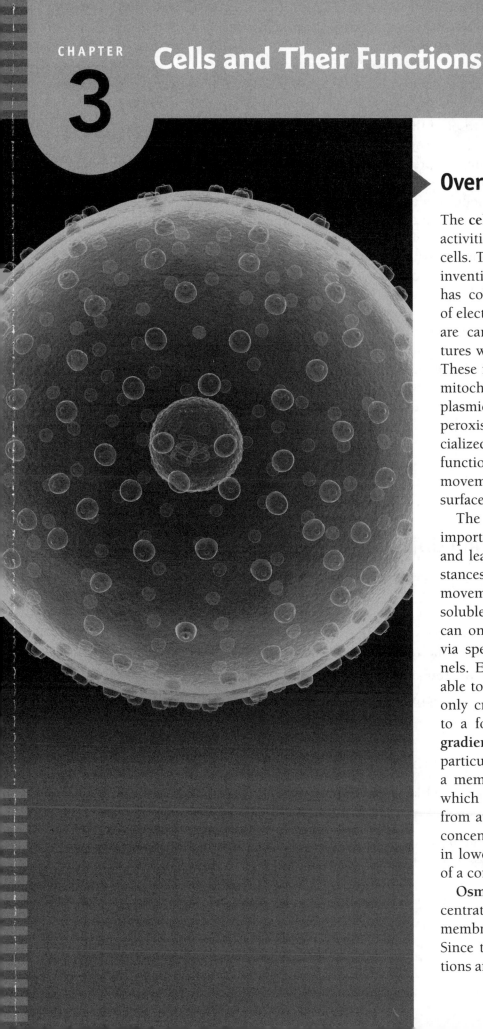

▶ Overview

The **cell** is the basic unit of life; all life activities result from the activities of cells. The study of cells began with the invention of the light microscope and has continued with the development of electron microscopes. Cell functions are carried out by specialized structures within the cell called **organelles**. These include the nucleus, ribosomes, mitochondria, Golgi apparatus, endoplasmic reticulum (ER), lysosomes, peroxisomes, and centrioles. Two specialized organelles, cilia and flagella, function in cell locomotion and the movement of materials across the cell surface.

The **plasma (cell) membrane** is important in regulating what enters and leaves the cell. Lipid-soluble substances can pass through freely, so their movement cannot be regulated. Water-soluble substances (including water) can only pass through the membrane via specific transporters or ion channels. Even if the membrane is permeable to a substance, the substance will only cross the membrane in response to a force. The force is frequently a **gradient**, which is a difference in a particular quality on the two sides of a membrane. For instance, **diffusion**, which is the movement of particles from an area where they are in higher concentration to an area where they are in lower concentration, uses the force of a concentration gradient.

Osmosis moves water down its concentration gradient with the aid of membrane proteins called *aquaporins*. Since the water and solute concentrations are inversely related, water moves from the area of *low* solute concentration to the region with a *high* solute concentration. Osmosis changes cell volume, so it must be prevented by keeping cells in solutions that have the same overall solute concentration as the cytosol. If the cell is placed in a solution of higher concentration, a **hypertonic solution**, it will shrink; in a solution of lower concentration, a **hypotonic solution**, it will swell and may burst. **Filtration** uses pressure gradients to move substances from an area of high pressure to one of lower pressure.

Other modes of transport use cellular energy to provide the force. **Active transport** uses ATP and specialized transporters to move substances against the concentration gradient (from low concentration to high). Energy is also used to move large particles and droplets of fluid across the plasma membrane by the processes of **endocytosis** and **exocytosis**. Concentration gradients and membrane permeability are not relevant to these vesicle-mediated processes.

An important cell function is the manufacture of proteins, including enzymes (organic catalysts). Protein manufacture is carried out by the ribosomes in the cytoplasm according to information coded in the deoxyribonucleic acid (DNA) of the nucleus. Specialized molecules of ribonucleic acid (RNA), called messenger RNA, play a key role in the process by carrying copies of the information in DNA to the ribosomes. DNA is also involved in the process of cell division or mitosis. Before cell division can occur, the DNA must double itself by the process of DNA replication, so each daughter cell produced by mitosis will have exactly the same kind and amount of DNA as the parent cell.

Any change in DNA sequence is called a **mutation**. Multiple mutations can cause a cell to multiply out of control, resulting in a tumor. A tumor that spreads to other parts of the body is termed **cancer**. Risk factors that influence the development of cancer include heredity, chemicals (carcinogens), ionizing radiation, physical irritation, diet, and certain viruses.

Addressing the Learning Objectives

1. LIST THREE TYPES OF MICROSCOPES USED TO STUDY CELLS.

EXERCISE 3-1

Write the appropriate term in each blank from the list below. Not all terms will be used.

compound light microscope	transmission electron microscope	
scanning electron microscope	micrometer	centimeter

1. 1/1,000 of a millimeter _____

2. Microscope that provides a three-dimensional view of an object _____

3. The most common microscope, which magnifies an object up to 1,000 times _____

4. A microscope that magnifies an object up to 1 million times _____

2. DESCRIBE THE COMPOSITION AND FUNCTIONS OF THE PLASMA MEMBRANE.

EXERCISE 3-2: Structure of the Plasma Membrane (Text Fig. 3-3)

1. Write the name of each labeled membrane component on the numbered lines in different colors. Choose a light color for component 3.
2. Color the different structures on the diagram with the corresponding color (except for structures 6 to 8). Color every example of components 1 through 5, not just those indicated by the leader lines. For instance, component 1 is found in three locations.

1. _____

2. _____

3. _____

4. _____

5. _____

6. _____

7. _____

8. _____

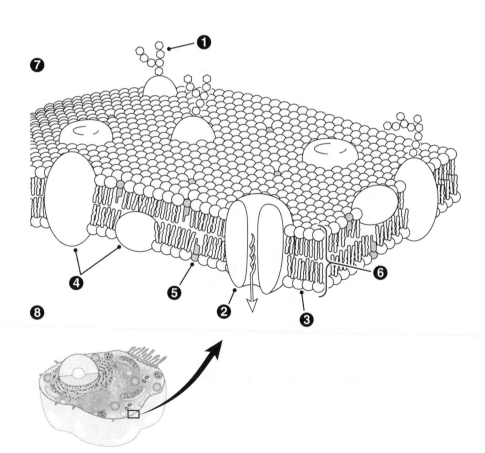

EXERCISE 3-3: Membrane Proteins (Table 3-1)

1. Write the appropriate membrane protein function in boxes 1 to 6 in different colors.
2. Color the protein in each diagram the corresponding color.

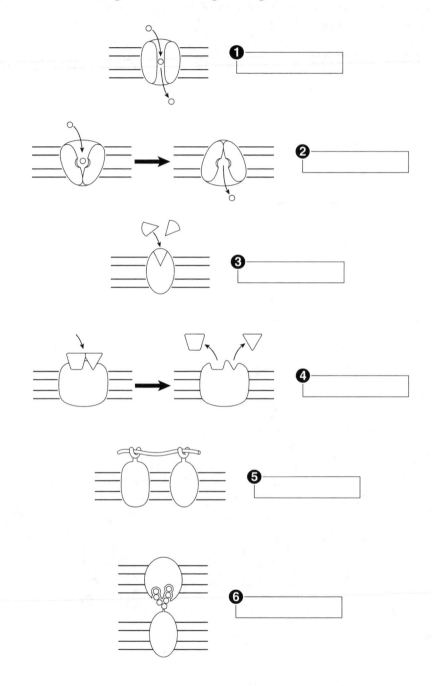

3. DESCRIBE THE CYTOPLASM OF THE CELL, AND CITE THE NAMES AND FUNCTIONS OF THE MAIN ORGANELLES.

EXERCISE 3-4: Typical Animal Cell Showing the Main Organelles (Text Fig. 3-2)

1. Write the name of each labeled part on the numbered lines in different colors. Make sure you use light colors for structures 1 and 7.
2. Color the different structures on the diagram with the corresponding colors.

   Note: Parts 12 and 15 have the same appearance in this diagram; write the name of one of the possible options in blank 12 and the other in blank 15.

1. _____

2. _____

3. _____

4. _____

5. _____

6. _____

7. _____

8. _____

9. _____

10. _____

11. _____

12. _____

13. _____

14. _____

15. _____

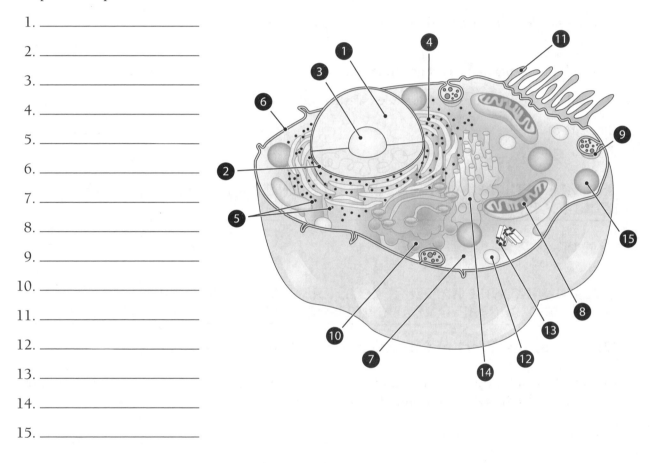

EXERCISE 3-5

Write the appropriate term in each blank from the list below. Not all terms will be used.

lysosome	nucleolus	cilia	ribosome
mitochondrion	nucleus	Golgi apparatus	vesicle

1. A structure that assembles ribosomes   _____

2. A structure that assembles amino acids into proteins   _____

3. A set of membranes involved in packaging proteins for export   _____

4. A small saclike structure used to transport substances within the cell   _____

5. A membranous organelle that generates ATP   _____

6. A small saclike structure that degrades waste products   _____

7. The site of DNA storage   _____

4. DESCRIBE METHODS BY WHICH SUBSTANCES ENTER AND LEAVE CELLS THAT DO NOT REQUIRE CELLULAR ENERGY.

See Exercise 3-7.

5. EXPLAIN WHAT WILL HAPPEN IF CELLS ARE PLACED IN SOLUTIONS WITH CONCENTRATIONS THE SAME AS OR DIFFERENT FROM THOSE OF THE CYTOPLASM.

EXERCISE 3-6: The Effect of Osmosis on Cells (Text Fig. 3-9)

Label each of the following solutions using the term that indicates the solute concentration in the solution relative to the solute concentration in the cell.

1. _____

2. _____

3. _____

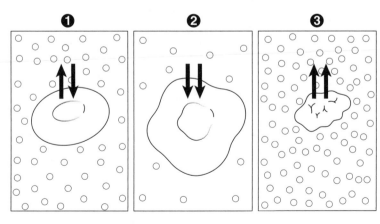

➡ Direction of osmotic water movement

6. DESCRIBE METHODS BY WHICH SUBSTANCES ENTER AND LEAVE CELLS THAT REQUIRE CELLULAR ENERGY.

EXERCISE 3-7

Write the appropriate term in each blank from the list below.

exocytosis   endocytosis   active transport   diffusion

osmosis   filtration   pinocytosis

1. The process that utilizes a carrier to move materials across the plasma membrane against the concentration gradient using ATP   _____

2. The movement of fluids through a membrane using a pressure gradient   _____

3. The movement of water down its concentration gradient   _____

4. The movement of a solute down its concentration gradient   _____

5. The process by which a cell takes in large particles   _____

6. The process by which materials are expelled from the cell using vesicles   _____

7. Small fluid droplets are brought into the cell using this method   _____

7. DESCRIBE THE COMPOSITION, LOCATION, AND FUNCTION OF THE DNA IN A CELL.

EXERCISE 3-8: Structure of DNA (Text Fig. 3-12C and D)

1. Write the name of each part of the DNA molecule on the numbered lines in contrasting colors.
2. Color the different parts on the diagram with the corresponding colors.

1. _____

2. _____

3. _____

4. _____

5. _____

6. _____

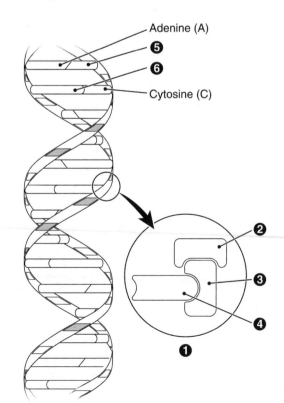

Adenine (A)

❺

❻

Cytosine (C)

❷

❸

❹

❶

8. COMPARE THE FUNCTIONS OF THREE TYPES OF RNA IN CELLS.

See Exercise 3-9.

9. EXPLAIN BRIEFLY HOW CELLS MAKE PROTEINS.

EXERCISE 3-9

Write the appropriate term in each blank from the list below. Not all terms will be used.

DNA   nucleotide   transcription   ribosomal RNA (rRNA)
transfer RNA (tRNA)   messenger RNA (mRNA)   translation

1. The process by which RNA is synthesized from the DNA   _____

2. A building block of DNA and RNA   _____

3. An important component of ribosomes   _____

4. The structure that carries amino acids to the ribosome   _____

5. The nucleic acid that carries information from the
   nucleus to the ribosomes   _____

6. The process by which amino acids are assembled into a protein   _____

10. NAME AND BRIEFLY DESCRIBE THE STAGES IN MITOSIS.

EXERCISE 3-10: Stages of Mitosis (Text Fig. 3-15)

Identify interphase and the indicated stages of mitosis. Find the DNA in each stage and color it.

1. _____

2. _____

3. _____

4. _____

5. _____

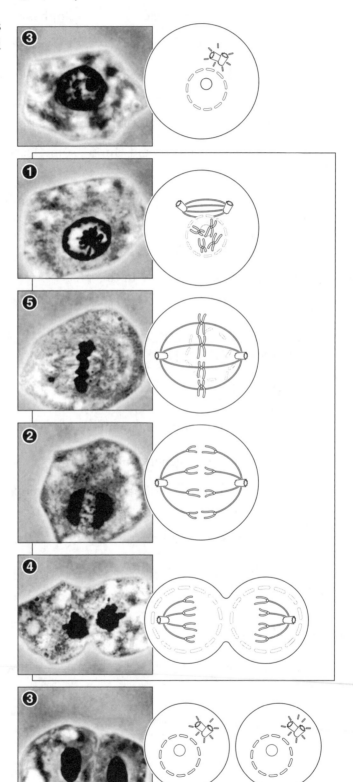

EXERCISE 3-11

Write the appropriate term in each blank from the list below. Not all terms will be used.

mitosis   anaphase   telophase   meiosis

metaphase   prophase   interphase

1. The process by which one cell divides into two identical daughter cells   _____

2. The nuclear membrane reforms during this phase   _____

3. A spindle begins to form and chromosomes condense during this phase   _____

4. The phase of mitosis when chromosomes are aligned in the middle of the cell   _____

5. DNA synthesis occurs during this phase   _____

6. The chromosomes separate in this phase   _____

11. DISCUSS THE CELLULAR CHANGES THAT MAY LEAD TO CANCER, AND LIST SEVERAL CANCER RISK FACTORS.

EXERCISE 3-12

Label each of the following statements as true (T) or false (F).

1. Cancers result from genetic mutations.   _____

2. Slower-growing cells are more likely to develop into cancers.   _____

3. The immune system often kills cancerous cells.   _____

4. Tumors that do not spread to other tissues are called cancers.   _____

EXERCISE 3-13

List seven risk factors for cancer in the spaces below.

1. _____   5. _____

2. _____   6. _____

3. _____   7. _____

4. _____

12. USE THE CASE STUDY TO DISCUSS THE IMPORTANCE OF THE PLASMA MEMBRANE TO THE FUNCTIONING OF THE BODY AS A WHOLE.

EXERCISE 3-14

Use the terms below to complete the paragraph. Not all terms will be used.

proteasomes	channel	osmosis	gene	mutation	DNA
diffusion	active transport	concentration	pressure	RNA	

Cystic fibrosis results from a small change, or (1) _____, in the substance that makes up the chromosomes, known as (2) _____. The change occurred in the small unit of heredity, or (3) _____, encoding a protein called the CFTR. The CFTR is a pore known as a (4)_____, that enables chloride to cross the plasma membrane. Chloride uses these pores to move down the (5) _____ gradient by the process of (6) _____. However, in cystic fibrosis, the CFTR is not inserted into the plasma membrane. Instead, it is degraded by small, barrel-shaped organelles called (7) _____. The lack of CFTRs in the plasma membrane decreases the movement of chloride into sweat gland cells. Normally, sodium follows chloride into sweat gland cells. So, in CF, sodium and water remain in the sweat gland and are secreted onto the skin surface, resulting in sweat that is saltier than usual.

In the lungs, the lack of the CFTR impedes water movement across plasma membranes into the lung passageways, a process called (8) _____. Without adequate water, thick ropy mucus can block passageways.

13. SHOW HOW WORD PARTS ARE USED TO BUILD WORDS RELATED TO CELLS AND THEIR FUNCTIONS.

EXERCISE 3-15

Complete the following table by writing the correct word part or meaning in the space provided. Write a word that contains each word part in the "Example" column.

Word Part	Meaning	Example
1. phag/o	_____	_____
2. _____	to drink	_____
3. -some	_____	_____
4. lys/o	_____	_____
5. _____	cell	_____
6. _____	above, over, excessive	_____
7. hem/o-	_____	_____
8. _____	same, equal	_____
9. hypo-	_____	_____
10. _____	in, within	_____

Making the Connections

The following concept map deals with the movement of materials through the plasma membrane. Complete the concept map by filling in the appropriate word or phrase that describes the indicated process.

1. Processes that move small quantities of material through the plasma membrane include...

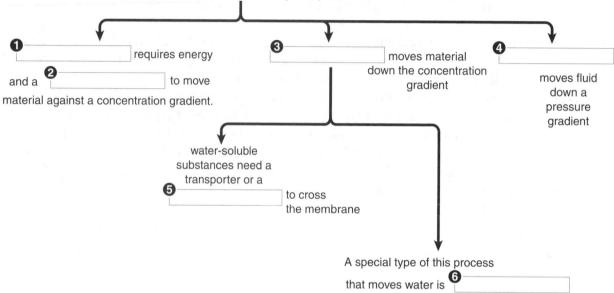

❶ _____ requires energy and a **❷** _____ to move material against a concentration gradient.

❸ _____ moves material down the concentration gradient

❹ _____ moves fluid down a pressure gradient

water-soluble substances need a transporter or a **❺** _____ to cross the membrane

A special type of this process that moves water is **❻** _____

2. Processes that move large quantities of material through the plasma membrane include...

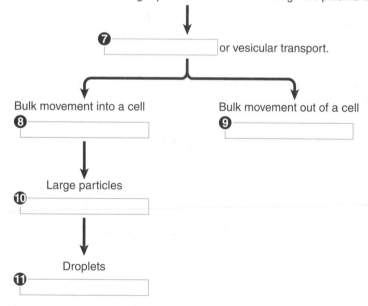

❼ _____ or vesicular transport.

Bulk movement into a cell
❽ _____

Bulk movement out of a cell
❾ _____

Large particles
❿ _____

Droplets
⓫ _____

Optional Exercise: Make your own concept map based on the components of the cell. Use the following terms and any others you would like to include: nucleus, mitochondria, cell membrane, protein, RNA, DNA, ATP, vesicle, ribosome, and ER.

Testing Your Knowledge

BUILDING UNDERSTANDING

I. MULTIPLE CHOICE

Select the best answer.

1. Which of the following organelles modifies proteins that have already been synthesized?
   a. mitochondrion
   b. rough endoplasmic reticulum
   c. ribosome
   d. lysosome

   1. _____

2. Which of these terms describes the process of programmed cell death?
   a. mitosis
   b. mutation
   c. apoptosis
   d. phagocytosis

   2. _____

3. Which of the following are required for active transport?
   a. vesicles and cilia
   b. transporters and ATP
   c. osmotic pressure and centrioles
   d. osmosis and lysosomes

   3. _____

4. During which stage of mitosis does chromatin condense into chromosomes?
   a. metaphase
   b. anaphase
   c. prophase
   d. telophase

   4. _____

5. Which of these processes enables large proteins to enter the cell?
   a. active transport
   b. osmosis
   c. endocytosis
   d. exocytosis

   5. _____

6. Which type of macromolecule makes up transporters, carriers, and enzymes in the plasma membrane?
   a. DNA
   b. protein
   c. carbohydrate
   d. phospholipid

   6. _____

7. Which of the following tools has the greatest magnification?
   a. scanning electron microscope
   b. transmission electron microscope
   c. light microscope
   d. magnifying glass

   7. _____

8. Which of these terms describes a solution that causes cells to shrink
   when they are placed in it?   8. _____
   a. isotonic
   b. hypotonic
   c. hypertonic
   d. osmotic

II. COMPLETION EXERCISE

Write the word or phrase that correctly completes each sentence.

1. Any chemical that causes cancer is called a(n) _____.

2. The plasma membrane contains two major kinds of lipids: cholesterol and

   _____.

3. The single small dark body within the nucleus is called the _____.

4. The four nitrogen bases found in DNA are A, T, G, and _____.

5. The four nitrogen bases found in RNA are A, C, G, and _____.

6. The assembly of an RNA strand is called _____.

7. The type of RNA that carries individual amino acids is called _____.

8. The chromosomes duplicate during the period between mitosis, which is called

   _____.

9. A cell has four chromosomes before entering the process of mitosis. After mitosis, the number
   of chromosomes in each daughter cell will be _____.

10. Movement of a fluid down a pressure gradient is called _____.

11. Droplets of water and dissolved substances are brought into the cell by the process of

    _____.

12. Water crosses plasma membranes through specialized channels called _____.

13. Bacteria are brought into the cell by the process of _____.

UNDERSTANDING CONCEPTS

I. TRUE/FALSE

For each statement, write T for true or F for false in the blank to the left of each number. If a statement is false, correct it by replacing the underlined term, and write the correct statement in the blank below the statement.

_____ 1. The nucleotide sequence "ACCTG" would be found in DNA.

_____

_____ 2. A living cell (with a tonicity equivalent to 0.9% NaCl) is placed in a solution containing 0.2% NaCl. This solution is <u>hypertonic</u>.

_____

_____ 3. Glucose is moving into a cell down its concentration gradient using a carrier protein. Glucose is traveling by <u>active transport</u>.

_____

_____ 4. A toxin has entered a cell. The cell is no longer capable of generating ATP. The most likely explanation for this effect is that the toxin has destroyed the <u>mitochondria</u>.

_____

_____ 5. The passage of water and dissolved materials through a membrane under pressure is called <u>osmosis</u>.

_____

_____ 6. It is impossible to count individual chromosomes during <u>interphase</u>.

_____

II. PRACTICAL APPLICATIONS

Study each discussion. Then write the appropriate word or phrase in the space provided. The following are observations you might make while working for the summer in a hospital laboratory.

1. A sample of breast tissue that was thought to be cancerous arrived at the pathology lab. Breast cancer is an example of a disease that occurs more often within certain families than in others,

   suggesting that one cancer risk factor is _____.

2. The tissue was in a liquid called normal saline so that the cells would neither shrink nor swell.

   Normal saline contains 0.9% salt and is thus considered to be _____.

3. The pathologist (Dr. C) sliced the tissue very thinly and placed it on a microscope slide, but he was unable to see anything. Unfortunately, he had forgotten to add a dye to the tissue. These

   special dyes are called _____.

4. Dr. C went back to his bench in order to get the necessary dye. He noticed that there were many impurities floating in the dye, and he decided to screen them out. He separated the solid particles from the liquid by forcing the liquid through a membrane, a process called

   _____.

5. Dr. C was clumsy and accidentally spilled the dye into a sink full of dishes. The water in the sink rapidly turned pink. The dye molecules had moved through the water by the process of

   _____.

6. Dr. C made up some new dye and treated the tissue. Finally, the tissue was ready for examination.

   The type of microscope that uses light to view stained tissues is called _____.

7. The pathologist looked at the tissue and noticed that the nuclei of many cells were in the process of dividing. This division process is called _____.

8. Some cells were in the stage of cell division called prophase. The DNA was condensed into structures called _____.

9. Dr. C sent a sample to a laboratory that specialized in identifying alterations in gene structure. These alterations are called _____.

10. The technicians in this laboratory identified a change in the sequence of a specific gene called the BRCA-1 gene. In place of an A nucleotide, there was the nucleotide that normally pairs with cytosine (C). This nucleotide is abbreviated as _____.

III. SHORT ESSAYS

1. Compare and contrast active transport and diffusion of water-soluble substances. List at least one similarity and at least one difference.

_____

_____

_____

_____

_____

2. You are in the hospital for a minor operation, and the technician is hooking you up to an IV. He knows you are a nursing student and jokingly asks if you would like a hypertonic, hypotonic, or isotonic solution to be placed in your IV. Which solution would you pick? Explain your answer.

_____

_____

_____

_____

_____

3. List the four stages of mitosis, and briefly describe each.

_____

_____

_____

_____

_____

CONCEPTUAL THINKING

1. Compare the structure of a cell to a factory or a city. Try to find cell structures that accomplish all of the different functions of the city or factory.

   _____

   _____

   _____

   _____

   _____

2. Your great aunt M is 96 years old and loves to hear about what you are learning in class. She recently attended an Elderhostel camp, where people were talking about this newfangled notion called DNA.
   a. She asks you to explain why DNA is so important. Explain the role of DNA in protein synthesis, using clear, uncomplicated language. You must define any term that your great aunt might not know. You can use an analogy if you like.

   _____

   _____

   _____

   _____

   _____

   b. Next, your great aunt wonders how the proteins get out of the cell. Explain the pathway a protein takes from the ribosome to the blood. You can use an illustration if you like.

   _____

   _____

   _____

   _____

3. You are a xenobiologist studying an alien cell isolated on Mars. Surprisingly, you notice that the cell contains some of the same substances as our cells. You quantify the concentration of these substances and determine that the cell contains 10% glucose and 0.3% calcium. The cell is placed in a solution containing 20% glucose and 0.1% calcium. The plasma membrane of this cell is very different than ours. It is permeable to glucose but not to calcium. That is, only glucose can cross the plasma membrane without using transporters. Use this information to answer the following questions:
   a. Will glucose move into the cell or out of the cell? Which transport mechanism will be involved?

   _____

   _____

b. Carrier proteins are present in the membrane that can transport calcium. If calcium moves down its concentration gradient, will calcium move into the cell or out of the cell? Which transport mechanism will be involved?

_____

_____

c. You place the cell in a new solution to study the process of osmosis. You know that sodium does not move across the alien cell membrane. You also know that the concentration of the intracellular fluid is equivalent to 1% sodium. The new solution contains 2% sodium.

   i. Is the 2% sodium solution hypertonic, isotonic, or hypotonic? _____

   ii. Will water flow into the cell or out of the cell? _____

   iii. What will be the effect of the water movement on cell volume? _____

Expanding Your Horizons

1. Carcinogens are a hot topic. Newspapers often carry large headlines proclaiming the newest carcinogen (milk, BBQ meat, mosquito repellent). You can find out the truth about these claims in the National Toxicology Program yearly report on carcinogens. The most recent report can be read online at http://ntp-server.niehs.nih.gov (look for "Report on Carcinogens"). The *Scientific American* article listed below can provide you with more information about the causes of cancer.

    ▪ Gibbs WW. Untangling the roots of cancer. Sci Am 2003;289:56–65.

2. Companies such as "23andMe" will screen your entire genome for about $100. All you have to do is spit into a tube and mail it. Within a few weeks, you will receive detailed information about your ancestry. In the past, they also provided individuals with information about genetic risk factors for various diseases (some of which were quite speculative), but in November 2013, they ceased providing health information pending regulatory review by the U.S. Food and Drug Administration (FDA). Regulatory agencies are justifiably concerned about the use of genome screening information. Will genetic screening soon be required to obtain health insurance? Will some parents use prenatal screening to select for a superbaby? How can we keep genetic information secure from computer hackers? See the website and articles listed below for more information.

    ▪ Aldhous P, Reilly M. How my genome was hacked. New Sci 2009;201:6–9.
    ▪ Hall SS. Revolution postponed. Sci Am 2010;303:60–67.
    ▪ Wong K. Finding my inner Neanderthal. Sci Am 2013. Available at: http://www.scientificamerican.com/blog/post/dna-finding-my-inner-neandertal/?page=1
    ▪ 23andme. Available at: https://www.23andme.com

Tissues, Glands, and Membranes

▶ Overview

The cell is the basic unit of life. Individual cells are grouped into **tissues** according to function. The four main groups of tissues include **epithelial tissue**, which forms glands, covers surfaces, and lines cavities; **connective tissue**, which gives support and form to the body; **muscle tissue**, which produces movement; and **nervous tissue**, which conducts electrical impulses.

All tissues derive from **stem cells**, actively dividing cells whose offspring can remain as stem cells or differentiate into specialized tissue cells. Small stem cell populations persist into adulthood so that tissues can renew and repair themselves. Glands produce substances used by other cells and tissues. **Exocrine glands** produce secretions that are released through ducts to nearby parts of the body. **Endocrine glands** produce hormones that are carried by the blood to all parts of the body.

The simplest combination of tissues is a **membrane**. Membranes serve several purposes, a few of which are mentioned here: they may serve as dividing partitions, may line hollow organs and cavities, and may anchor various organs. Membranes that have epithelial cells on the surface are referred to as **epithelial membranes**. Two types of epithelial membranes are **serous membranes**, which line body cavities and cover the internal organs, and **mucous membranes**, which line passageways leading to the outside.

Connective tissue membranes cover or enclose organs, providing protection and support. These membranes include the fascia around muscles, the meninges around the brain and spinal cord, and the tissues around the heart, bones, and cartilage.

If the normal pattern of cell growth is disrupted by the formation of cells that multiply out of control, the result is a **tumor**. A tumor that is confined locally and does not spread is called a benign tumor; a tumor that spreads from its original site to other parts of the body, a process termed **metastasis**, is called a malignant tumor. The general term for any type of malignant tumor is **cancer**. Tissue biopsy, radiography, ultrasound, computed tomography, and magnetic resonance imaging are the techniques most frequently used to diagnose cancer. Most benign tumors can be removed surgically; malignant tumors are usually treated by surgery, radiation, chemotherapy, or by a combination of these methods. **Biological therapy** uses cancer vaccines to activate the immune system against a particular tumor cell, much like regular vaccines activate the immune system against a particular bacterium or virus.

The study of tissues, known as **histology**, requires much memorization. In particular, you may be challenged to learn the different types of epithelial and connective tissue as well as the classification scheme of epithelial and connective membranes. Learning the structure of these different tissues and membranes will help you understand the amazing properties of the body, such as how we can jump from great heights, swim without becoming waterlogged, and fold our ears over without breaking them. Also, the study of histology is necessary for understanding tissue diseases such as cancer.

Addressing the Learning Objectives

1. DEFINE STEM CELLS, AND DESCRIBE THEIR ROLE IN DEVELOPMENT AND REPAIR OF TISSUE.

EXERCISE 4-1

Label each of the following statements as true (T) or false (F).

1. Mature, fully differentiated body cells frequently undergo mitosis to replace damaged cells   _____

2. Tissues that regenerate easily, such as the intestinal lining, contain more stem cells than do tissues that regenerate poorly, such as the brain   _____

3. Stem cells spend more time in interphase than do mature body cells   _____

2. NAME THE FOUR MAIN GROUPS OF TISSUES, AND GIVE THE LOCATION AND GENERAL CHARACTERISTICS OF EACH.

EXERCISE 4-2: Four Types of Epithelium (Text Figs. 4-1 and 4-2)

Label each of the following types of epithelium.

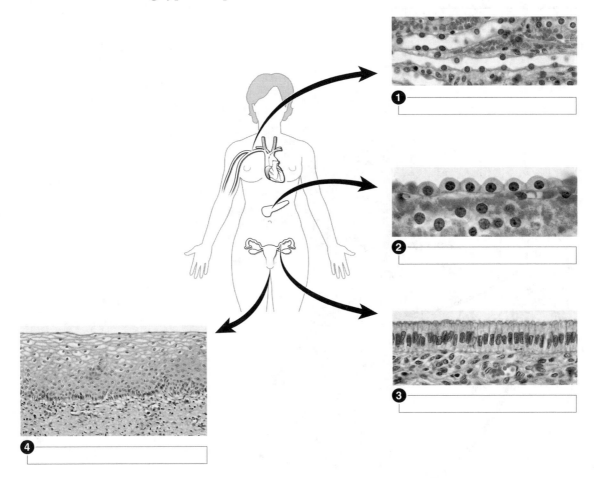

EXERCISE 4-3: Muscle Tissue (Text Fig. 4-6)

Write the names of the three types of muscle tissue in the appropriate blanks in different colors. Color some of the muscle cells the corresponding colors. Look for the nuclei, and color them a different color.

1. _____

2. _____

3. _____

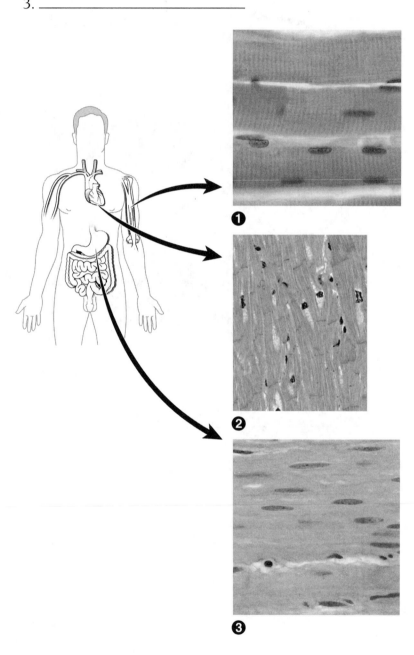

EXERCISE 4-4: Nervous Tissue (Text Fig. 4-7)

1. Write the names of each tissue in boxes 5 to 7.
2. Label each of the following neural structures and tissues.

1. _____

2. _____

3. _____

4. _____

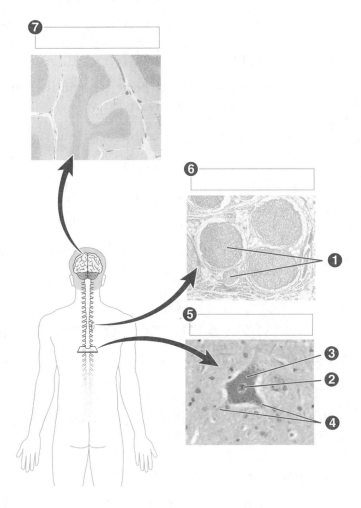

EXERCISE 4-5

Write the appropriate term in each blank from the list below. Not all terms will be used.

stratified   squamous   columnar   transitional

simple   tissue   cuboidal

1. A group of cells similar in structure and function   _____

2. Term that describes flat, irregular epithelial cells   _____

3. A term that means "*in layers*"   _____

4. Term that describes long and narrow epithelial cells   _____

5. Term that describes square epithelial cells   _____

6. Cells arranged in a single layer   _____

EXERCISE 4-6

Write the appropriate term in each blank from the list below. Not all terms will be used.

bone   myocardium   voluntary muscle   epithelial tissue

neuron   smooth muscle   neuroglia   connective tissue

1. The thigh muscle, for example   _____

2. Tissue that forms when cartilage gradually becomes
   impregnated with calcium salts   _____

3. The thick, muscular layer of the heart wall   _____

4. A type of tissue found in membrane and glands   _____

5. Also known as visceral muscle   _____

6. A cell that carries nerve impulses is called a(n)   _____

7. A tissue in which cells are separated by large amounts of
   acellular material called matrix   _____

3. DESCRIBE THE DIFFERENCE BETWEEN EXOCRINE AND ENDOCRINE GLANDS, AND GIVE EXAMPLES OF EACH.

EXERCISE 4-7

Fill in the blank after each statement—does it apply to exocrine glands (EX), endocrine glands (EN), or both (B)?

1. A gland that secretes into the blood _____

2. A gland that secretes through ducts _____

3. A gland that secretes onto the body surface _____

4. The pituitary gland, for example _____

5. A group of cells that produces substances for use by other parts
   of the body _____

6. Salivary glands, for example _____

4. CLASSIFY THE DIFFERENT TYPES OF CONNECTIVE TISSUE.

EXERCISE 4-8: Connective Tissue (Text Figs. 4-4 and 4-5)

Write the names of the seven examples of connective tissue in the appropriate boxes in different colors. Color some of the cells of each tissue type with the corresponding colors. You can also label the cells if you would like.

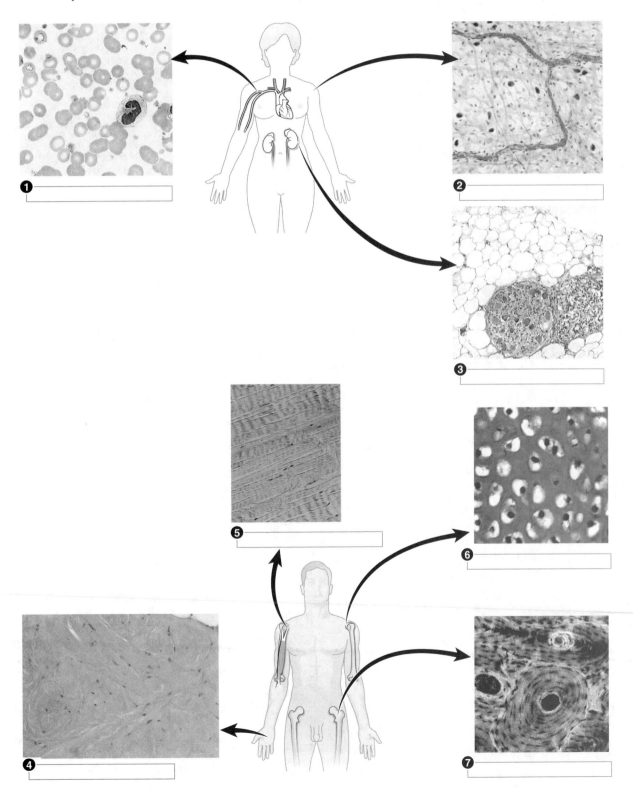

EXERCISE 4-9

Write the appropriate term in each blank from the list below. Not all terms will be used.

ligament   tendon   collagen   chondrocyte

capsule   hyaline cartilage   elastic cartilage   fibrocartilage

1. A cord of connective tissue that connects a muscle to a bone   _____

2. A tough membranous connective tissue that encloses an organ   _____

3. The cartilage found between the bones of the spine   _____

4. A fiber found in most connective tissues   _____

5. A cell that synthesizes cartilage   _____

6. A strong, gristly cartilage that reinforces the trachea   _____

5. DESCRIBE THREE TYPES OF EPITHELIAL MEMBRANES.

EXERCISE 4-10

Write the appropriate term in each blank from the list below. Not all terms will be used.

mesothelium   serous membrane   cutaneous membrane   parietal layer

visceral layer   mucous membrane   peritoneum   serous pericardium

1. An epithelial membrane that lines a body cavity or covers an internal organ   _____

2. The epithelial membrane also known as the skin   _____

3. A membrane that lines a space open to the outside of the body   _____

4. The portion of a serous membrane attached to an organ   _____

5. The portion of a serous membrane attached to the body wall   _____

6. The epithelial portion of serous membranes   _____

7. The serous membrane covering the heart   _____

6. LIST SIX TYPES OF CONNECTIVE TISSUE MEMBRANES.

EXERCISE 4-11

Write the appropriate term in each blank from the list below. Not all terms will be used.

periosteum	fibrous pericardium	perichondrium	peritoneum
synovial membrane	superficial fascia	deep fascia	pleura

1. The sheet of tissue that underlies the skin   _____

2. The connective tissue membrane that lines joint cavities   _____

3. A tough membrane composed entirely of connective tissue
    that serves to anchor and support an organ or to cover a muscle   _____

4. A layer of fibrous connective tissue around a bone   _____

5. The membrane that covers cartilage   _____

6. The serous membrane covering a lung   _____

7. The serous membrane covering the abdominal organs   _____

7. EXPLAIN THE DIFFERENCE BETWEEN BENIGN AND MALIGNANT TUMORS, AND GIVE SEVERAL EXAMPLES OF EACH TYPE.

EXERCISE 4-12

Write the appropriate term in each blank from the list below. Not all terms will be used.

lymphoma	leukemia	adenoma	sarcoma
osteoma	papilloma	angioma	chondroma
nevus	malignant	benign	meningioma

1. Term for a tumor that does not spread   _____

2. Term for a tumor that spreads to other tissues   _____

3. The general term for a tumor composed of blood or lymphatic
    vessels   _____

4. A benign connective tissue tumor that originates in a bone   _____

5. A benign epithelial tumor that originates in a gland   _____

6. A malignant tumor originating in connective tissue   _____

7. A tumor that originates in cartilage   _____

8. A small skin tumor that may become malignant   _____

9. A wart is an example of this type of benign tumor   _____

10. A malignant neoplasm of lymphatic tissue   _____

11. A cancer of white blood cells   _____

8. IDENTIFY THE MOST COMMON METHODS OF DIAGNOSING AND TREATING CANCER.

EXERCISE 4-13

In the blanks below, list some symptoms and signs of cancer.

1. _____

2. _____

3. _____

4. _____

5. _____

6. _____

EXERCISE 4-14

Write the appropriate term in each blank from the list below.

biopsy   biological therapy   ultrasound

positron emission tomography   tumor markers   computed tomography

magnetic resonance imaging   radiography

1. Mammography is an example of this technique   _____

2. The use of magnetic fields and radio waves to diagnose soft
   tissue cancers   _____

3. Technique using high-frequency sound waves   _____

4. The removal of living tissue for microscopic examination   _____

5. Technique that diagnoses tumors based on their abnormally
   high activity levels   _____

6. Use of many x-rays to produce a cross-sectional image of
   body parts   _____

7. Some tests screen for these substances, which are produced
   in high quantities by malignant cells   _____

8. The use of a vaccine to stimulate an immune response against
   malignant cells   _____

EXERCISE 4-15

List the three traditional cancer treatment methods in the blanks below. Briefly describe why each treatment works.

1. _____

   _____

   _____

2. _____

   _____

   _____

3. _____

   _____

   _____

9. USING THE CASE STUDY AND INFORMATION IN THE TEXT, DESCRIBE THE WARNING SIGNS OF CANCER.

EXERCISE 4-16

Use the terms below to complete the paragraph. Not all terms will be used.

chemotherapy	lump	abnormal bleeding	DNA
RNA	ultraviolet	x-rays	stratified
simple	squamous	cuboidal	carcinoma
epithelium	sarcoma	cutaneous	serous
dense irregular	dense regular	collagen	keratin

In Paul's case, repeated exposure to the (1) _____ rays in sunlight caused damage to the genetic material, known as (2) _____. This damage resulted in uncontrolled growth of cells in the connective tissue membrane that covers the body, known as the (3) _____ membrane. This membrane consists of an outer layer of (4) _____, in which the cells are tightly packed together. The cells are flat, so they are described as (5) _____ and organized in multiple layers, so the tissue is described as (6) _____. The inner layer of the membrane consists of (7) _____ _____ connective tissue. This connective tissue contains bundles of a protein called (8) _____ in a seemingly random arrange-ment. In Paul's case, the only warning sign for his cancer was a (9) _____ on his nose. The dermatologist found a malignant tumor of epithelial cells, known as a(n) (10) _____. The tumor was successfully excised surgically, and he did not require additional drug treatment, known as (11) _____, or the administration of (12) _____, known as radiation.

10. SHOW HOW WORD PARTS ARE USED TO BUILD WORDS RELATED TO TISSUES, GLANDS, AND MEMBRANES.

EXERCISE 4-17

Complete the following table by writing the correct word part or meaning in the space provided. Write a word that contains each word part in the "Example" column.

Word Part	Meaning	Example
1. _____	cartilage	_____
2. _____	inflammation	_____
3. onc/o	_____	_____
4. _____	tumor, swelling	_____
5. oss-, osse/o	_____	_____
6. blast/o	_____	_____
7. peri-	_____	_____
8. _____	muscle	_____
9. _____	false	_____
10. hist/o	_____	_____
11. neo-	_____	_____
12. _____	vessel	_____
13. _____	side, rib	_____
14. leuk/o	_____	_____
15. neur/o	_____	_____

Making the Connections

The following concept map deals with the classification of tissues. Complete the concept map by filling in the appropriate word or phrase that classifies or describes the tissue.

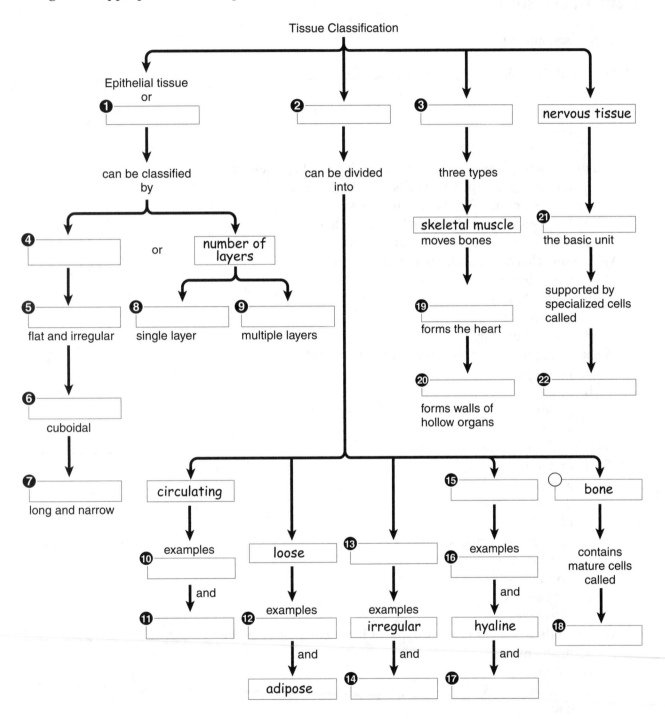

Optional Exercise: Make your own concept map summarizing the classification of membranes. Use the following terms and any others you would like to include: epithelial tissue membrane, connective tissue membrane, serous, mucous, cutaneous, peritoneum, fibrous pericardium, serous pericardium, parietal layer, visceral layer, and fascia.

Testing Your Knowledge

BUILDING UNDERSTANDING

I. MULTIPLE CHOICE

Select the best answer.

1. Which of the following is a type of connective tissue?
    a. transitional
    b. squamous
    c. cuboidal
    d. areolar

1. _____

2. Which of these phrases describes simple cuboidal epithelium?
    a. flat, irregular, epithelial cells in a single layer
    b. square epithelial cells in a single layer
    c. long, narrow, epithelial cells in a single layer
    d. square epithelial cells in many layers

2. _____

3. Which muscle type is under voluntary control?
    a. smooth muscle
    b. skeletal muscle
    c. cardiac muscle
    d. visceral muscle

3. _____

4. What is the proper scientific name for a nerve cell?
    a. neuroglia
    b. nevus
    c. neuron
    d. axon

4. _____

5. Which cell type secretes mucus?
    a. endocrine
    b. goblet
    c. areolar
    d. fibroblast

5. _____

6. Which cell type produces cartilage?
    a. chondrocytes
    b. fibroblasts
    c. osteoblasts
    d. osteocytes

6. _____

7. Which of these tissues is an example of loose connective tissue?
    a. ligament
    b. adipose
    c. tendon
    d. bone

7. _____

8. What is the tough connective tissue membrane that covers most parts of all bones?
    a. perichondrium
    b. periosteum
    c. fascia
    d. ligament

8. _____

9. Which of these tumors is composed primarily of vessels?   9. _____
   a. myoma
   b. angioma
   c. chondroma
   d. glioma
10. Which of these phrases describes the membrane surrounding skeletal
    muscles?   10. _____
    a. deep fascia
    b. fibrous pericardium
    c. superficial fascia
    d. perichondrium

II. COMPLETION EXERCISE

Write the word or phrase that correctly completes each sentence.

1. The technique that uses high-frequency sound waves to visualize tumors is
   called _____.

2. Cells that form bone are called _____.

3. The cartilage found at the ends of long bones is called _____.

4. The connective tissue membrane lining joint cavities is called the _____.

5. The study of tissues is called _____.

6. The epithelial membrane that lines the walls of the abdominal cavity is called
   the _____.

7. A mucous membrane can also be called the _____.

8. A single layer of flat, irregular epithelial cells is called _____.

9. The microscopic, hairlike projections found in the cells lining most of the respiratory tract
   are called _____.

10. The process of tumor cell spread is called _____.

11. A malignant tumor that originates in connective tissue is called a(n) _____.

12. The extent of tumor spread is determined by a process called _____.

UNDERSTANDING CONCEPTS

I. TRUE/FALSE

For each statement, write T for true or F for false in the blank to the left of each number. If a statement is false, correct it by replacing the underlined term, and write the correct statement in the blanks below the statement.

_____ 1. The peritoneum is an example of a <u>mucous membrane</u>.

_____

_____ 2. Stimulating one's own immune system to fight cancer is called <u>chemotherapy</u>.

_____

_____ 3. Tumors that undergo metastasis are called <u>benign</u> tumors.

_____

_____ 4. The strip of tissue connecting the kneecap to the thigh muscle is an example of a <u>tendon</u>.

_____

_____ 5. The periosteum is a <u>connective tissue membrane surrounding bone</u>.

_____

_____ 6. The pituitary gland releases prolactin into the bloodstream. The pituitary gland is thus an <u>exocrine</u> gland.

_____

_____ 7. The <u>parietal</u> layer of the peritoneum is in contact with the stomach.

_____

_____ 8. Ligaments are classified as <u>irregular loose connective tissue</u>.

_____

II. PRACTICAL APPLICATIONS

Study each discussion. Then write the appropriate word or phrase in the space provided.

➤ Group A: Cancer

You are spending some time in an oncology clinic as a volunteer. The nurse has asked you to perform an initial interview of a patient.

1. Ms. M was your first patient. You noted that she had a large wart on her chin. The scientific name for a wart is _____.

2. Ms. M complained of headaches and blurred vision. You suspect a tumor of the neurons or of the cells that that support the neurons. The support cells are called _____.

3. Ms. M was subjected to a test that uses magnetic fields to show changes in soft tissues. This test is called _____.

4. The test confirmed your diagnosis, showing a small tumor just beneath the membranes that surround the brain. These membranes are called the _____.

5. Since they lack epithelial tissue, these membranes are classified as _____.

6. The tumor appeared to be encapsulated, and other studies did not reveal any secondary tumors. Tumors that are confined to a single area are called _____.

7. The oncologist did not see the need for any further treatment. Ms. M was relieved because she had always feared chemotherapy. Chemotherapy, in the context of cancer, involves treatment with a particular class of drugs called _____.

➤ Group B: Tissues and Membranes

You are working as a sports therapist for a wrestling team. At a particularly brutal competition, you are asked to evaluate a number of injuries.

1. Mr. K suffered a crushing injury to the lower leg yesterday in a sumo wrestling match when his opponent fell on him. Initially, he had little pain. Now, he complains of numbness and pain in the foot and leg. This type of injury is made worse by the tight, fibrous covering of the muscles, known as the _____.

2. Mr. K is also complaining of pain in the knee. You suspect an injury to the membrane that lines the joint cavity, a membrane called the _____.

3. You note that Mr. K has a significant amount of fat. Fat is contained in a type of connective tissue called _____.

4. Based on its consistency, this tissue is classified as _____.

5. Ms. J suffered a painful bump on her ankle. The swelling involved the superficial tissues and the fibrous covering of the bone, or the _____.

6. Mr. S was involved in a closely fought match when his opponent bent his ear back. Thankfully, the cartilage in his ear was able to spring back into shape. This kind of cartilage is called _____.

7. Later, Mr. S suffered a penetrating wound to his abdomen when his opponent accidentally threw him into the seating area. You fear that the wound may have penetrated the membrane that lines his abdomen, called the _____.

8. The wrestling coach comes over to talk to you during a break in the match. He has a question about his favorite shampoo. The advertisement stated that it contained collagen. He asks you which cells in the body synthesize collagen. These cells are called _____.

III. SHORT ESSAYS

1. What are antineoplastic agents, and what is the rationale for using them to treat cancer?

_____

_____

_____

_____

_____

2. Compare and contrast benign and malignant tumors. List at least one similarity and two differences.

_____

_____

_____

_____

_____

3. Name two methods of cancer treatment other than chemotherapy, radiation, and surgery, and briefly describe how they work.

_____

_____

_____

_____

_____

CONCEPTUAL THINKING

1. Why are bone and blood both considered to be connective tissue? Define connective tissue in your answer.

   _____

   _____

   _____

   _____

2. Which tissue, epithelial or connective, would be best suited to the following functions?
   a. cushioning the kidneys against a blow

   _____

   b. creating a virtually waterproof barrier between the body and the environment

   _____

   c. preventing toxins from entering the blood from the gastrointestinal tract

   _____

3. A number of the components of a joint are listed below. Name the tissue type and/or membrane for each. Provide as many descriptive terms as you can. For instance, the skin can be described as both the cutaneous membrane and an epithelial membrane.
   a. the ends of the bones are covered with tough, translucent cartilage

   _____

   b. a fluid-secreting membrane secretes synovial fluid

   _____

   c. small fat pads cushion the joint

   _____

   d. tendons pass over the joint and contribute to the strength of the joint

   _____

   e. ligaments bind the bones together

   _____

Expanding Your Horizons

We are constantly reading headlines in the news about "New Treatments for Cancer!" Many of these claims are based on successful trials of drugs in mice that do not necessarily prove to be effective in humans. Investigate ongoing trials of different cancer treatments at the National Cancer Institute website (http://www.cancer.gov). The website of the National Center for Complementary and Alternative Medicine (http://nccam.nih.gov) discusses the efficacy of alternative treatments such as massage, acupuncture, and herbal remedies.

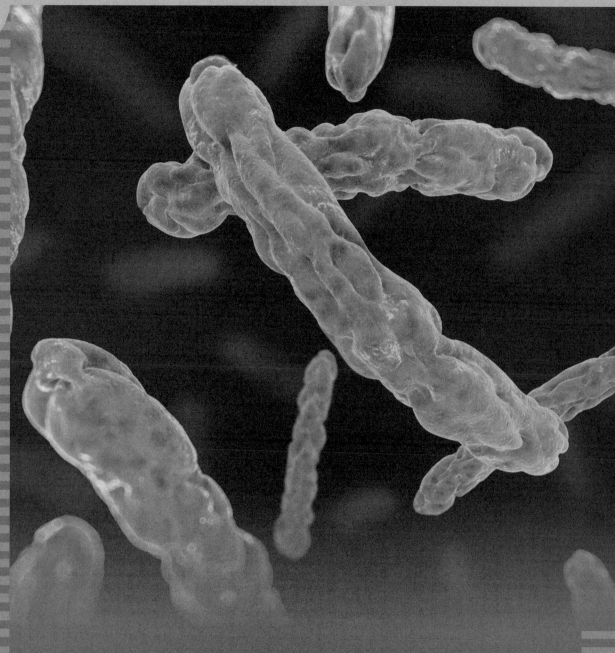

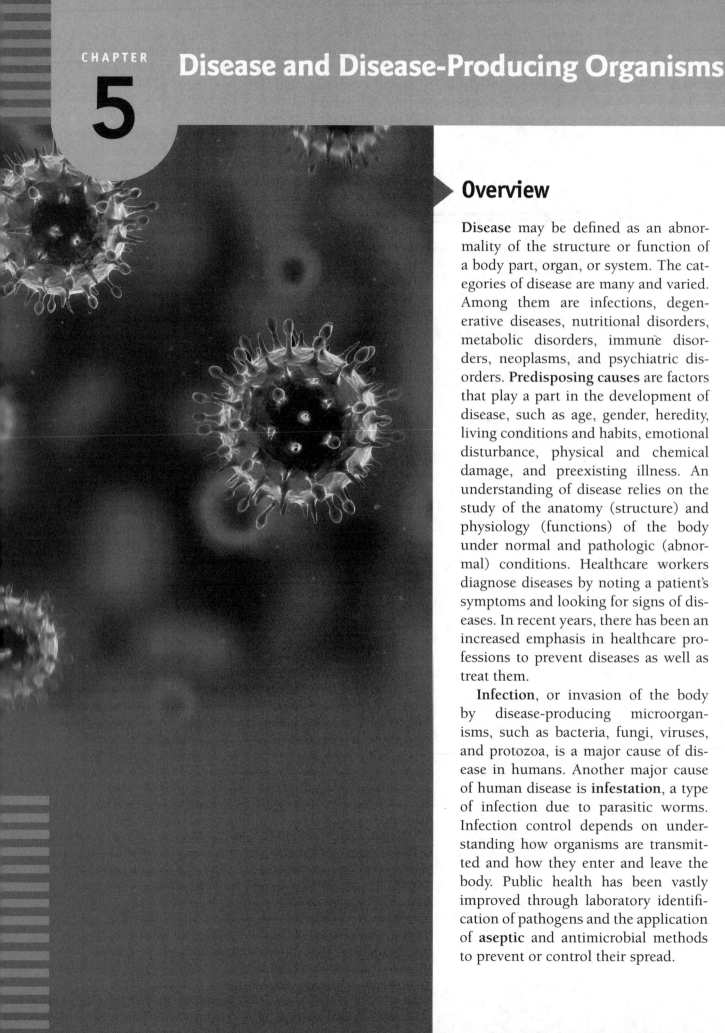

Disease and Disease-Producing Organisms

Overview

Disease may be defined as an abnormality of the structure or function of a body part, organ, or system. The categories of disease are many and varied. Among them are infections, degenerative diseases, nutritional disorders, metabolic disorders, immune disorders, neoplasms, and psychiatric disorders. **Predisposing causes** are factors that play a part in the development of disease, such as age, gender, heredity, living conditions and habits, emotional disturbance, physical and chemical damage, and preexisting illness. An understanding of disease relies on the study of the anatomy (structure) and physiology (functions) of the body under normal and pathologic (abnormal) conditions. Healthcare workers diagnose diseases by noting a patient's symptoms and looking for signs of diseases. In recent years, there has been an increased emphasis in healthcare professions to prevent diseases as well as treat them.

Infection, or invasion of the body by disease-producing microorganisms, such as bacteria, fungi, viruses, and protozoa, is a major cause of disease in humans. Another major cause of human disease is **infestation**, a type of infection due to parasitic worms. Infection control depends on understanding how organisms are transmitted and how they enter and leave the body. Public health has been vastly improved through laboratory identification of pathogens and the application of **aseptic** and antimicrobial methods to prevent or control their spread.

This chapter does not contain any difficult concepts, but it is rich in detail. You may find it useful to construct a chart that summarizes the physical characteristics, diseases, and treatment methods for the different microorganisms and multicellular parasites. The concept maps will also help you to master the content.

Addressing the Learning Objectives

1. DEFINE *DISEASE*, AND LIST SEVEN CATEGORIES OF DISEASE.

EXERCISE 5-1

Write the appropriate term in each blank from the list below.

neoplasm   infectious disease ✓   psychiatric disorder ✓   metabolic disorder ✓

nutritional disorder ✓   immune disorder   degenerative disorder ✓

1. A disease that can be transmitted between individuals   *Infectious disease*

2. A disorder involving tissue breakdown, such as that associated with aging   *degenerative disor*

3. A cancer or other tumor   *Neoplasm*

4. Disorders involving a disruption in the chemical reactions of the body   *metabolic disorder*

5. Rickets is an example of this type of disorder (vit D deficiency in children)   *nutritional disorder*
   Inperfect Calcifel

6. Multiple sclerosis is an example of this type of disorder   *immune disorder (autoimmune*
   (potential disability of brain and spinal cord)

7. Another term for mental disorder   *psychiatric disorder*

2. EXPLAIN THE MEANING OF A *PREDISPOSING* CAUSE OF DISEASE, AND LIST SEVEN SUCH CAUSES.

EXERCISE 5-2

List seven predisposing causes of disease in the spaces below.

1. *Age*
2. *gender*
3. *life style*
4. *previous condit'*
5. *heredity*
6. *environment*
7. *emotional disturbance*

3. DESCRIBE THE SCIENCE OF EPIDEMIOLOGY, AND CITE SOME TYPES OF STUDIES DONE IN THAT FIELD.

See Exercises 5-3 and 5-4.

4. DEFINE TERMS USED IN THE DIAGNOSIS AND TREATMENT OF DISEASE.

EXERCISE 5-3

Write the appropriate term in each blank from the list below. Not all terms will be used.

chronic   acute ✓   endemic ✓   idiopathic   epidemiology

iatrogenic ✓   communicable   pandemic   epidemic ✓

1. Term for a disease that is the result of adverse effects of medical treatment _____ Iatrogenic

2. Describing a relatively severe disorder of short duration _____ Acute

3. A disease that is found continuously, but to a low extent in a population _____ endemic

4. Term for a disease that persists over a long period _____ Chronic

5. A disease that can be transmitted between individuals _____ epidemic

6. A disease prevalent throughout an entire country or continent _____ pandemic

7. A disease of unknown cause _____ Idiopathic

8. The study of the geographic distribution of a disease _____ epidemiology

EXERCISE 5-4

Write the appropriate term in each blank from the list below. Not all terms will be used.

therapy   syndrome ✓   etiology ✓   symptom ✓   prevalence rate ✓

sign ✓   prognosis ✓   diagnosis ✓   incidence rate

1. A group of signs or symptoms that occur together _____ Syndrome

2. The study of the cause of a disorder _____ etiology

3. A condition of disease that is experienced by the patient _____ Symptom

4. The process of determining the nature of an illness _____ diagnostic

5. Evidence of disease noted by a healthcare worker _____ Sign

6. A prediction of the probable outcome of a disease _____ prognostic

7. A course of treatment _____ therapy

8. The overall frequency of a disease in a given group _____ prevalence

5. DEFINE *COMPLEMENTARY AND ALTERNATIVE MEDICINE*; CITE FOUR ALTERNATIVE OR COMPLEMENTARY FIELDS OF PRACTICE.

EXERCISE 5-5

Label each of the following statements as true (T) or false (F).

1. Complementary and alternative medicine refers to practices that cannot be combined with traditional modern medicine practices. _____

2. Naturopathy uses manipulation to correct misalignment for musculoskeletal disorders. _____

3. Yoga and meditation are examples of complementary medical practices. _____

4. Complementary and alternative medical practices are evaluated by the same institute that evaluates traditional modern medical practices, the NIH. _____

5. Echinacea and other botanicals are classified as drugs and are tightly regulated by the U.S. Food and Drug Administration (FDA). _____

6. EXPLAIN METHODS BY WHICH MICROORGANISMS CAN BE TRANSMITTED FROM ONE HOST TO ANOTHER.

EXERCISE 5-6

Write the appropriate term in each blank from the list below. Not all terms will be used.

systemic	local	opportunistic	pathogen	communicable
parasite	host	vector	portal of entry	

1. Term for an infection that affects the whole body _____

2. Term for an infection that takes hold in a weakened host _____

3. Any organism that lives on or within another organism at that organism's expense _____

4. The avenue by which a microorganism invades the body _____

5. An organism that causes disease _____

6. An animal that transmits a disease-causing organism from one host to another _____

7. NAME FOUR TYPES OF MICROORGANISMS, AND GIVE THE CHARACTERISTICS OF EACH.

EXERCISE 5-7

Fill in the blank after each statement—does it apply to bacteria (B), viruses (V), protozoa (P), or fungi (F)?

1. Single-cell organisms with a plasma membrane but without a true nucleus
   _____

2. Group of microbes that includes the ciliates _____

3. Obligate intracellular parasites unaffected by antibiotics _____

4. Microbes composed of nucleic acid and a protein coat _____

5. Single-cell forms are called yeasts _____

6. Filamentous forms are called molds _____

7. Single-cell, animal-like microbes _____

8. Some of these microbes produce endospores _____

EXERCISE 5-8: Bacteria (Text Figs. 5-6, 5-7, and 5-8)

1. Write the name of each type of bacteria on the appropriate numbered line in a particular color.
2. Color the bacteria on the diagram with the corresponding colors.

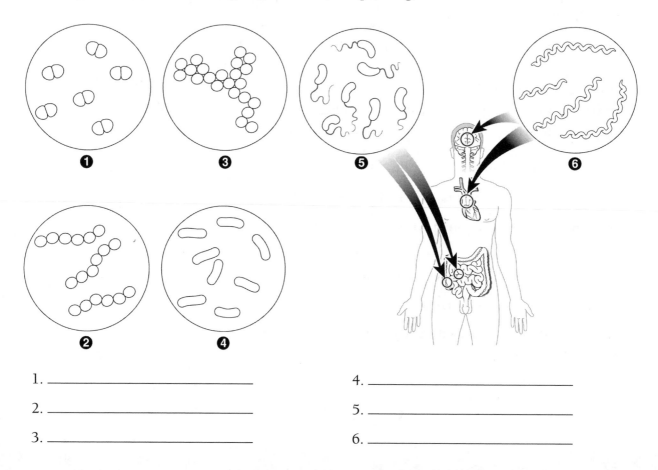

1. _____   4. _____

2. _____   5. _____

3. _____   6. _____

EXERCISE 5-9

Write the appropriate term in each blank from the list below. Not all terms will be used.

diplococci   staphylococci   streptococci   vibrios   chlamydiae

bacilli   spirilla   spirochetes   rickettsiae

1. Spherical bacteria organized in clusters   _____

2. Spherical bacteria arranged in chains   _____

3. Long, wavelike cells resembling corkscrews but not capable of waving motions   _____

4. Bacteria shaped like straight rods that may form endospores   _____

5. Short rods with a small curvature   _____

6. A group of bacteria smaller than the rickettsiae   _____

7. Spherical bacteria occurring in pairs   _____

8. Long wavelike cells capable of waving and twisting motions   _____

EXERCISE 5-10: Pathogenic Protozoa (Text Fig. 5-12)

1. Write the name of each type of protozoon on the appropriate numbered lines 1 to 3 in a particular color. Use light colors, and do not use red.
2. Color the protozoa on the diagram with the corresponding colors.
3. Use black to write the names of the different forms of protozoa in lines 4 to 7.
4. Color the red blood cells red.

1. _____

2. _____

3. _____

4. _____

5. _____

6. _____

7. _____

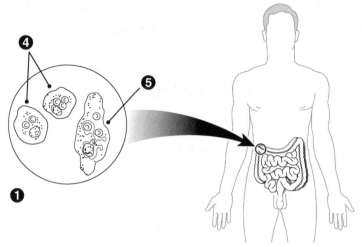

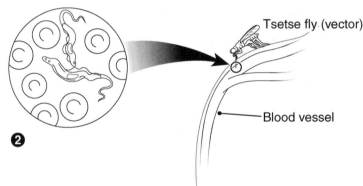

Tsetse fly (vector)

Blood vessel

Anopheles mosquito (vector)

Blood vessel

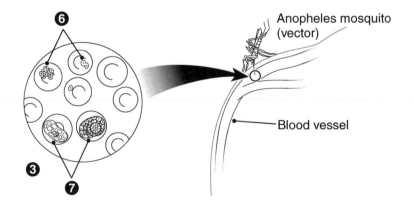

8. DEFINE *NORMAL FLORA,* AND EXPLAIN THE VALUE OF IT.

EXERCISE 5-11

Address this objective by writing an answer in the lines below. Use your own words; avoid copying from the textbook.

_____

_____

_____

_____

_____

_____

9. LIST SOME DISEASES CAUSED BY EACH TYPE OF MICROORGANISM.

EXERCISE 5-12

For each of the following diseases, state if they are bacterial (B), viral (V), protozoan (P), or fungal (F).

1. African sleeping sickness   _____

2. AIDS   _____

3. *Pneumocystis* pneumonia (PCP)   _____

4. Malaria   _____

5. Thrush   _____

6. Rocky Mountain spotted fever   _____

7. Diphtheria   _____

8. Influenza   _____

9. Chickenpox   _____

10. Cholera   _____

11. Tinea   _____

10. NAME TWO CATEGORIES OF PARASITIC WORMS, AND GIVE EXAMPLES OF EACH.

EXERCISE 5-13

Write the appropriate term in each blank from the list below.

infestation   pinworm   ascaris   hookworm

filaria   fluke   vermifuge

1. *Enterobius vermicularis*, a small roundworm that infests the
   large intestine   _____

2. The most common form of roundworm that infests
   the intestines or lungs   _____

3. A general term describing an invasion by a parasitic worm   _____

4. An anthelmintic agent   _____

5. The worm that can cause elephantiasis   _____

6. A leaf-shaped worm that can invade the blood, lungs, or
   intestine   _____

7. A small worm that can cause anemia   _____

11. EXPLAIN THE ROLE OF PARASITIC ARTHROPODS IN CAUSING DISEASE.

EXERCISE 5-14

Complete the following table by placing the characteristics below in the column pertaining to either head lice or scabies.

a parasitic animal that exclusively colonizes the scalp
lays eggs on hair
lays eggs in burrows within the skin
a member of the spider family
an insect
the host's immune response against its saliva causes itching
the host's immune response against its feces and eggs causes itching

Head Lice (pediculosis capitis)	Scabies

12. GIVE FOUR REASONS FOR THE EMERGENCE AND SPREAD OF MICROORGANISMS TODAY.

EXERCISE 5-15

List four reasons why infectious diseases are on the rise.

1. _____
_____

2. _____
_____

3. _____
_____

4. _____
_____

13. DESCRIBE FOUR PUBLIC HEALTH MEASURES TAKEN TO PREVENT THE SPREAD OF DISEASE.

EXERCISE 5-16

List four common public measures that improve human health.

1. _____
_____

2. _____
_____

3. _____
_____

4. _____
_____

14. DIFFERENTIATE BETWEEN *STERILIZATION* AND *DISINFECTION*, AND GIVE THREE EXAMPLES OF EACH.

EXERCISE 5-17

Fill in the blank after each situation—does it apply to sterilization (S), disinfection (D), or antiseptic (A)?

1. A father applies iodine solution to a nasty scrape on his child's knee. _____

2. A scrub nurse uses ethylene oxide gas to treat delicate instruments. _____

3. A woman uses hydrogen peroxide to clean earrings before inserting them. _____

4. A student applies bleach solution to the countertop after completing his experiment. _____

5. A lab technician treats an agar solution in an autoclave before using it to prepare agar plates. _____

15. DESCRIBE FOUR MEASURES INCLUDED AS PART OF BODY SUBSTANCE PRECAUTIONS.

EXERCISE 5-18

Identify which techniques are appropriate (G: good) and which might contribute to the spread of disease or are otherwise harmful (B: bad).

1. Changing gloves every two or three patients to avoid excess garbage _____

2. Recapping needles to avoid accidental injection _____

3. Using barriers (such as a mask) for every patient, not just potentially infective patients _____

4. Washing hands after gloves are removed _____

5. Using bleach to cleanse hands _____

6. Using hydrogen peroxide to cleanse hands _____

16. NAME TWO TYPES OF ANTIMICROBIAL AGENTS, AND GIVE THREE EXAMPLES OF EACH.

EXERCISE 5-19

Use the terms below to complete the paragraph.

protein coat   antibiotic   protease inhibitor

nosocomial   nucleic acid

Bacterial infections are commonly treated with a(n) (1) _____, but overuse and abuse of these agents have led to an increase in drug-resistant (2) _____ infections in hospitals. Effective treatments for viral infections are sparse. A (3) _____ such as indinavir blocks a specific enzyme that viruses need to spread to other cells. Other drugs block the removal of the (4) _____ of the virus or inhibit the production of the viral (5) _____ by the host cell.

17. DESCRIBE THREE METHODS USED TO IDENTIFY MICROORGANISMS IN THE LABORATORY.

EXERCISE 5-20

Use the terms below to complete the paragraph. Not all terms will be used.

PCR   Gram stain   acid-fast stain   red

blue   bluish-purple   Gram-negative   Gram-positive

The most common staining procedure used to identify bacteria is the (1) _____. Bacteria that appear bluish purple after staining are described as (2) _____. Bacteria that do not retain the stain are described as (3) _____. The acid-fast staining procedure uses a different dye. Acid-fast bacteria that retain the dye become (4) _____ in color. Bacteria that are not acid-fast do not retain the dye; they are usually visualized by counterstaining with another dye to make them (5) _____ in color. Viruses cannot be identified by microscopy; instead, genetic techniques such as (6) _____ must be used.

18. USING THE CASE STUDY, DESCRIBE A VIRUS'S MECHANISM OF INFECTION AND THE HOST'S RESPONSE.

EXERCISE 5-21

Place the following events in order by writing the appropriate numbers in the blanks. The first one has been done for you.

_____ a. Epithelial cells die.

_____ b. Endocytosis brings the virus into the epithelial cell.

_____ c. Viral RNA is transcribed into mRNA.

_____ d. New viruses are exported from the epithelial cell.

___1___ e. Droplets carry influenza viruses into Maria's throat.

_____ f. The viral RNA travels to the cell nucleus.

_____ g. Ribosomes translate viral mRNA into viral proteins.

_____ h. Proteins of the virus coat attach to proteins on epithelial cell membranes.

_____ i. The virus sheds its protein coat.

_____ j. New viral RNA and proteins combine to make new viruses

19. SHOW HOW WORD PARTS ARE USED TO BUILD WORDS RELATED TO DISEASE.

EXERCISE 5-22

Complete the following table by writing the correct word part or meaning in the space provided. Write a word that contains each word part in the "Example" column.

Word Part	Meaning	Example
1. pan-	_____	_____
2. py/o	_____	_____
3. _____	self, separate, distinct	_____
4. aer/o	_____	_____
5. _____	fungus	_____
6. _____	together	_____
7. _____	chain	_____
8. an-	_____	_____
9. _____	grapelike cluster	_____
10. iatro	_____	_____

Making the Connections

The following concept map deals with disease transmission. Complete the concept map by filling in the blanks with the appropriate word or phrase.

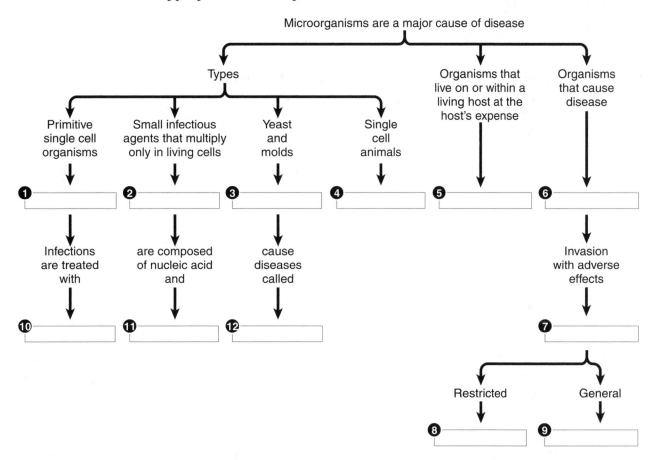

Optional Exercise: Make your own concept map of terms relating to multicellular parasites using the following terms and any others you would like to include: worms, infestation, vermifuge, flatworms, roundworms, flukes, tapeworms, ascaris, hookworms, pinworms, helminthology, trichina, arthropods, lice, mites, and scabies.

Testing Your Knowledge

BUILDING UNDERSTANDING

I. MULTIPLE CHOICE

Select the best answer.

1. How would a Gram-positive coccus appear after Gram staining?   1. _____
    a. red rod
    b. purple rod
    c. red sphere
    d. purple sphere
2. Which term describes an infection picked up in the hospital?   2. _____
    a. aseptic
    b. nosocomial
    c. local
    d. systemic
3. What is the study of fungi called?   3. _____
    a. mycology
    b. virology
    c. helminthology
    d. protozoology
4. Which term describes the short, curved rods responsible for cholera?   4. _____
    a. streptococci
    b. spirochetes
    c. vibrios
    d. staphylococci
5. Which type of pathogen would NOT be affected by antibacterial agents?   5. _____
    a. viruses
    b. spirochetes
    c. diplococci
    d. rickettsiae
6. Which type of infectious agent causes athlete's foot, tinea capitis, thrush, and *Candida* infections?   6. _____
    a. viruses
    b. fungi
    c. rickettsiae
    d. filariae
7. Which of the following is a nutritional disorder?   7. _____
    a. osteoporosis
    b. multiple sclerosis
    c. filariasis
    d. scurvy

8. Which type of infectious agent are pinworms, hookworms,
   trichina, and filaria? 8. _____
   a. flatworms
   b. flagellates
   c. flukes
   d. roundworms
9. Which type of infectious agent consists solely of a protein? 9. _____
   a. viruses
   b. bacteria
   c. rickettsiae
   d. prions

II. COMPLETION EXERCISE

Write the word or phrase that correctly completes each sentence.

1. Short hairlike structures that help bacteria glide along solid surfaces and help them anchor to solid surfaces are _____.

2. An insect or other animal that transmits a disease-causing organism from one host to another is a(n) _____.

3. An obligate intracellular parasite composed of nucleic acid and a protein coat is a(n) _____.

4. Legionnaire disease, tuberculosis, and tetanus (lockjaw) are caused by rod-shaped microorganisms known as _____.

5. One type of flatworm, composed of many segments (proglottids), may grow to a length of 50 ft—hence the name _____.

6. The science of studying diseases in populations is called _____.

7. An alternative medical practice based on the philosophy of helping people develop healthy lifestyles is called _____.

8. The group of protozoa that consists of all obligate parasites is _____.

9. Bacteria that are pathogenic may cause injury and even death by the action of poisons referred to as _____.

10. The number of new disease cases appearing in a particular population during a specific time period is known as the _____.

11. A device that sterilizes equipment using steam under pressure is a(n) _____.

12. Tetanus is caused by an organism that exists in two forms. One of these is the growing vegetative form. The other is a resting and resistant form, the _____.

13. The fuzzy, filamentous, multicellular forms of fungi are called _____.

14. Infections without a known cause are described as _____.

15. The large class of multicellular animals that includes insects and spiders is the _____.

UNDERSTANDING CONCEPTS

I. TRUE/FALSE

For each statement, write T for true or F for false in the blank to the left of each number. If a statement is false, correct it by replacing the underlined term, and write the correct statement in the blank below the statement.

_____ 1. Ms. J has severe anemia as a result of chemotherapy. Her anemia can be termed an <u>idiopathic</u> disease.

_____

_____ 2. A large number of influenza infections were reported in cold regions of the United States last winter. During this period, influenza was <u>pandemic</u>.

_____

_____ 3. An <u>aerobic</u> organism requires oxygen for growth.

_____

_____ 4. Protozoa covered with tiny hairs are called <u>ciliates</u>.

_____

_____ 5. <u>Gram-negative</u> organisms appear bluish-purple under the microscope after staining with the Gram stain.

_____

_____ 6. <u>Disinfection</u> kills every living organism on or in an object.

_____

II. PRACTICAL APPLICATIONS

Ms. L, age 10, has been brought to the clinic by her father. As a nursing student, you are responsible for her primary evaluation. Write the appropriate word or phrase in each space provided.

1. First, you evaluate Ms. L's signs and symptoms. Ms. L has been complaining of fatigue. Is fatigue a sign or a symptom? _____

2. You take a blood sample, and determine that Ms. L has an abnormally low number of red blood cells (anemia). Is red blood cell number a sign or a symptom? _____

3. You take a stool sample and observe large numbers of round worms in her feces. The presence of worms or other multicellular parasites in or on the body is called a(n) _____.

4. Based on the worm's round appearance and the fact that Ms. L has anemia, you suspect that the worms responsible for Ms. L's condition are called _____.

5. The likely portal of entry for the organisms causing her illness is the _____.

6. Ms. L will require a medication to kill the worms; the general term for this type of drug is _____

7. Ms. L's worm problem has rendered her more susceptible to other pathogens, and she has a *Candida* infection in her mouth. An infection that only affects weakened hosts is termed _____.

8. The *Candida* infection is caused by the group of microorganisms called _____.

9. Ms. L's school was notified, and a letter was sent home to the parents. The letter also reminded parents to periodically check their child's head for pediculosis capitis, commonly known as _____.

10. The other students were also examined. Out of 100 students, only five students were afflicted by the same disorder. Epidemiologists would describe this statistic as the _____.

III. SHORT ESSAYS

1. Define the term *predisposing cause* in relation to disease, and give several examples of predisposing causes.

_____

_____

_____

2. What is the science of pathophysiology, and why is it important in medicine?

_____

_____

_____

3. Describe some disadvantages of antibiotic therapy.

_____

_____

_____

4. Compare and contrast bacteria and protozoa.

_____

_____

_____

5. Why are opportunistic infections more prevalent now than in the past?

_____

_____

_____

6. Have immunizations and antibiotics successfully conquered infectious disease? Explain.

_____

_____

_____

CONCEPTUAL THINKING

1. Mr. S engaged in unsafe sexual practices. One week later, he experienced painful urination and an abnormal discharge from his penis. Microscopic analysis of the discharge revealed the presence of small round cells organized in pairs.
   a. Based on the microscopic findings, which type of microorganism is responsible for Mr. S's symptoms? Provide as much information as you can.

   _____

   _____

   b. What is the most probable diagnosis?

   _____

   c. What would be the most appropriate treatment?

   _____

2. Young Ms. Y has been suffering from nausea, a stuffy nose, coughing, and a fever for several days. She has been diagnosed with influenza, and her mother requests that she be placed on antibiotics. Will the doctor fulfill her request? Why or why not?

   _____

   _____

   _____

3. Write a sentence using the words *etiology* and *idiopathic*.

   _____

   _____

   _____

4. Based on your knowledge of word parts, what is a *Streptobacillus*? Give as much information as you can.

_____

_____

_____

5. Reread the case study introduction and summary, and answer the following questions.
   a. Does Maria have any predisposing causes?

   _____

   b. How was Maria's disease transmitted?

   _____

   c. How could Maria's infectious agent have been identified?

   _____

   d. What is Maria's diagnosis and prognosis?

   _____

Expanding Your Horizons

The SARS scare in spring 2003 and the H1N1 scare in 2010 were reminders of how vulnerable we are to infectious diseases. Thankfully, neither epidemic/pandemic was as severe as many had feared. Read the article below for a "worst-case scenario" of what might have happened.

- Wolfe N. Preventing the next pandemic. Sci Am 2009;300:76–81.

The Integumentary System

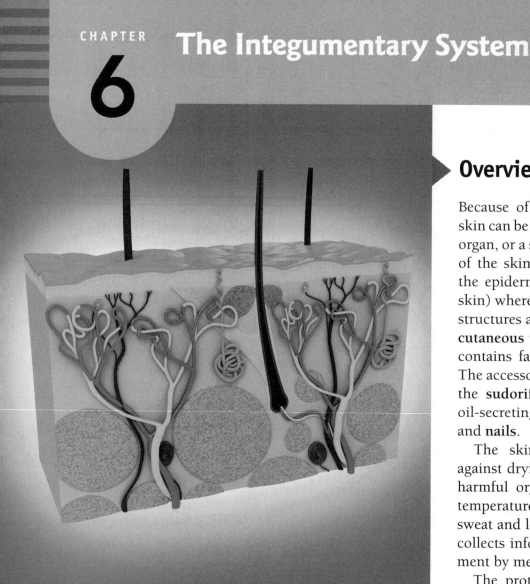

▶ Overview

Because of its various properties, the skin can be classified as a membrane, an organ, or a system. The outermost layer of the skin is the **epidermis**. Beneath the epidermis is the **dermis** (the true skin) where glands and other accessory structures are mainly located. The **subcutaneous tissue** underlies the skin. It contains fat that serves as insulation. The accessory structures of the skin are the **sudoriferous** (sweat) **glands**, the oil-secreting **sebaceous glands**, **hair**, and **nails**.

The skin protects deeper tissues against drying and against invasion by harmful organisms. It regulates body temperature through evaporation of sweat and loss of heat at the surface. It collects information from the environment by means of sensory receptors.

The protein **keratin** in the epidermis thickens and protects the skin and makes up hair and nails. **Melanin** is the main pigment that gives the skin its color. It functions to filter out harmful ultraviolet radiation from the sun. Skin color is also influenced by the quantity and oxygen content of blood circulating in the surface blood vessels.

Much can be learned about the condition of the skin by observing for discoloration, injury, or lesions. Aging, exposure to sunlight, and the health of other body systems also have a bearing on the skin's condition and appearance. The skin is subject to numerous diseases, common forms of which are atopic dermatitis (eczema), acne, psoriasis, infections, and cancer.

This chapter does not contain any particularly difficult material. However, you must be familiar with the different tissue types discussed in Chapter 4 in order to understand the structure and function of the skin and associated glands. Familiarity with the terminology used in describing and treating disease will help you learn about skin diseases.

Addressing the Learning Objectives

1. NAME AND DESCRIBE THE LAYERS OF THE SKIN.

EXERCISE 6-1

Write the appropriate term in each blank from the list below. Not all terms will be used.

melanocyte	integument	keratin
dermis	epidermis	stratum corneum
dermal papillae	stratum basale	subcutaneous layer

1. A pigment-producing cell that becomes more active in the presence of ultraviolet light _____

2. The protein in the epidermis that thickens and protects the skin _____

3. The deeper of the two major layers of skin _____

4. The uppermost epidermal layer, consisting of flat, keratin-filled cells _____

5. Another name for the skin as a whole _____

6. Projections of the dermis responsible for fingerprints _____

7. The deepest layer of the epidermis, which contains living, dividing cells _____

8. Another name for the superficial fascia _____

2. DESCRIBE THE SUBCUTANEOUS LAYER.

EXERCISE 6-2

Match the structures in the list below with their functions by writing the appropriate term in each blank.

adipose tissue   elastic fibers   blood vessels   nerves

1. Connect the subcutaneous tissue with the dermis _____

2. Insulates the body and acts as an energy reserve _____

3. Carry sensory information from the skin to the brain _____

4. Supply skin with nutrients and oxygen _____

3. GIVE THE LOCATIONS AND FUNCTIONS OF THE ACCESSORY STRUCTURES OF THE INTEGUMENTARY SYSTEM.

EXERCISE 6-3

Write the appropriate term in each blank from the list below. Not all terms will be used.

apocrine   eccrine   ceruminous   sudoriferous   wax

vernix caseosa   sebaceous   sebum   meibomian

1. A general term for any gland that produces sweat   _____

2. Sweat glands found throughout the skin that help cool the body   _____

3. Glands that are found only in the ear canal   _____

4. Excess activity of these glands contributes to acne vulgaris   _____

5. The product of ceruminous glands   _____

6. Sweat glands in the armpits and groin that become active at puberty   _____

7. Glands that lubricate the eye   _____

8. A secretion of sebaceous glands that covers newborn babies   _____

EXERCISE 6-4: The Skin (Text Fig. 6-1)

1. Write the names of the three skin layers in the numbered boxes 1 to 3.
2. Write the name of each labeled part on the numbered lines in different colors. Use a light color for structures 4 and 12. Use the same color for structures 15 and 16, for structures 13 and 14, and for structures 8 and 9.
3. Color the different structures on the diagram with the corresponding colors. Try to color every structure in the figure with the appropriate color. For instance, structure 8 is found in three locations.

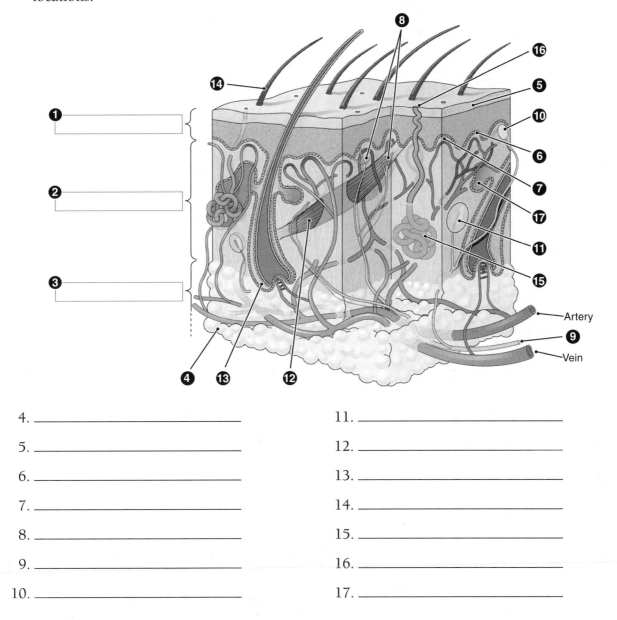

4. _____
5. _____
6. _____
7. _____
8. _____
9. _____
10. _____

11. _____
12. _____
13. _____
14. _____
15. _____
16. _____
17. _____

EXERCISE 6-5: The Nail (Text Fig. 6-5B)

1. In the middle column of the table, write the name of the labeled structure.
2. In the right-hand column of the table, write a brief description of the structure.

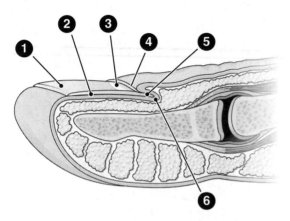

Number	Structure	Description
1.		
2.		
3.		
4.		
5.		
6.		

4. LIST THE MAIN FUNCTIONS OF THE INTEGUMENTARY SYSTEM.

EXERCISE 6-6

Fill in the numbered blanks in the table below.

Function	Structures Involved
(1)	Stratum corneum, shedding skin cells
Protection against dehydration	(2)
Regulation of body temperature	(3)
(4)	Nerve endings, specialized receptors

5. SUMMARIZE THE INFORMATION TO BE GAINED BY OBSERVATION OF THE SKIN.

EXERCISE 6-7

Write the appropriate term in each blank from the list below. Not all terms will be used.

carotenemia   albinism   cyanosis   jaundice   pallor   flushing

1. A condition in which the skin takes on a bluish coloration

   _____

2. A condition in which the skin takes on a yellowish discoloration due to excess consumption of deeply colored vegetables

   _____

3. Paleness of the skin

   _____

4. Redness of the skin, often related to fever

   _____

5. A condition in which the skin takes on a yellowish discoloration due to excess bile pigments

   _____

EXERCISE 6-8: Some Common Skin Lesions (Text Fig. 6-7)

Identify each of the types of skin lesions pictured in the figure.

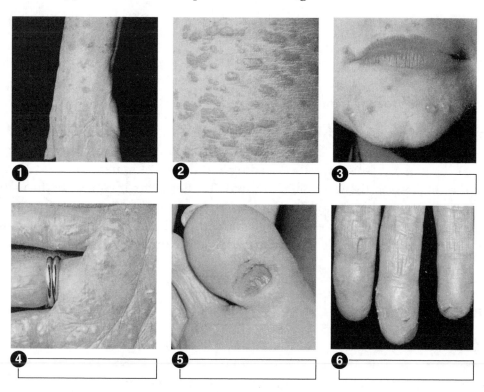

1. _____
2. _____
3. _____
4. _____
5. _____
6. _____

EXERCISE 6-9

Write the appropriate term in each blank from the list below. Not all terms will be used.

pustule   macule   vesicle   ulcer

fissure   excoriation   laceration

1. A spot that is neither raised nor depressed   _____

2. A scratch into the skin   _____

3. A crack in the skin   _____

4. A vesicle filled with pus   _____

5. A small sac that contains fluid; a blister   _____

6. A sore associated with disintegration and death of tissue   _____

6. CITE THE STEPS IN REPAIR OF SKIN WOUNDS AND THE FACTORS THAT AFFECT HEALING.

EXERCISE 6-10

Use the terms below to complete the paragraph.

cicatrix   blood clot   fibroblasts   stratum basale

capillaries   inflammatory response   collagen

The repair of any tissue requires actively dividing cells. In skin, these cells are found in the deepest layer of the epidermis, the (1) _____. After a skin wound, the first responses are the formation of a(n) (2) _____ to stanch the bleeding and the development of a(n) (3) _____ _____ that makes the tissue swollen and red. Tissue repair involves the formation of new (4) _____ to bring blood to the new tissue and the production of a protein called (5) _____ by connective tissue cells called (6) _____. Sometimes, the damage is too extensive for complete repair, so the tissue is replaced by connective tissue that forms a scar, or (7) _____.

7. DESCRIBE HOW THE SKIN CHANGES WITH AGE.

EXERCISE 6-11

List five age-related changes that occur in the integumentary system.

1. _____

2. _____

3. _____

4. _____

5. _____

8. LIST THE MAIN DISORDERS OF THE INTEGUMENTARY SYSTEM.

EXERCISE 6-12

Write the appropriate term in each blank from the list below. Not all terms will be used.

scleroderma   erythema   dermatosis   dermatitis

pruritus   atopic dermatitis   psoriasis   urticaria

acne vulgaris   impetigo

1. Any skin disease   _____

2. Redness of the skin   _____

3. Any inflammation of the skin   _____

4. Another term for itching   _____

5. A disease resulting from excess collagen production   _____

6. A skin disease resulting from infection with staphylococcal
   bacteria   _____

7. An allergic reaction resulting in the temporary appearance of
   raised red patches; also known as *hives*   _____

8. A disease associated with raised, red, flat areas resulting
   from overgrowth of the epidermis   _____

9. USING INFORMATION IN THE CASE STUDY AND THE TEXT, DESCRIBE THE CAUSES, CLASSIFICATION, AND DANGERS OF A BURN.

EXERCISE 6-13

Complete the following table by filling in the missing information. Note that the burn types are not listed in order of increasing damage.

Burn Classification	Skin Layers Involved	Description	Example
Deep partial thickness			
	Epidermis only		
	Full skin and potentially underlying tissues		Major flame exposure
Superficial partial thickness			Severe sunburn or mild scalding

10. SHOW HOW WORD PARTS ARE USED TO BUILD WORDS RELATED TO THE INTEGUMENTARY SYSTEM.

EXERCISE 6-14

Complete the following table by writing the correct word part or meaning in the space provided. Write a word that contains each word part in the "Example" column.

Word Part	Meaning	Example
1. cyan/o	_____	_____
2. _____	dark, black	_____
3. _____	hard	_____
4. eryth	_____	_____
5. dermat/o	_____	_____
6. _____	cornified, keratinized	_____
7. ap/o-	_____	_____
8. _____	state of	_____
9. _____	hair	_____
10. -emia	_____	_____

Making the Connections

The following concept map deals with the structural features of the skin. Complete the concept map by filling in the appropriate word or phrase in each box.

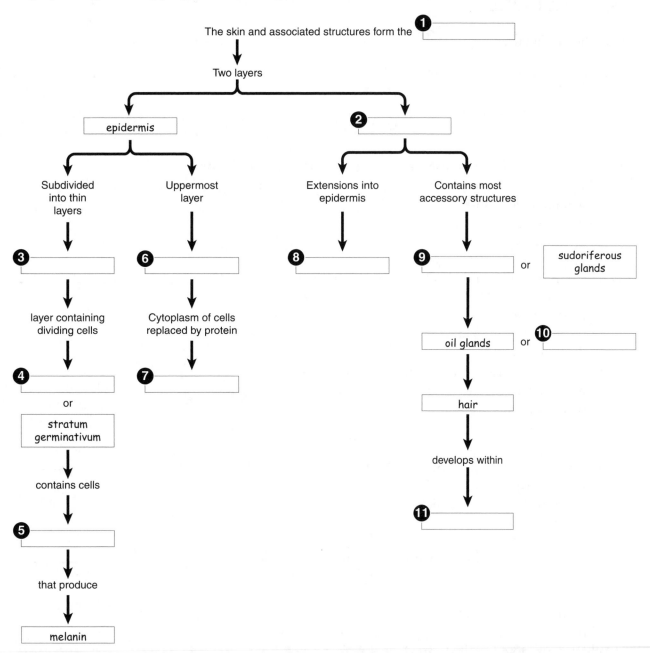

Optional Exercise: Make your own concept map based on the observation and pathology of skin. Use the following terms and any others you would like to include: surface lesion, macule, papule, nodule, vesicle, pustule, deeper lesion, excoriation, laceration, ulcer, and fissure.

Testing Your Knowledge

BUILDING UNDERSTANDING

I. MULTIPLE CHOICE

Select the best answer.

1. Where are new epidermal cells produced?
   a. dermis
   b. stratum corneum
   c. stratum basale
   d. subcutaneous layer

2. Which of the following diseases results from the accumulation of bile pigments in blood?
   a. atopic dermatitis
   b. impetigo
   c. alopecia
   d. jaundice

3. Which of the following glands is NOT a modified sweat gland?
   a. mammary gland
   b. sebaceous gland
   c. ceruminous gland
   d. ciliary gland

4. Which term describes a narrowing of a blood vessel?
   a. dilation
   b. constriction
   c. closure
   d. merger

5. Which of the following signs most commonly results from dietary causes?
   a. jaundice
   b. cyanosis
   c. liver spots
   d. carotenemia

6. What is the white half-moon at the base of the nail called?
   a. root
   b. matrix
   c. lunula
   d. cuticle

7. Which term describes bed sores?
   a. laceration
   b. decubitus ulcer
   c. psoriasis
   d. herpes

8. Which term describes a burn involving the epidermis and the entire dermis, such as a severe scald burn?
   a. full thickness
   b. superficial partial thickness
   c. superficial thickness
   d. deep partial thickness

1. _____

2. _____

3. _____

4. _____

5. _____

6. _____

7. _____

8. _____

9. What is the name of the muscle connected to hair follicles?   9. _____
   a. ciliary muscle
   b. arrector pili
   c. sebaceous muscle
   d. vernix
10. Which of the following skin disorders could be treated with
    antifungal agents?   10. _____
    a. ringworm
    b. acne vulgaris
    c. impetigo
    d. shingles

II. COMPLETION EXERCISE

Write the word or phrase that correctly completes each sentence.

1. The outer layer of the epidermis, which contains flat, keratin-filled cells, is called the

   _____.

2. Fingerprints are created by extensions of the dermis into the epidermis. These extensions are

   called _____.

3. The main pigment of the skin is _____.

4. The cells that secrete collagen to help close a wound are called _____.

5. Another name for a wart is a(n) _____.

6. The subcutaneous layer is also called the hypodermis or the _____.

7. Baldness is very common in elderly men. Absence of hair from any area where it is normally

   present is called _____.

8. A spot that is neither raised nor depressed, like a freckle, is known as a(n)

   _____.

9. The most malignant form of skin cancer is called _____.

10. Hair and nails are composed mainly of a protein named _____.

11. A crack in the skin, such as that observed with athlete's foot, is called a(n)

    _____.

12. Very painful burns involving the epidermis and part of the dermis are classified as

    _____.

13. Overactivity of the sebaceous glands during adolescence may play a part in the common skin

    disease called _____.

14. The body surface area involved in a burn may be estimated using the rule of

    _____.

UNDERSTANDING CONCEPTS

I. TRUE/FALSE

For each statement, write T for true or F for false in the blank to the left of each number. If a statement is false, correct it by replacing the underlined term, and write the correct statement in the blanks below the statement.

_____ 1. The nail cuticle, which seals the space between the nail plate and the skin above the nail root, is an extension of the <u>stratum basale</u>.

_____

_____ 2. In cold weather, the blood vessels in the skin <u>constrict</u> in order to conserve heat.

_____

_____ 3. The skin produces <u>vitamin A</u> under the influence of ultraviolet light.

_____

_____ 4. Sebum is produced by <u>sudoriferous glands</u>.

_____

_____ 5. The part of the hair below the skin surface is the <u>follicle</u>.

_____

_____ 6. The <u>stratum corneum</u> is the deepest layer of the epidermis.

_____

II. PRACTICAL APPLICATIONS

Study each discussion. Then write the appropriate word or phrase in the space provided.

➤ Group A

Mr. B has suffered a fall in a downhill mountain biking competition. Working as a first-aid volunteer, you are the first person arriving at the scene of the accident.

1. Mr. B has light scratches on his left cheek. The medical term for a scratch is

   _____.

2. A branch tore a long jagged wound in his right arm. This wound is called a(n)

   _____.

3. The skin of Mr. B's nose is very red, a symptom that is described as _____.

4. The redness is accompanied by blisters. A blister is also called a bulla, or

   _____.

5. You determine that the redness and blisters are due to sunburn and that the damage is confined to the epidermis. This type of burn is categorized as _____.

6. When you examine Mr. B, you see numerous healed wounds. These wounds were healed through the actions of cells called _____.

7. The new tissue that healed the wound is called a scar, or a(n) _____.

8. Some scars contain sharply raised areas resulting from excess collagen production. These areas are called _____.

➤ Group B

As a volunteer with Doctors Without Borders, you are working at a clinic in Malawi. Mr. L has brought his two children, ages 1 and 6, to the clinic.

1. You notice that the baby has small, pimple-like protrusions on her cheeks. These protrusions are called _____.

2. The baby also has scaly, crusty areas in the folds of her elbows and knees and has been scratching incessantly. Severe itching is also called _____.

3. Based on her symptoms, you suspect that the baby has a noncontagious skin disorder called _____.

4. The older child has a number of blisters on his hands that contain pus. Microscopic examination reveals the presence of staphylococci. This contagious skin disease is called _____.

5. You also examined the father, who had small facial lesions and complained of pain, sensitivity, and itching. You noted that the lesions followed the path of nerves, leading to the diagnosis of _____.

6. The father asks if antibiotics will help his condition. You reply that the best medication for his disorder is a(n) _____.

III. SHORT ESSAYS

1. Explain why the skin is valuable in diagnosis.

   _____

   _____

   _____

2. Compare and contrast eccrine and apocrine sweat (sudoriferous) glands.

   _____

   _____

   _____

3. Describe the role that the skin plays in the regulation of body temperature.

   _____

   _____

   _____

CONCEPTUAL THINKING

1. Describe the location and structure of the different tissue types (epithelial, muscle, nervous, connective) present in the integumentary system.

_____

_____

_____

_____

2. Chapter 5 discussed predisposing causes of disease. Briefly describe four predisposing causes of disease that are associated with alopecia.

_____

_____

_____

_____

3. Discuss the functional implications of the following structural changes in skin associated with aging.

   a. less subcutaneous fat and decreased circulation in the dermis

   _____

   _____

   b. decreased melanin production in hair and skin

   _____

   _____

   c. decreased cell proliferation in the hair follicle

   _____

   _____

   d. reduced collagen and elastic production in the dermis

   _____

   _____

   e. decreased activity of the eccrine sudoriferous glands

   _____

   _____

Expanding Your Horizons

Why did different skin tones evolve? It is often thought that darker pigmentation (more melanin) has evolved to protect humans from skin cancer. However, since skin cancer occurs later in life (usually postreproductive age), it cannot exert much evolutionary pressure. The advantages and disadvantages of darker skin tone are discussed in these *Scientific American* articles:

- Jablonski NG, Chaplin G. Skin deep. Sci Am 2002;287:74–81.
- Tavera-Mendoza LE, White JH. Cell defenses and the sunshine vitamin. *Sci Am* 2007;297:62–72.

Movement and Support

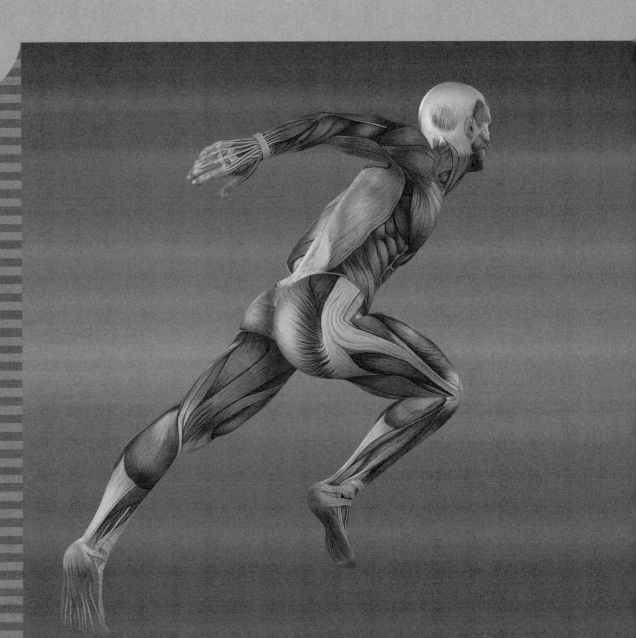

The Skeleton: Bones and Joints

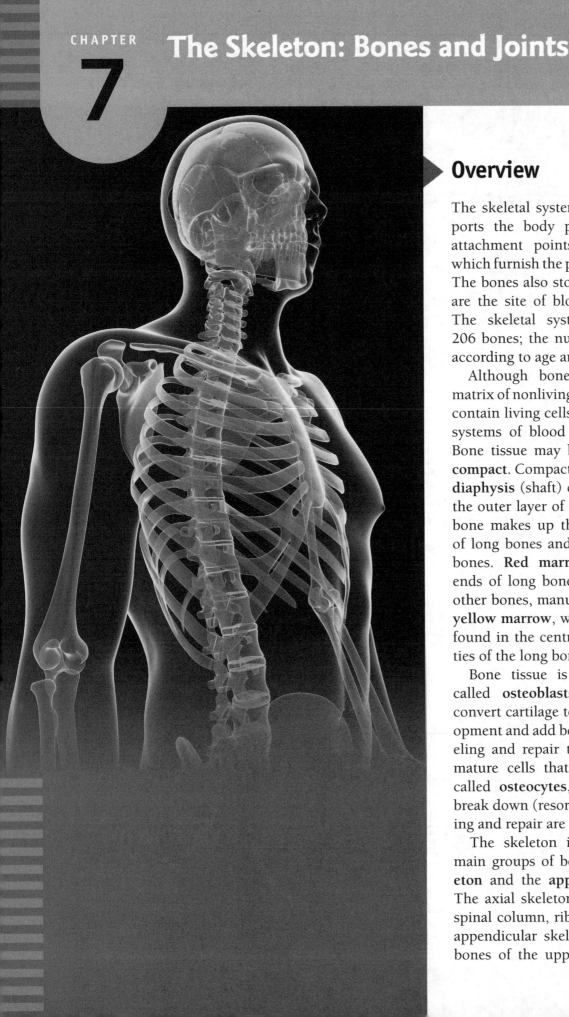

Overview

The skeletal system protects and supports the body parts and serves as attachment points for the muscles, which furnish the power for movement. The bones also store calcium salts and are the site of blood cell production. The skeletal system includes some 206 bones; the number varies slightly according to age and the individual.

Although bone tissue contains a matrix of nonliving material, bones also contain living cells and have their own systems of blood vessels and nerves. Bone tissue may be either **spongy** or **compact**. Compact bone is found in the **diaphysis** (shaft) of long bones and in the outer layer of other bones. Spongy bone makes up the **epiphyses** (ends) of long bones and the center of other bones. **Red marrow**, present at the ends of long bones and the center of other bones, manufactures blood cells; **yellow marrow**, which is largely fat, is found in the central (medullary) cavities of the long bones.

Bone tissue is produced by cells called **osteoblasts**, which gradually convert cartilage to bone during development and add bone tissue for remodeling and repair throughout life. The mature cells that maintain bone are called **osteocytes**, and the cells that break down (resorb) bone for remodeling and repair are the **osteoclasts**.

The skeleton is divided into two main groups of bones, the **axial skeleton** and the **appendicular skeleton**. The axial skeleton includes the skull, spinal column, ribs, and sternum. The appendicular skeleton consists of the bones of the upper and lower limbs, the shoulder girdle, and the pelvic girdle. Disorders of bones include metabolic disorders, tumors, infection, structural disorders, and fractures.

A **joint** is the region of union of two or more bones. Joints are classified into three main types on the degree of movement permitted. In order of increasing movement, the types are synarthroses, amphiarthroses, and diarthroses. **Diarthroses** are also called **synovial joints** because the material between the bones is synovial fluid, which is secreted by the synovial membrane lining the joint cavity. The bones in synovial joints are connected by ligaments. Synovial joints show the greatest degree of movement, and the six types of synovial joints allow for a variety of movements in different directions.

Addressing the Learning Objectives

1. LIST THE FUNCTIONS OF BONES.

EXERCISE 7-1

List five functions of bones in the spaces provided.

1. _____

2. _____

3. _____

4. _____

5. _____

2. DESCRIBE THE STRUCTURE OF A LONG BONE.

(Also see Exercises 7-4 through 7-6.)

EXERCISE 7-2: Structure of a Long Bone (Text Fig. 7-2)

1. Write the names of the three parts of a long bone in the numbered boxes 1 to 3.
2. Write the name of each labeled part on the numbered lines in different colors. Use a dark color for structure 5.
3. Color the different structures on the diagram with the corresponding color.

4. _____

5. _____

6. _____

7. _____

8. _____

9. _____

10. _____

11. _____

12. _____

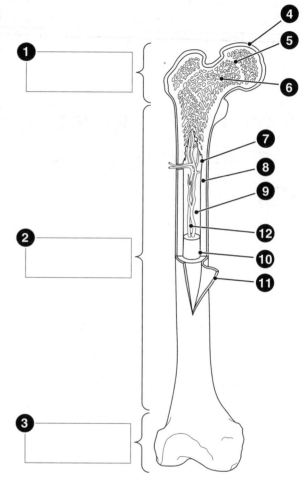

3. NAME THE THREE DIFFERENT TYPES OF CELLS IN BONE, AND DESCRIBE THE FUNCTIONS OF EACH.

EXERCISE 7-3

Write the name of the appropriate bone cell in each blank from the list below.

osteoblast   osteocyte   osteoclast

1. A cell that resorbs bone matrix   _____

2. A mature bone cell that is completely surrounded by hard bone tissue   _____

3. A cell that builds bone tissue   _____

4. DIFFERENTIATE BETWEEN COMPACT BONE AND SPONGY BONE WITH RESPECT TO STRUCTURE AND LOCATION.

EXERCISE 7-4

Fill in the blank after each statement—does it apply to compact bone (C) or spongy bone (S)?

1. Makes up the interior of the epiphyses of long bones _____

2. Makes up the center of short bones _____

3. Makes up the shaft of a long bone _____

4. A meshwork of small, bony plates _____

5. Very hard bone with few spaces _____

EXERCISE 7-5

Fill in the blank after each statement—does it apply to red marrow (R) or yellow marrow (Y)?

1. Found in the spaces of spongy bone _____

2. Composed largely of fat _____

3. Site of blood cell synthesis _____

4. Found in the shaft of a long bone _____

EXERCISE 7-6

Write the appropriate term in each blank from the list below. Not all terms will be used.

diaphysis   epiphysis   medullary cavity   central canal

periosteum   endosteum   spongy bone   osteon

1. The shaft of a long bone   _____

2. The tough connective tissue membrane that covers bones   _____

3. The end of a long bone   _____

4. The type of bone tissue found at the end of long bones   _____

5. The thin membrane that lines the central cavity of long bones _____

6. The hollow portion of a long bone containing yellow marrow   _____

7. The longitudinal canal in the middle of each osteon   _____

5. EXPLAIN HOW A LONG BONE GROWS.

EXERCISE 7-7

Label each of the following statements as true (T) or false (F).

1. Long bones grow in length by producing new bone tissue in the middle of the diaphysis.

    _____

2. Once bone growth is complete, the epiphyseal plate turns into the epiphyseal line.

    _____

3. Long bones elongate by converting cartilage in the bone ends into bone tissue.

    _____

4. Osteoclasts and osteoblasts stop working once bone growth is complete.

    _____

5. As a bone lengthens, the medullary cavity becomes larger.

    _____

6. NAME AND DESCRIBE NINE MARKINGS FOUND ON BONES.

EXERCISE 7-8

Write the appropriate term in each blank from the list below. Not all terms will be used.

crest   condyle   head   process

foramen   fossa   sinus   meatus

1. A short channel or passageway

    _____

2. An air space found in some skull bones

    _____

3. A rounded knoblike end separated by a slender region from the rest of the bone

    _____

4. A rounded projection

    _____

5. A distinct border or ridge

    _____

6. A depression on a bone surface

    _____

7. A hole that permits the passage of a vessel or nerve

    _____

7. NAME, LOCATE, AND DESCRIBE THE BONES IN THE AXIAL SKELETON.

EXERCISE 7-9: The Skull (Text Fig. 7-5)

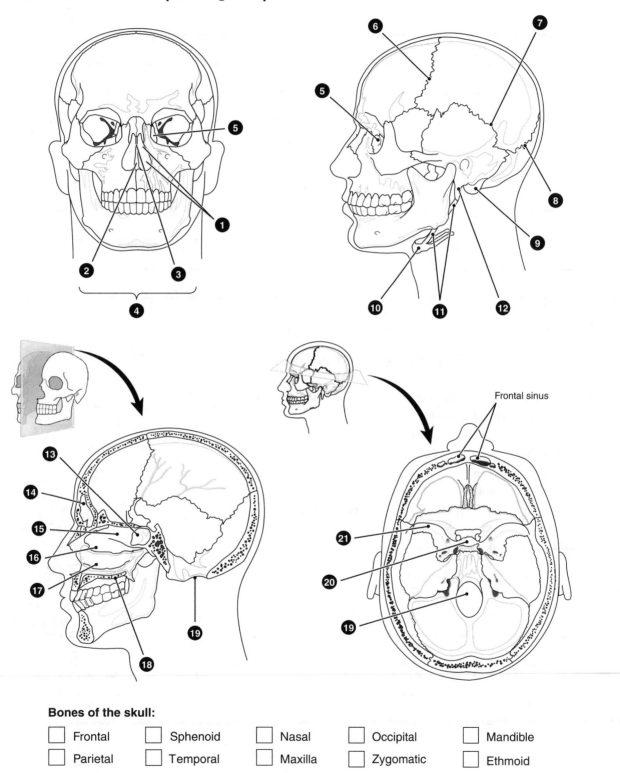

Frontal sinus

Bones of the skull:

☐ Frontal   ☐ Sphenoid   ☐ Nasal   ☐ Occipital   ☐ Mandible

☐ Parietal   ☐ Temporal   ☐ Maxilla   ☐ Zygomatic   ☐ Ethmoid

1. Color the boxes next to the names of the skull bones in different, light colors.
2. Color the skull bones with the corresponding colors.
3. Label each of the following numbered bones and bone features (write their names in black).

1. _____   12. _____

2. _____   13. _____

3. _____   14. _____

4. _____   15. _____

5. _____   16. _____

6. _____   17. _____

7. _____   18. _____

8. _____   19. _____

9. _____   20. _____

10. _____   21. _____

11. _____

EXERCISE 7-10: Vertebral Column (Text Fig. 7-7A)

1. Color each of the following bones the indicated color.
   a. cervical vertebrae—blue
   b. thoracic vertebrae—red
   c. lumbar vertebrae—green
   d. sacrum—yellow
   e. coccyx—violet
2. Label each of the indicated bones and bone parts.

1. _____

2. _____

3. _____

4. _____

5. _____

6. _____

7. _____

EXERCISE 7-11: The Vertebrae (Text Fig. 7-7B and C)

1. Look at the top left ⟨...⟩ writing the names of the labeled parts on ⟨...⟩
2. Identify these parts in ⟨...⟩ in each empty bullet.
3. Write the names of p⟨...⟩me color for parts 6 and 7. Color structu⟨...⟩

Cer⟨...⟩
vert⟨...⟩

Tho⟨...⟩
vert⟨...⟩

Pos⟨...⟩

Ant⟨...⟩

Tra⟨...⟩
for a⟨...⟩

1. _____
2. _____
3. _____
4. _____

8. _____

EXERCISE 7-12: Bones of the Thorax (Text Fig. 7-8)

1. Write the name of each labeled part on the lines beside the bullets in different colors. Structures 1, 3, 6, and 7 will not be colored, so write their names in black.
2. Color the different structures on the diagram with the corresponding colors.

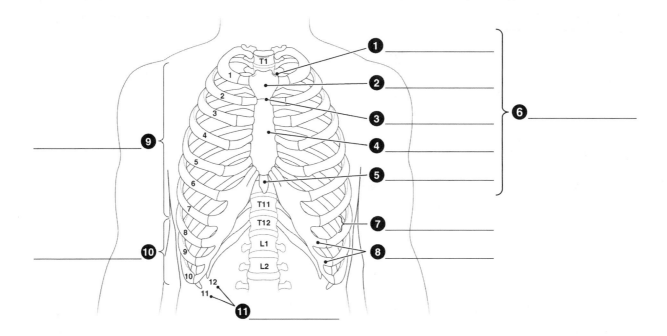

EXERCISE 7-13

Write the appropriate term in each blank from the list below. Not all terms will be used.

parietal bone   temporal bone   frontal bone   hyoid bone

nasal bone   occipital bone   maxilla   mandible

sphenoid bone   zygomatic bone

1. The only movable bone of the skull   _____

2. A bone of the upper jaw   _____

3. The U-shaped bone lying just below the mandible   _____

4. The bone that articulates with the parietal and temporal
   bones and forms the posterior inferior part of the cranium   _____

5. The bone that forms the forehead   _____

6. One of two slender bones that form the bridge of the nose   _____

7. One of two large bones that articulate with the frontal bone
   and form the superior lateral portions of the cranium   _____

8. The anatomic name for the cheekbone   _____

EXERCISE 7-14

Write the appropriate term in each blank from the list below. Not all terms will be used.

floating ribs   true ribs   fontanel   costal

foramina   xiphoid process   manubrium   clavicular notch

1. The T-shaped, superior portion of the sternum   _____

2. The portion of the sternum that is made of cartilage in children   _____

3. An adjective that refers to the ribs   _____

4. A soft spot in the infant skull that later closes   _____

5. The last two pairs of ribs, which are very short and do not
   extend to the front of the body   _____

6. The point of articulation between the sternum and the collarbone   _____

7. Ribs that attach to the sternum by individual cartilages   _____

8. DESCRIBE THE NORMAL CURVES OF THE SPINE, AND EXPLAIN THEIR PURPOSE.

EXERCISE 7-15

Write the appropriate term in each blank from the list below.

cervical region   thoracic region   lumbar region   coccyx

thoracic curve   lumbar curve   cervical curve

1. A primary curve of the spine   _____

2. The second part of the vertebral column, made up of
   12 vertebrae   _____

3. The spinal curve that appears when the infant holds
   his or her head up   _____

4. The spinal curve that appears when the infant begins to walk   _____

5. The most inferior part of the vertebral column   _____

6. The region of the spine that contains the largest, strongest
   vertebrae   _____

7. The region of the vertebral column made up of the first seven
   vertebrae   _____

9. NAME, LOCATE, AND DESCRIBE THE BONES IN THE APPENDICULAR SKELETON.

EXERCISE 7-16: The Skeleton (Text Fig. 7-1)

1. Write the name of each labeled part on the lines beside the bullets in different colors. Use the same color for structures 23 to 25 and for structures 19 and 20.
2. Color the different structures on the diagram with the corresponding colors. Try to color every structure in the figure with the appropriate color. For instance, structure number 3 is found in two locations.

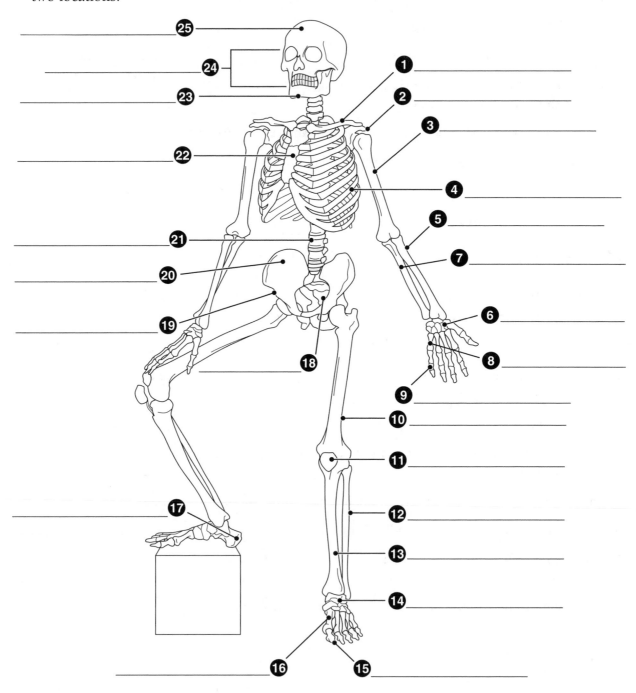

EXERCISE 7-17: Bones of the Shoulder Girdle (Text Fig. 7-9)

1. Use three contrasting colors to write the names of the illustrated bones in lines 1, 4, and 5. Color the bones the corresponding colors.
2. Write the names of the bone features in black on the other numbered lines.

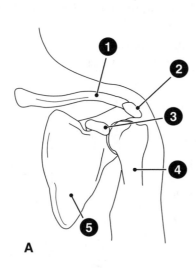

A

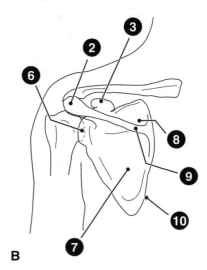

B

1. _____

2. _____

3. _____

4. _____

5. _____

6. _____

7. _____

8. _____

9. _____

10. _____

EXERCISE 7-18: Upper Extremity (Text Fig. 7-10)

1. Color the large bone in part A with a bright color. Use this color to write the name of the bone in the first line, and all of its features on lines 2 to 6.
2. Write the name of the joint indicated by bullet 7 on line 7.
3. Use a bright color to shade and write the name of the bone indicated by bullet 8. Use the same color to write the name of this bone's features (bullets 9 to 10).
4. Use a contrasting color to shade and write the name of the bone indicated by bullet 11. Use the same color to write the name of this bone's features (bullets 12 to 16).

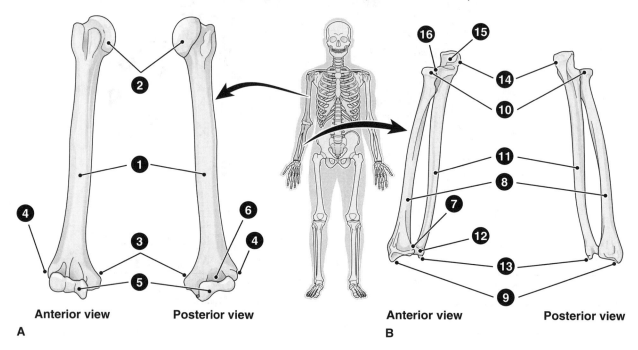

1. _____   9. _____

2. _____   10. _____

3. _____   11. _____

4. _____   12. _____

5. _____   13. _____

6. _____   14. _____

7. _____   15. _____

8. _____   16. _____

EXERCISE 7-19: Pelvic Bones (Text Fig. 7-14)

1. Color the boxes next to the names of the pelvic bones in different, light colors.
2. Color the pelvic bones in parts A and B of the diagram with the corresponding color.
3. Label each of the following numbered bones and bone features.

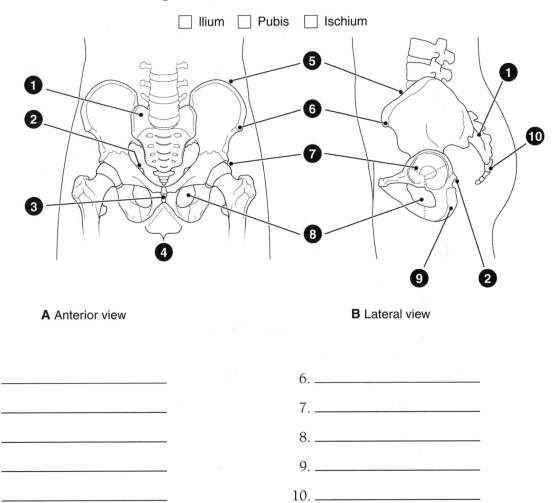

☐ Ilium   ☐ Pubis   ☐ Ischium

A Anterior view   **B** Lateral view

1. _____   6. _____

2. _____   7. _____

3. _____   8. _____

4. _____   9. _____

5. _____   10. _____

EXERCISE 7-20: Lower Extremity (Text Fig. 7-16)

1. Color the large bone in part A with a bright color. Use this color to write the name of the bone in the first line, and all of its features on lines 2 to 10.
2. Use a bright color to shade and write the name of the bone indicated by bullet 11. Use the same color to write the name of this bone's features (bullets 12 to 16).
3. Use a contrasting color to shade and write the name of the bone indicated by bullet 17. Use the same color to write the name of this bone's features (bullets 18 to 19).
4. Use black to write the name of the joints indicated by bullets 20 and 21 on the appropriate lines.

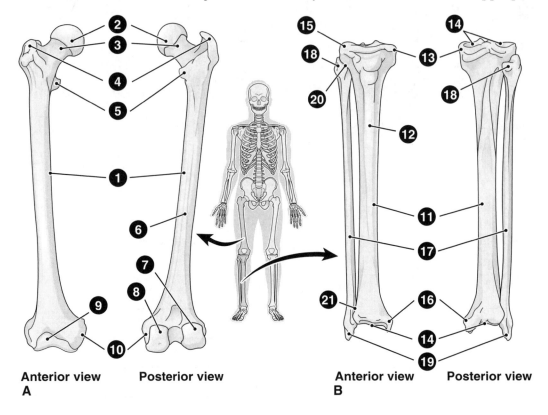

Anterior view   **Posterior view**   **Anterior view**   **Posterior view**
A   **B**

1. _____   12. _____

2. _____   13. _____

3. _____   14. _____

4. _____   15. _____

5. _____   16. _____

6. _____   17. _____

7. _____   18. _____

8. _____   19. _____

9. _____   20. _____

10. _____   21. _____

11. _____

EXERCISE 7-21

Write the appropriate term in each blank from the list below. Not all terms will be used.

olecranon   carpal bones   clavicle   ulna   radius

metacarpal bones   phalanges   scapula   humerus

1. The anatomic name for the collarbone   _____

2. The five bones in the palm of the hand   _____

3. The medial forearm bone (in the anatomic position)   _____

4. The upper part of the ulna, which forms the point of the elbow   _____

5. The 14 small bones that form the framework of the fingers
   on each hand   _____

6. The bone located on the thumb side of the forearm   _____

7. The bone containing the supraspinous and infraspinous fossae   _____

EXERCISE 7-22

Write the appropriate term in each blank from the list below. Not all terms will be used.

greater trochanter   patella   tibia   calcaneus   pubis

fibula   ilium   ischium acetabulum

1. The deep socket in the hip bone that holds the head of
   the femur   _____

2. The most inferior bone in the pelvis   _____

3. The lateral bone of the leg   _____

4. A bone that is wider and more flared in females   _____

5. The scientific name for the kneecap   _____

6. The largest of the tarsal bones; the heel bone   _____

7. The large, rounded projection at the upper and lateral
   portion of the femur   _____

EXERCISE 7-23

Write "male" or "female" in the spaces below to make each statement true.

1. The pelvic outlet is narrower in the _____ than in the

   _____.

2. The angle of the pubic arch is broader in the _____ than in the

   _____.

3. The sacrum and coccyx are shorter and less curved in the _____ than in

   the _____.

4. The ilia are narrower in the _____ than in the _____.

10. DESCRIBE FIVE TYPES OF BONE DISORDERS.

EXERCISE 7-24

Write the appropriate term in each blank from the list below. Not all terms will be used. In addition, write down the type of each bone disorder in the same blank (e.g., structural).

osteomyelitis	scoliosis	osteoporosis	chondrosarcoma
kyphosis	lordosis	osteopenia	osteosarcoma

1. An excessive concave curvature of the thoracic spine   _____

2. A lateral curvature of the vertebral column   _____

3. A mild reduction in bone density levels   _____

4. A bone infection caused by pus-producing bacteria   _____

5. A bone disorder common in older women that may lead
   to fracture   _____

6. A malignant tumor originating in cartilage   _____

7. An excessive lumbar curve   _____

11. NAME AND DESCRIBE EIGHT TYPES OF FRACTURES.

EXERCISE 7-25: Types of Fractures (Text Fig. 7-21)

Label each of the types of fractures pictured below.

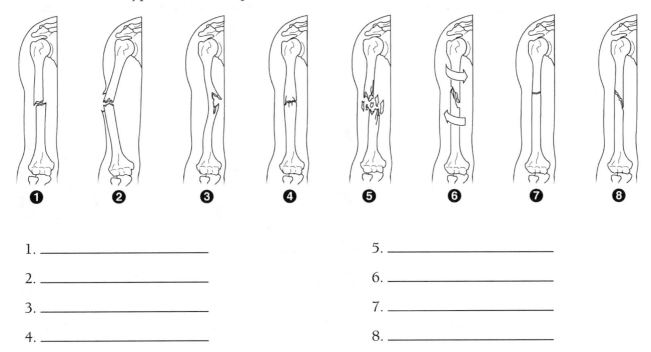

❶	❷	❸	❹	❺	❻	❼	❽

1. _____

2. _____

3. _____

4. _____

5. _____

6. _____

7. _____

8. _____

12. DESCRIBE THREE CATEGORIES OF JOINTS BASED ON DEGREE OF MOVEMENT, AND GIVE EXAMPLES OF EACH.

EXERCISE 7-26

Write the appropriate term in each blank from the list below.

cartilage	articulation	diarthrosis	amphiarthrosis
synarthrosis	fibrous connective tissue	synovial fluid	

1. The region where two or more bones unite; a joint   _____

2. A slightly moveable joint   _____

3. A freely moveable joint   _____

4. An immovable joint   _____

5. The material joining the bones of most synarthroses   _____

6. The material joining the bones of most amphiarthroses   _____

7. The material between the bones of diarthroses   _____

EXERCISE 7-27: The Knee Joint (Text Fig. 7-23)

1. Write the name of each labeled part on the numbered lines in different colors. Use a dark color for part 7, which can be outlined.
2. Color the different structures on the diagram with the corresponding colors. Try to color every structure in the figure with the appropriate color. For instance, structure number 2 is found in two locations.

1. _____
2. _____
3. _____
4. _____
5. _____
6. _____
7. _____
8. _____
9. _____
10. _____
11. _____
12. _____

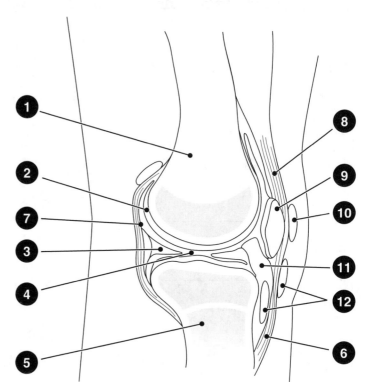

13. NAME SIX TYPES OF SYNOVIAL JOINTS, AND DEMONSTRATE THE MOVEMENTS THAT OCCUR AT EACH.

EXERCISE 7-28: Movements at Synovial Joints (Text Fig. 7-24)

Label each of the illustrated motions with the correct terms.

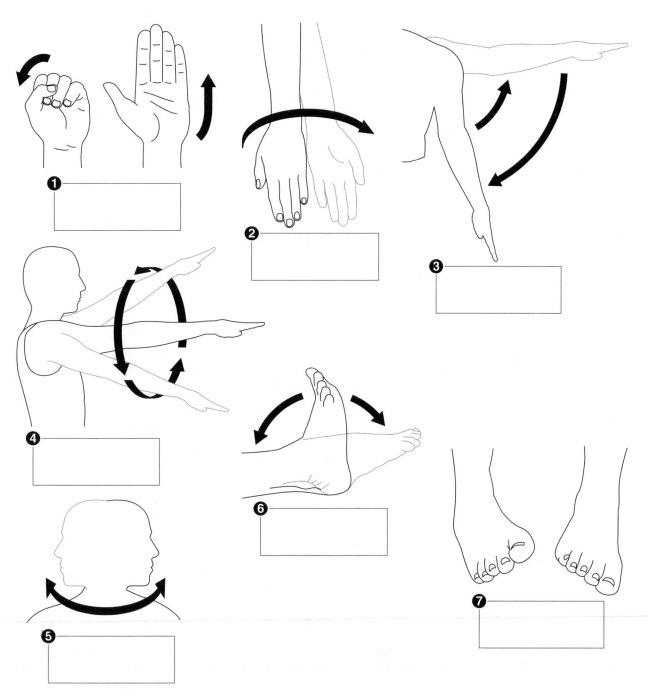

EXERCISE 7-29

Write the appropriate term in each blank from the list below. Not all terms will be used.

flexion   rotation   abduction   extension   adduction
supination circumduction dorsiflexion plantar flexion pronation

1. A movement that increases the angle between two bones   _____

2. Movement away from the midline of the body   _____

3. Motion around a central axis   _____

4. A bending motion that decreases the angle between two parts   _____

5. Movement toward the midline of the body   _____

6. The act of turning the palm up or forward   _____

7. The act of pointing the toes downward   _____

EXERCISE 7-30: Types of Synovial Joints (Text Table 7-3)

Label each of the different types of synovial joints.

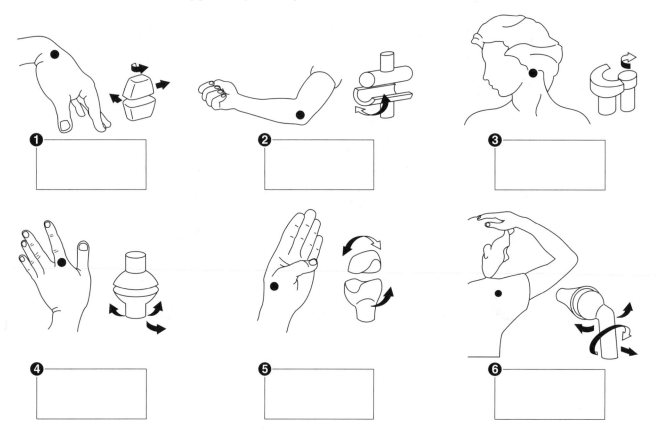

14. DESCRIBE THREE TYPES OF JOINT DISORDERS.

EXERCISE 7-31

Write the appropriate term in each blank from the list below. Not all terms will be used.

rheumatoid arthritis   osteoarthritis   sprain   septic arthritis

bursitis   bunion   gout

1. Another name for degenerative joint disease   _____

2. An injury to the ligaments of a joint   _____

3. Arthritis caused by overproduction of uric acid   _____

4. A crippling autoimmune disease of joints   _____

5. Joint inflammation caused by bacteria   _____

6. Inflammation of a small sac filled with synovial fluid   _____

EXERCISE 7-32

Which of the following are common causes of backache? Circle all that apply.
a. intervertebral disk disorders
b. osteoarthritis
c. lifting a large weight by holding it close to the body
d. using the legs, not the back, to lift a heavy load

15. DESCRIBE METHODS USED TO CORRECT DISEASED JOINTS.

EXERCISE 7-33

Write the appropriate term in each blank from the list below.

arthroplasty   arthroscope   arthrocentesis

1. The removal of excess fluid from the joint cavity   _____

2. An instrument to identify and repair joint problems   _____

3. Joint replacement   _____

16. DESCRIBE HOW THE SKELETAL SYSTEM CHANGES WITH AGE.

EXERCISE 7-34

Use the terms below to complete the paragraph.

protein   intervertebral disks   collagen   calcium   intercostal cartilages

In older adults, bones are weaker because of a loss in (1) _____ salts and a general decline in the manufacture of (2) _____. Height may be reduced because the (3) _____ become thinner. The chest may become smaller because the (4) _____ become calcified and less flexible. The reduction in levels of the protein (5) _____ in tendons and ligaments makes movement more difficult.

17. USING THE CASE STUDY, DISCUSS HOW FRACTURES HEAL.

EXERCISE 7-35

The following five steps detail how Reggie's femur fracture healed, but they are out of order. Put them in order by numbering the steps 1 to 5 in the order in which they occurred.

_____ a. Osteoblasts deposit spongy bone, forming a hard callus.

_____ b. A blood clot forms at the fracture site.

_____ c. Osteoclasts and osteoblasts remodel hard callus into normal bone.

_____ d. Blood vessels from the periosteum invade the blood clot.

_____ e. Fibroblasts secrete collagen, and chondrocytes produce cartilage to make a soft callus.

18. SHOW HOW WORD PARTS ARE USED TO BUILD WORDS RELATED TO THE SKELETON.

EXERCISE 7-36

Complete the following table by writing the correct word part or meaning in the space provided. Write a word that contains each word part in the "Example" column.

Word Part	Meaning	Example
1. _____	lack of	_____
2. -clast	_____	_____
3. _____	rib	_____
4. amphi-	_____	_____
5. arthr/o	_____	_____
6. _____	away from	_____
7. _____	around	_____
8. _____	toward, added to	_____
9. dia-	_____	_____
10. pariet/o	_____	_____

Making the Connections

The following concept map deals with bone structure. Complete the concept map by filling in the appropriate term or phrase that describes the indicated structure or process.

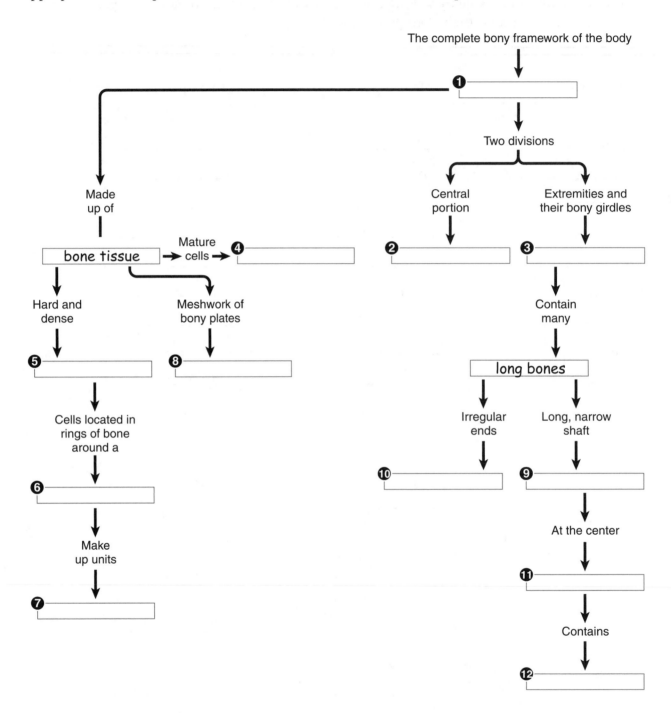

Optional Exercise: Make your own concept map based on the different bone markings. Use the following terms and any others you would like to include: bone markings, projections, depressions, head, process, condyle, crest, spine, foramen, sinus, fossa, meatus, sella turcica, mastoid sinus, foramen magnum, acromion, intervertebral foramina, supraspinous fossa, and scapula spine. Try to find an example of each bone marking.

Testing Your Knowledge

BUILDING UNDERSTANDING

I. MULTIPLE CHOICE

Select the best answer.

1. What cells dissolve bone matrix?   1. _____
   a. osteoblasts
   b. osteocytes
   c. osteoclasts
   d. osteons
2. Which of the following terms best describes sutures?   2. _____
   a. synovial joint
   b. diarthrosis
   c. synarthrosis
   d. amphiarthrosis
3. Which of the following is a projection?   3. _____
   a. process
   b. fossa
   c. foramen
   d. sinus
4. What bone makes up the posterior portion of the hard palate?   4. _____
   a. vomer bone
   b. palatine bone
   c. hyoid bone
   d. mandible
5. The os coxa is a fused bone consisting of the ilium, ischium,
   and a third bone. What is the name of this third bone?   5. _____
   a. femur
   b. acetabulum
   c. sacrum
   d. pubis
6. Which of the following terms best describes the patella?   6. _____
   a. sesamoid
   b. axial
   c. tarsal
   d. symphysis
7. Which of the following bones is part of the shoulder girdle?   7. _____
   a. sternum
   b. tibia
   c. scapula
   d. os coxae
8. Which rib type attaches to the sternum by individual costal cartilages? 8. _____
   a. false ribs
   b. floating ribs
   c. xiphoid ribs
   d. true ribs

9. Which of the following describes the foramen magnum?   9. _____
   a. a large hole in a hip bone near the symphysis pubis
   b. the curved rim along the top of the hip bone
   c. a hole between vertebrae that allows for passage of a spinal nerve
   d. a large opening at the base of the skull through which the spinal cord passes

10. Which of the following joint types is only capable of flexion and extension?   10. _____
    a. pivot
    b. ball and socket
    c. hinge
    d. gliding

II. COMPLETION EXERCISE

Write the word or phrase that correctly completes each sentence.

1. The first cervical vertebra is called the _____.

2. The bat-shaped bone that extends behind the eyes and also forms part of the base of the skull is called the _____.

3. The bone located between the eyes that extends into the nasal cavity, eye sockets, and cranial floor is called the _____.

4. The hard bone matrix is composed mainly of salts of the element _____.

5. The greater and lesser trochanters are found on the _____.

6. The small fluid-filled sacs near some joints are called _____.

7. Pivot, hinge, and gliding joints are examples of freely movable joints, also called
   _____.

8. Swimming the front crawl requires a broad circular movement at the shoulder that is a combination of simpler movements. This combined motion is called _____.

9. When you bend your foot upward to walk on your heels, the position of the foot is technically called _____.

10. In the embryo, most of the developing bones are made of _____.

11. The type of bone tissue that makes up the shaft of a long bone is called
    _____.

12. A malignant tumor of bone tissue is called a(n) _____.

UNDERSTANDING CONCEPTS

I. TRUE/FALSE

For each statement, write T for true or F for false in the blank to the left of each number. If a statement is false, correct it by replacing the <u>underlined</u> term, and write the correct statement in the blanks below the statement.

_____ 1. The shaft of a long bone contains <u>yellow</u> marrow.

_____

_____ 2. The ethmoid bone is in the <u>axial</u> skeleton.

_____

_____ 3. Moving a bone toward the midline is <u>abduction</u>.

_____

_____ 4. The pointed projection of the ulna on the posterior surface of the elbow is called the <u>axis</u>.

_____

_____ 5. Increasing the angle at a joint is <u>extension</u>.

_____

_____ 6. There are <u>six</u> pairs of false ribs.

_____

_____ 7. A mature bone cell is an <u>osteocyte</u>.

_____

_____ 8. <u>Immovable</u> joints are called amphiarthroses.

_____

_____ 9. The ends of a long bone are composed mainly of <u>spongy</u> bone.

_____

_____ 10. Excessive lumbar curve of the spine is called <u>scoliosis</u>.

_____

_____ 11. The medial malleolus is found at the distal end of the <u>fibula</u>.

_____

_____ 12. A malignant bone tumor arising in cartilage is called a <u>chondrosarcoma</u>.

_____

II. PRACTICAL APPLICATIONS

Ms. M, aged 67, suffered a serious fall at a recent bowling tournament. As a physician assistant trainee, you are responsible for her preliminary evaluation.

1. Her right forearm is bent at a peculiar angle. You suspect a fracture to the radius or to the
   _____.

2. The broken bone does not project through the skin. Fractures without an open wound are called _____.

3. An x-ray reveals a single break at an angle across the bone. This type of fracture is called a(n)
   _____.

4. The arm x-ray also reveals a number of fractures in the wrist bones, which are also called the
   _____.

5. The wrist bones are fractured in many places. This type of fracture is called a(n)
   _____.

6. Ms. M also reports pain in the hip region. The hip joint consists of the femur and a deep socket called the _____.

7. An x-ray reveals a crack in one of the "sitting bones" that support the weight of the trunk when sitting. This bone is called the _____.

8. The large number of fractures Ms. M suffered suggests that she may have a bone disorder. Changes in her bone mass can be detected using a test called a(n) _____.

9. The test confirms that she has a significant reduction in bone mass and bone protein, leading to the diagnosis of _____.

10. The physician prescribes a new medication designed to increase the activity of cells that synthesize new bone tissue. These cells are called _____.

III. SHORT ESSAYS

1. What is the function of the fontanels?

   _____

   _____

   _____

2. Describe the four curves of the adult spine, and explain the purpose of these curves.

   _____

   _____

   _____

3. What are the differences among true ribs, false ribs, and floating ribs?

_____

_____

_____

4. List the bones that make up the elbow joint, and describe three different articulations between these three bones.

_____

_____

_____

CONCEPTUAL THINKING

1. The following questions relate to the knee joint.

   a. Classify the knee joint in terms of the degree of movement permitted.

   _____

   b. List the types of movement that can occur at the knee joint.

   _____

   c. Based on your answer to part b, classify the knee joint based on the types of movement permitted.

   _____

   d. Classify the knee joint in terms of the material between the adjoining bones.

   _____

   e. List the bones that articulate within the capsule of the knee joint.

   _____

2. List, in order, the movements (e.g., abduction) that must occur in order to accomplish the following actions:

   a. A child brings her leg far behind her and then kicks the ball, bringing her leg in front of her.

   _____

   b. You hear your friend shouting, and turn your head to the right in the direction of the sound.

   _____

   c. The person in the car next to you bends his arm at the elbow to scratch his nose and then straightens his arm again.

   _____

3. Using the text's Figure 7-5 for reference, list all of the bones that make up the eye socket.

_____

_____

_____

Expanding Your Horizons

The human skeleton has evolved from that of four-legged animals. Unfortunately, the adaptation is far from perfect; our upright posture can cause problems like backache and knee injuries. If you could design the human skeleton from scratch, what would you change? A *Scientific American* article suggests some improvements.

- Olshansky JS, Carnes BA, Butler RN. If humans were built to last. Sci Am 2001;284:50–55.

The Muscular System

▶ Overview

There are three basic types of muscle tissue: skeletal, smooth, and cardiac. This chapter focuses on **skeletal muscle**, which is usually attached to bones. Skeletal muscle is also called **voluntary muscle**, because it is normally under conscious control. The muscular system is composed of more than 650 individual muscles.

Skeletal muscles are activated by electrical impulses from the nervous system. A neuron (nerve cell) makes contact with a muscle cell at the **neuromuscular junction**. The neurotransmitter **acetylcholine** transmits the signal from the neuron to the muscle cell by producing an electrical change called the **action potential** in the muscle cell membrane. The action potential causes the release of **calcium** from specialized endoplasmic reticulum (known as **sarcoplasmic reticulum**) into the muscle cell cytoplasm (known as the **sarcoplasm**). Calcium enables two types of intracellular filaments inside the muscle cell, made of **actin** and **myosin**, to contact each other. The myosin filaments pull the actin filaments closer together, resulting in muscle shortening.

ATP is the direct source of energy for the contraction, and it is made on demand by muscle cells. Only a small amount of ATP can be synthesized without oxygen (**anaerobically**), from creatine phosphate and glucose. Most ATP is synthesized from glucose and fatty acids by **oxidation**, a process that requires adequate amounts of oxygen and mitochondria. A reserve supply of glucose is stored in muscle cells in the form of **glycogen**. Additional oxygen is stored by a muscle cell pigment called **myoglobin**.

Muscles usually work in groups to execute a body movement. The muscle that produces a given movement is called the **prime mover**; the muscle that produces the opposite action is the **antagonist**. Muscles that assist the prime mover are called **synergists**; the prime mover and synergists are collectively known as **agonists**.

Muscles act with the bones of the skeleton as lever systems, in which the joint is the pivot point or fulcrum. Exercise and proper body mechanics help maintain muscle health and effectiveness. Continued activity delays the undesirable effects of aging.

This chapter contains some challenging concepts, particularly in respect to the mechanism of muscle contraction, and many muscles to memorize. Try to learn the muscle names and actions by using your own body. You should be familiar with the different movements and the anatomy of joints from Chapter 7 before you tackle this chapter.

Addressing the Learning Objectives

1. COMPARE THE THREE TYPES OF MUSCLE TISSUE.

EXERCISE 8-1

Fill in the blank after each statement—does it apply to skeletal muscle (SK), smooth muscle (SM), or cardiac muscle (CA)? Note that some characteristics may apply to more than one muscle type.

1. Usually attached to bone   _____

2. Only found in the heart   _____

3. Found in the walls of hollow organs   _____

4. Regulates the diameter of an opening, such as a blood vessel   _____

5. Involuntary   _____

6. Voluntary   _____

7. Contains intercalated disks   _____

8. Contains striations   _____

9. Does not contain striations   _____

10. Cells have single nuclei   _____

11. Cells have multiple nuclei   _____

12. Controlled by the somatic nervous system   _____

2. DESCRIBE THREE FUNCTIONS OF SKELETAL MUSCLE.

EXERCISE 8-2

List three functions of skeletal muscle in the spaces below.

1. _____

2. _____

3. _____

3. DESCRIBE THE STRUCTURE OF A SKELETAL MUSCLE TO THE LEVEL OF INDIVIDUAL CELLS.

EXERCISE 8-3: Structure of a Skeletal Muscle (Text Fig. 8-1)

Label each of the indicated parts. Hint: Parts 3, 4, and 5 are membranes.

1. _____

2. _____

3. _____

4. _____

5. _____

6. _____

7. _____

8. _____

9. _____

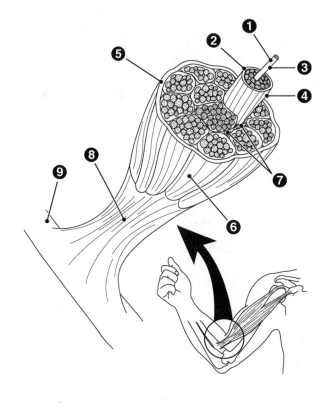

4. OUTLINE THE STEPS IN SKELETAL MUSCLE CONTRACTION.

EXERCISE 8-4: Neuromuscular Junction (Text Fig. 8-2B and C)

Label each of the indicated parts. Hint: Part 8 is a chemical.

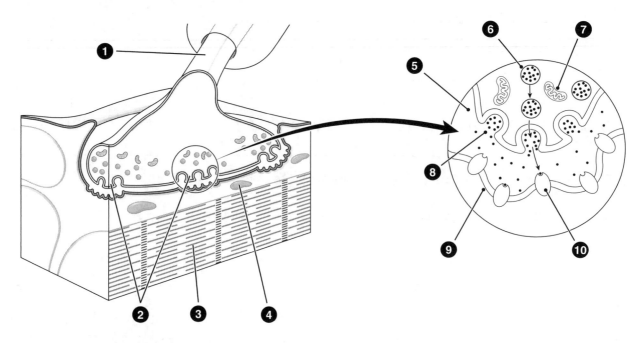

1. _____

2. _____

3. _____

4. _____

5. _____

6. _____

7. _____

8. _____

9. _____

10. _____

EXERCISE 8-5

Write the appropriate term in each blank from the list below. Not all terms will be used.

synaptic cleft   motor end plate   motor unit   actin   sarcoplasmic reticulum

myosin   troponin   sarcomere   tropomyosin

1. The protein that makes up muscle's lighter, thin filaments   _____

2. The protein that interacts with actin to form cross-bridges   _____

3. The membrane of the muscle cell that binds ACh   _____

4. The space between the neuron and the muscle cell   _____

5. A single neuron and all of the muscle fibers it stimulates   _____

6. A protein that binds calcium during muscle contraction   _____

7. The organelle that stores calcium in resting muscle cells   _____

8. A contracting subunit of skeletal muscle   _____

EXERCISE 8-6

In the blanks, write the name of the substance that is accomplishing each action. Each term may be used more than once.

ATP   calcium   acetylcholine

myoglobin   creatine phosphate   glycogen

1. Substance released into the synaptic cleft   _____

2. The immediate source of energy for muscle contraction   _____

3. Binds to troponin when muscle contracts   _____

4. Used to detach the myosin head   _____

5. Pumped back into the ER when muscle relaxes   _____

6. Causes an action potential when it binds the motor end plate   _____

7. A compound similar to ATP that can be used to generate ATP   _____

8. A polysaccharide that can be used to generate glucose   _____

9. A compound that stores oxygen within muscle cells   _____

EXERCISE 8-7

The events of muscle contraction are listed below, but they are out of order. Place the following events in order by writing the appropriate numbers in the blanks. The first one has been done for you.

___1___ a. Acetylcholine (ACh) is released from an axon terminal into the synaptic cleft at the neuromuscular junction.

_____ b. Using stored energy, myosin heads pull actin filaments together within the sarcomeres, and the cell shortens.

_____ c. Myosin heads bind to actin forming cross-bridges.

_____ d. ACh binds to receptors in the muscle's motor end plate and produces an action potential.

_____ e. New ATP is used to detach myosin heads and move them back to position for another "power stroke."

_____ f. The action potential travels to the sarcoplasmic reticulum (SR).

_____ g. Calcium shifts troponin and tropomyosin so that binding sites on actin are exposed.

_____ h. Muscle relaxes when stimulation ends and the calcium is pumped back into the sarcoplasmic reticulum.

_____ i. The sarcoplasmic reticulum releases calcium into the cytoplasm.

5. LIST COMPOUNDS STORED IN MUSCLE CELLS THAT ARE USED TO GENERATE ENERGY.

See Exercises 8-6 and 8-8.

6. EXPLAIN WHAT HAPPENS IN MUSCLE CELLS CONTRACTING ANAEROBICALLY.

EXERCISE 8-8

Circle all answers that are correct (there are two).
   In muscle cells contracting anaerobically:

   a. Creatine phosphate can generate ATP.
   b. Mitochondria break down fatty acids for energy.
   c. Glycolysis can generate ATP by partially breaking down glucose.
   d. Muscle fatigue results when cells run out of ATP.

7. CITE THE EFFECTS OF EXERCISE ON MUSCLES.

EXERCISE 8-9

Label each of the following statements as true (T) or false (F).

1. Resistance exercise causes muscle hypertrophy. _____

2. Blood vessels constrict in actively contracting muscles. _____

3. Weight lifting is the most efficient way to improve endurance. _____

4. Regular exercise increases the number of capillaries in muscles. _____

5. Regular exercise decreases the number of mitochondria in muscles. _____

6. Short bursts of high-intensity exercise appear to be more effective at promoting health than do longer periods of low-intensity exercise. _____

7. Static stretching before a workout can improve muscle strength and reduce the risk of injury. _____

8. COMPARE ISOTONIC AND ISOMETRIC CONTRACTIONS.

See Exercise 8-10.

9. EXPLAIN HOW MUSCLES WORK TOGETHER TO PRODUCE MOVEMENT.

EXERCISE 8-10

Write the appropriate term in each blank from the list below.

origin   prime mover   antagonist   synergist
isotonic   isometric   insertion

1. A muscle acting as a helper to accomplish a particular movement _____

2. The muscle attachment joined to the more moveable part _____

3. The muscle attachment joined to the less moveable part _____

4. The muscle that produces a given movement _____

5. A muscle that relaxes during a given movement _____

6. A contraction in which the muscle shortens but muscle tension remains the same _____

7. A contraction in which muscle tension increases but muscle length is unchanged _____

EXERCISE 8-11: Muscle Attachment to Bones (Text Fig. 8-6)

1. Write the names of the labeled muscles in the blanks 1 to 4. Beside each muscle name, state whether the muscle acts as an antagonist, synergist, or prime mover in elbow flexion.
2. Identify the origin and the insertion of muscle 1 by writing the correct terms in blanks 5 and 6.

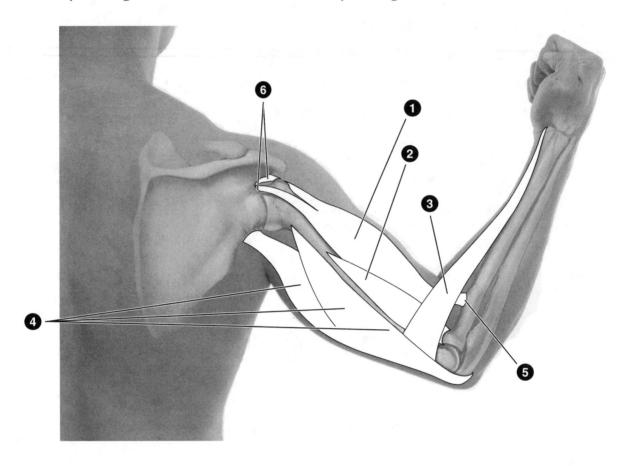

1. _____

2. _____

3. _____

4. _____

5. _____

6. _____

10. COMPARE THE WORKINGS OF MUSCLES AND BONES TO LEVER SYSTEMS.

EXERCISE 8-12

Fill in the blank after each muscle action—does it apply to a first (F), second (S), or third (T)-class lever?

1. Nodding the head   _____

2. Performing a biceps curl   _____

3. Biting an apple with your incisors (front teeth)   _____

11. EXPLAIN HOW MUSCLES ARE NAMED.

EXERCISE 8-13

For each muscle name, write the characteristic(s) used for the name. Choose between the following six options: location, size, shape, direction of fibers, number of heads, and/or action. The number of blanks indicates how many characteristics apply to each muscle. Note that femoris means thigh, brachii means arm, and teres means rounded.

1. trapezius _____

2. quadriceps femoris _____ _____

3. rectus abdominis _____ _____

4. flexor carpi _____ _____

5. teres minor _____ _____

12. NAME SOME OF THE MAJOR MUSCLES IN EACH MUSCLE GROUP, AND DESCRIBE THE LOCATIONS AND FUNCTIONS OF EACH.

EXERCISE 8-14: Superficial Muscles: Anterior View (Text Fig. 8-8)

1. Write the name of each labeled muscle on the numbered lines in different colors.
2. Color the different muscles on the diagram with the corresponding colors.

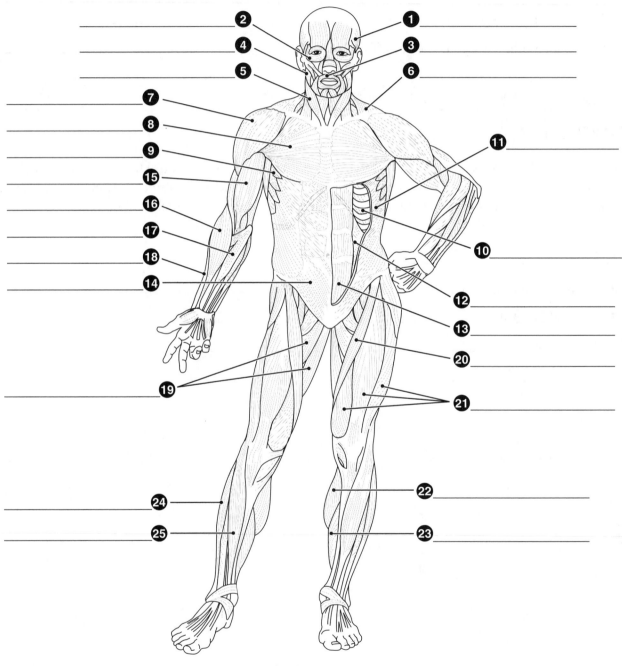

Anterior view

EXERCISE 8-15: Superficial Muscles: Posterior View (Text Fig. 8-9)

1. Write the name of each labeled muscle or tendon on the numbered lines in different colors (part 16 is a bone feature). If possible, use the same colors you used for the muscles in Exercise 8-14. Note: You may need to consult Figure 8-15 in the textbook in order to label all of the muscles.
2. Color the different muscles and tendons on the diagram with the corresponding colors.

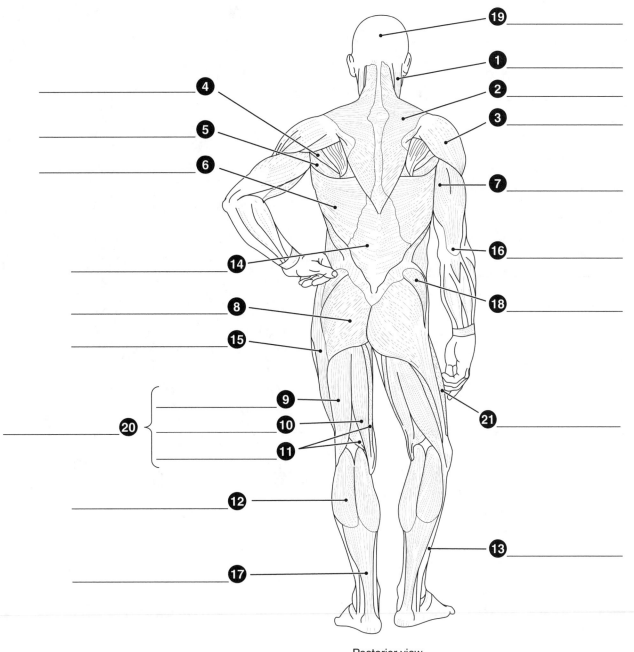

Posterior view

EXERCISE 8-16: Muscles of the Head (Text Fig. 8-10)

Write the name of each labeled muscle or tendon on the numbered lines in different colors. If possible, use the same colors you used for the muscles in Exercises 8-14 and 8-15. Color the different muscles and tendons on the diagram with the corresponding colors.

1. _____
2. _____
3. _____
4. _____
5. _____
6. _____
7. _____
8. _____
9. _____
10. _____
11. _____
12. _____
13. _____

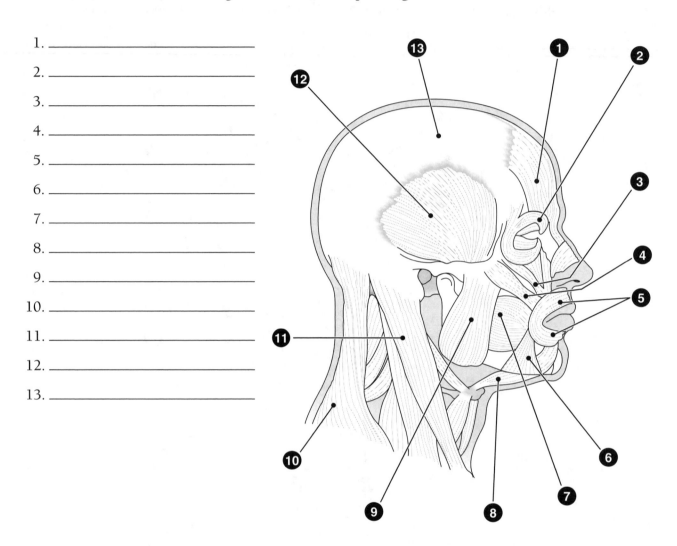

EXERCISE 8-17: Muscles of the Thigh: Anterior View (Text Fig. 8-15A)

Write the name of each labeled muscle, tendon, or bone on the numbered lines in different colors. If possible, use the same colors you used for the muscles in Exercise 8-14. Color the different structures on the diagram with the corresponding colors.

1. _____

2. _____

3. _____

4. _____

5. _____

6. _____

7. _____

8. _____

9. _____

10. _____

11. _____

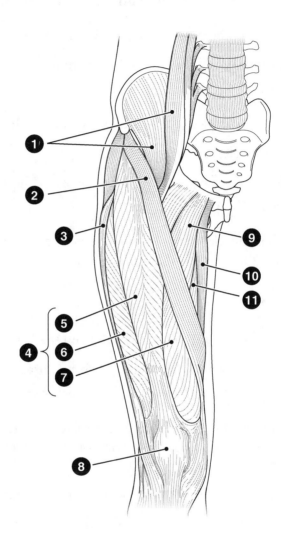

EXERCISE 8-18

Write the appropriate term in each blank from the list below. Not all terms will be used.

sternocleidomastoid   buccinator   masseter   trapezius   orbicularis oris

deltoid   rotator cuff   orbicularis oculi   brachialis

1. The muscle capping the shoulder and upper arm   _____

2. A deep muscle group that supports the shoulder joint   _____

3. A muscle that closes the eye   _____

4. The muscle that makes up the fleshy part of the cheek   _____

5. A muscle that closes the jaw   _____

6. A muscle on the side of the neck that flexes the head   _____

7. A main flexor of the forearm   _____

8. A triangular muscle on the back of the neck and the upper back that extends the head   _____

EXERCISE 8-19

Write the appropriate term in each blank from the list below. Not all terms will be used.

triceps brachii   serratus anterior   brachioradialis   biceps brachii

intercostals   levator ani   erector spinae   latissimus dorsi

1. A large muscle of the middle and lower back that inserts in the humerus and extends the arm at the shoulder behind the back   _____

2. The muscle in the pelvic floor that aids in defecation   _____

3. A muscle on the front of the arm that flexes the elbow and supinates the hand   _____

4. The large muscle on the back of the arm that extends the elbow   _____

5. A chest muscle inferior to the axilla that moves the scapula forward   _____

6. A deep muscle that extends the vertebral column   _____

7. Muscles between the ribs that can enlarge the thoracic cavity   _____

EXERCISE 8-20

Write the appropriate term in each blank from the list below. Not all terms will be used.

rectus abdominis   transversus abdominis   gluteus maximus

gluteus medius   iliopsoas   adductor longus

gracilis   biceps femoris   rectus femoris

1. Member of the quadriceps femoris muscle group _____

2. The muscle that forms much of the fleshy part
   of the buttock _____

3. A deep muscle of the buttock that abducts the thigh _____

4. A vertical muscle covering the anterior surface of
   the abdomen _____

5. A muscle that aids in pressing the thighs together _____

6. A muscle extending from the pubic bone to the tibia
   that adducts the thigh at the hip _____

7. A powerful flexor of the thigh that arises from the ilium _____

EXERCISE 8-21

Write the appropriate term in each blank from the list below. Not all terms will be used.

sartorius   gastrocnemius   soleus   fibularis longus

tibialis anterior   quadriceps femoris   semimembranosus   flexor digitorum group

1. The thin muscle that travels down and across the medial
   surface of the thigh _____

2. The chief muscle of the calf of the leg _____

3. The muscle that inverts and dorsiflexes the foot _____

4. Muscles that flex the toes _____

5. The muscle that everts the foot _____

6. A deep muscle that plantar flexes the foot at the ankle _____

7. A group of four muscles forming the bulk of the
   anterior thigh _____

13. DESCRIBE HOW MUSCLES CHANGE WITH AGE.

EXERCISE 8-22

Your great aunt suffers a minor fall, so you take her to the hospital to be checked. A healthcare worker records her height as 5 ft 2 in. Your great aunt asks you why she is falling so much lately and why her height has decreased in the past two years. Which changes in her muscular system could account for frequent falls and decreased height?

_____

_____

_____

14. LIST THE MAJOR DISORDERS OF MUSCLES AND THEIR ASSOCIATED STRUCTURES.

EXERCISE 8-23

Write the appropriate term in each blank from the list below. Not all terms will be used.

tendinitis	myalgia	rhabdomyolysis	muscular dystrophy
strain	myasthenia gravis	fibromyositis	fibromyalgia syndrome

1. Inflammation of muscle and connective tissue _____

2. A muscle injury resulting from overuse or overstretching _____

3. A group of disorders seen more frequently in male children that cause progressive weakness and paralysis _____

4. Necrosis of muscle cells subsequent to aggressive overtraining _____

5. A disorder associated with muscle aches, tenderness, and stiffness that result from autoimmune reactions _____

6. A general term that means muscle pain _____

7. Plantar fasciitis is a form of this inflammatory problem _____

15. DESCRIBE SOME OF THE DIAGNOSTIC SIGNS OF MUSCULAR DYSTROPHY BASED ON THE CASE STUDY.

EXERCISE 8-24

In the lines below, list the diagnostic findings used by Dr. Schroder to diagnose Shane with muscular dystrophy.

a. Physical exam:

_____

_____

b. Blood test:

_____

_____

c. Muscle biopsy:

_____

_____

d. Genetic testing:

_____

_____

16. SHOW HOW WORD PARTS ARE USED TO BUILD WORDS RELATED TO THE MUSCULAR SYSTEM.

EXERCISE 8-25

Complete the following table by writing the correct word part or meaning in the space provided. Write a word that contains each word part in the "Example" column.

Word Part	Meaning	Example
1. _____	muscle	_____
2. brachi/o	_____	_____
3. _____	strength	_____
4. erg/o	_____	_____
5. -algia	_____	_____
6. _____	four	_____
7. _____	absent, lack of	_____
8. _____	flesh	_____
9. vas/o	_____	_____
10. iso-		

Making the Connections

The following concept map deals with substances and structures required for muscle contraction. Each pair of terms is linked together by a connecting phrase into a sentence. The sentence should be read in the direction of the arrow. Complete the concept map by filling in the appropriate term or phrase. There is one right answer for each term. However, there are many correct answers for the connecting phrases.

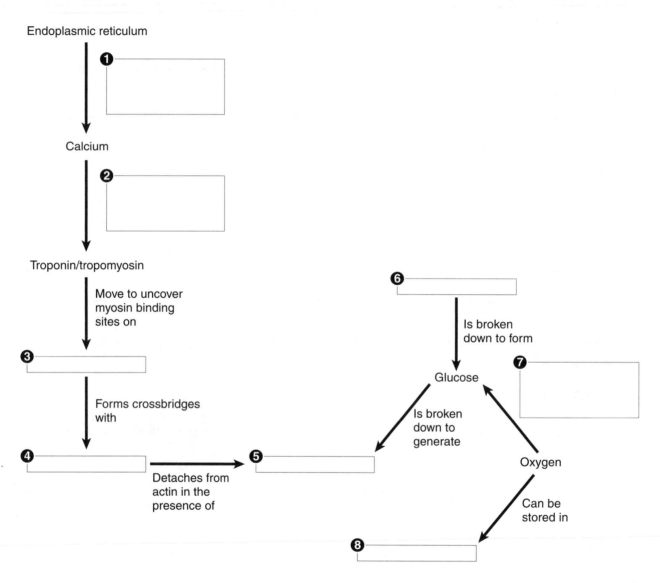

Optional Exercise: Make your own concept map based on the events of muscle contraction. Use the following terms and any others you would like to include: neuron, acetylcholine, neuromuscular junction, synaptic cleft, motor end plate, myosin, actin, endoplasmic reticulum, calcium, sarcomere, troponin/tropomyosin, and ATP.

Testing Your Knowledge

BUILDING UNDERSTANDING

I. MULTIPLE CHOICE

Select the best answer.

1. Which of the following statements is NOT true of skeletal muscle?   1. _____C_____
   a. The cells are long and threadlike.
   b. It is described as striated.
   c. It is involuntary.
   d. The cells are multinucleated.

2. When muscles and bones act together in the body as a lever system, what is the pivot point or fulcrum of the system?   2. _____
   a. joint
   b. tendon
   c. ligament
   d. myoglobin

3. Which of the following is an action of the quadriceps femoris?   3. _____
   a. flex the thigh
   b. extend the leg
   c. adduct the leg
   d. abduct the thigh

4. Which of these terms describes muscle soreness resulting from overuse?   4. _____
   a. fibromyalgia syndrome
   b. fibrositis
   c. strain
   d. tendinitis

5. Which of the following is NOT a muscle of the hamstring group?   5. _____
   a. biceps femoris
   b. rectus femoris
   c. semimembranosus
   d. semitendinosus

6. What is the name of the connective tissue layer around individual muscle fibers?   6. _____
   a. epimysium
   b. perimysium
   c. superficial fascia
   d. endomysium

7. Where does ATP attach during muscle contraction?   7. _____
   a. tropomyosin
   b. myosin
   c. actin
   d. troponin

8. Which of the following processes requires oxygen?   8. _____
    a. ATP generation from fatty acids
    b. glycolysis
    c. creatine phosphate breakdown
    d. lactic acid synthesis
9. Which of these terms describes a broad tendon sheet?   9. _____
    a. ligament
    b. aponeurosis
    c. bursa
    d. meniscus

II. COMPLETION EXERCISE

Write the word or phrase that correctly completes each sentence.

1. Normally, muscles are in a partially contracted state, even when not in use. This state of mild constant tension is called _____.

2. The hemoglobin-like compound that stores oxygen in muscle is _____.

3. The muscle attachment that is usually relatively fixed is called its _____.

4. A contraction in which the muscle lengthens as it exerts force is described as _____.

5. The band of connective tissue that attaches the gastrocnemius muscle to the heel is the _____.

6. The muscles of the pelvic floor together form the _____.

7. A muscle that must relax during a given movement is called the _____.

8. A term that means muscle pain is _____.

9. The muscular partition between the thoracic and abdominal cavities is the _____.

10. A superficial muscle of the neck and upper back acts at the shoulder. This muscle is the _____.

11. The large muscle of the upper chest that flexes the arm across the body is the _____.

12. A spasm of the visceral muscles is known as _____.

13. The muscle responsible for dorsiflexion and inversion of the foot is the _____.

UNDERSTANDING CONCEPTS

I. TRUE/FALSE

For each statement, write T for true or F for false in the blank to the left of each number. If a statement is false, correct it by replacing the underlined term, and write the correct statement in the blanks below the statement.

_____ 1. The triceps brachii <u>flexes</u> the arm at the elbow.

_____

_____ 2. Muscle fatigue can result when <u>phosphate</u> accumulates in the cell.

_____

_____ 3. In an <u>isometric</u> contraction, muscle tension increases but the muscle does not shorten.

_____

_____ 4. A muscle that stabilizes a body part during a movement is called a(n) <u>synergist</u>.

_____

_____ 5. The neurotransmitter used at the neuromuscular junction is <u>norepinephrine</u>.

_____

_____ 6. A contracting subunit of skeletal muscle is called a(n) <u>cross-bridge</u>.

_____

_____ 7. The storage form of glucose is called <u>creatine phosphate</u>.

_____

_____ 8. The element that binds to the troponin and tropomyosin complex is <u>calcium</u>.

_____

II. PRACTICAL APPLICATIONS

Study each discussion. Then write the appropriate word or phrase in the space provided.

➤ **Group A**

Ms. J is sitting at her desk studying for her anatomy final exam.

1. She has excellent posture with her back straight. The deep back muscle responsible for her erect posture is the _____.

2. Despite her excellent posture, Ms. J has developed a muscle spasm that has fixed her head in a flexed, rotated position. This condition is called wryneck, or _____.

3. Ms. J has a cheerful disposition, and she likes to whistle while she works. The cheek muscle involved in whistling is the _____.

4. Ms. J is furiously writing notes, and her hand is flexed around the pen. The muscle groups that flex the hand are called the _____.

5. Every few pages, Ms. J flexes her arm. This action requires contraction of the anterior portion of both the pectoralis major and the _____.

6. After several hours of intense studying, Ms. J takes a break. Stretching, she straightens her leg at the knee joint. The muscle that accomplishes this action is the _____.

➤ Group B

Ms. K, age 29, joined a gym that offered an aggressive form of cross-training.

1. They begin the workout with a 20-minute run. Endurance training increases the quantity of a specific organelle that generates ATP aerobically. This organelle is the _____.

2. Next, they begin on the resistance training portion of the workout. Ms. K asks if the resistance training will give her bigger muscles. She is told that her muscle cells may increase in size, a change called _____.

3. Ms. K's muscles ache the day after her first workout, probably reflecting microtears in her muscle. This stiffness and pain is known as _____.

4. Ms. K enjoyed her workout so much that she started training every day. The day after a particularly vigorous workout, she noticed swelling in her legs and tea-colored urine. She was told that her muscle cells had been overworked and were undergoing necrosis, and that the muscle cell contents were damaging her kidneys. This disorder is known as _____.

III. SHORT ESSAYS

1. For each of the following word pairs, write a sentence explaining their roles in muscle contraction.
   a. glycogen and glucose

   _____

   _____

   _____

   b. calcium and oxygen

   _____

   _____

   _____

   c. acetylcholine and ATP

   _____

   _____

   _____

2. Ms. L has suffered an embarrassing (but relatively painless) fall. Mr. L is staring at her with his mouth open. Ms. L asks him to activate two muscles to close his jaw. Name these two muscles.

_____

_____

_____

CONCEPTUAL THINKING

1. Your great uncle has decided that it's time to get in shape, so he has started jogging four times weekly. He complains that he gets tired easily because of all of the lactic acid accumulating in his muscles. Do you think he's right? Discuss the causes of muscle fatigue in your answer.

_____

_____

_____

2. After two weeks your great uncle wants to quit his exercise program, claiming that it isn't doing him any good and makes him too tired to watch late-night hockey game reruns. What would you tell him about the benefits of endurance exercise?

_____

_____

_____

3. During his physical exam, Shane would have been asked to perform various actions. For each of these actions, state the term that describes the movement and which muscles accomplished the action.

	Movement (e.g., flexion)	Muscle(s) involved
Bend arm at the elbow		
Straighten arm at the elbow		
Stand on tiptoe		
Straighten leg at knee		

Expanding Your Horizons

Are world-class athletes born or made? It is no coincidence that athletic performance tends to run in families. Genetic influences on muscular function are discussed in an article in *Scientific American.*

- Andersen JL, Schjerling P, Saltin B. Muscle, genes and athletic performance. Sci Am 2000;283(3):48–55.

This information could be used to screen for elite athletes—perhaps children of the future will know in which sports they can excel based on their genetic profiles. This possibility is discussed in a special *Scientific American* issue entitled "Building the Elite Athlete."

- Taubes G. Toward Molecular Talent Scouting. Scientific American Presents: Building the Elite Athlete 2000;11:26–31.

Coordination and Control

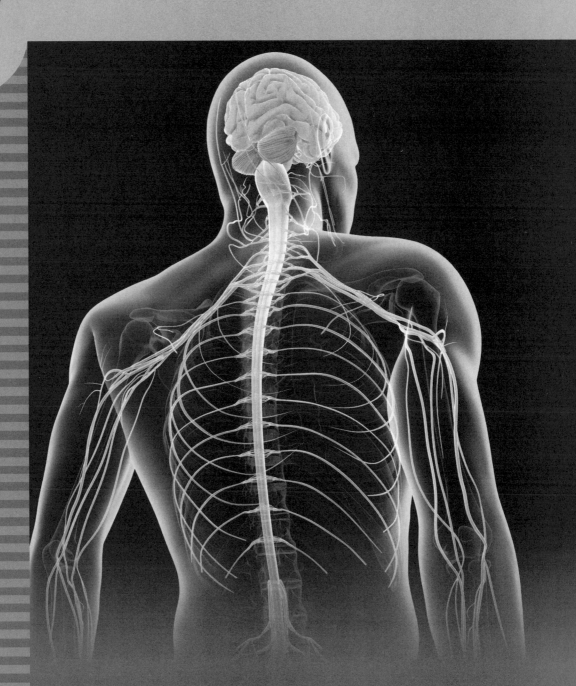

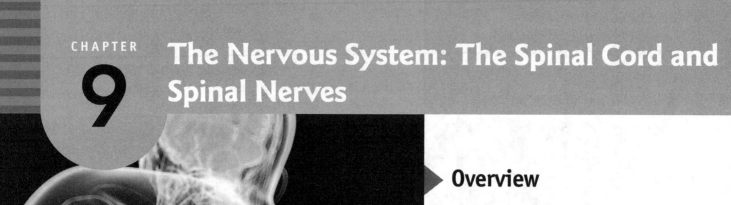

9

The Nervous System: The Spinal Cord and Spinal Nerves

▶ Overview

The nervous system is the body's coordinating system, receiving, sorting, and controlling responses to both internal and external changes (stimuli). The nervous system as a whole is divided structurally into the **central nervous system (CNS)**, made up of the brain and the spinal cord, and the **peripheral nervous system (PNS)**, made up of the cranial and spinal nerves. The PNS connects all parts of the body with the CNS. The brain and cranial nerves are the subject of Chapter 10. Functionally, the peripheral nervous system is divided into the **somatic (voluntary) system** and the **autonomic (involuntary) system**.

The nervous system functions by means of the **nerve impulse**, an electrical current or **action potential** that spreads along the membrane of the **neuron** (nerve cell). Each neuron is composed of a cell body and fibers, which are threadlike extensions from the cell body. A **dendrite** is a fiber that carries impulses toward the cell body, and an **axon** is a fiber that carries impulses away from the cell body. Some axons are covered with a sheath of fatty material called **myelin**, which insulates the fiber and speeds conduction along the fiber. In the PNS, neuron fibers are collected in bundles to form **nerves**. Bundles of fibers in the CNS are called **tracts**. Nerve cells make contact at a junction called a **synapse**. Here, a nerve impulse travels across a very narrow cleft between the cells by means of a chemical referred to as a **neurotransmitter**. Neurotransmitters are released from axons of presynaptic cells to be picked up by receptors in the membranes of responding cells, the postsynaptic cells.

A neuron may be classified as a **sensory (afferent)** type, which carries impulses toward the CNS, or a **motor (efferent)** type, which carries impulses away from the CNS. **Interneurons** are connecting neurons within the CNS.

The basic functional pathway of the nervous system is the **reflex arc**, in which an impulse travels from a receptor, along a sensory neuron to a synapse or synapses in the CNS, and then along a motor neuron to an effector organ that carries out a response.

The spinal cord carries impulses to and from the brain. It is also a center for simple reflex activities in which responses are coordinated within the cord.

The **autonomic nervous system** controls unconscious activities. This system regulates the actions of glands, smooth muscle, and the heart muscle. The autonomic nervous system has two divisions, the **sympathetic nervous system** and the **parasympathetic nervous system**, which generally have opposite effects on a given organ.

Addressing the Learning Objectives

1. OUTLINE THE ORGANIZATION OF THE NERVOUS SYSTEM ACCORDING TO STRUCTURE AND FUNCTION.

EXERCISE 9-1: Anatomic Divisions of the Nervous System (Text Fig. 9-1)

1. Label the structures of the nervous system by writing the correct term in spaces 1 to 4. Label the two anatomic divisions of the nervous system by writing the correct term in boxes 5 and 6.

1. _____

2. _____

3. _____

4. _____

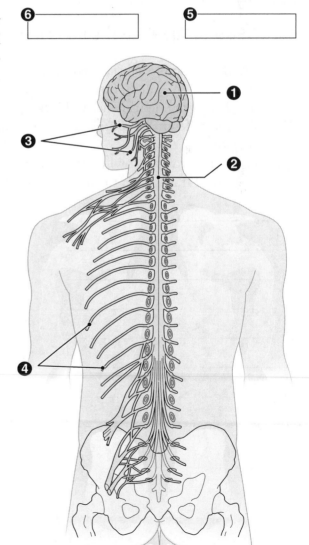

EXERCISE 9-2

Write the appropriate term in each blank from the list below.

central nervous system   peripheral nervous system   somatic nervous system

sympathetic nervous system   parasympathetic nervous system   autonomic nervous system

1. The functional division that controls the heart   _____

2. The functional division that controls skeletal muscles   _____

3. The system that promotes the fight-or-flight response   _____

4. The system that stimulates the activity of the digestive tract   _____

5. The structural division that includes the brain   _____

6. The structural division that includes the cranial nerves   _____

2. DESCRIBE THE STRUCTURE OF A NEURON.

Also see Exercise 9-4.

EXERCISE 9-3: The Motor Neuron (Text Fig. 9-2)

1. Write the name of each labeled part on the numbered lines in different colors. Structures 4 and 5 will not be colored, so write their names in black.
2. Color the different structures on the diagram with the corresponding colors.
3. Add large arrows showing the direction the nerve impulse will travel from the dendrites to the muscle.

1. _____

2. _____

3. _____

4. _____

5. _____

6. _____

7. _____

8. _____

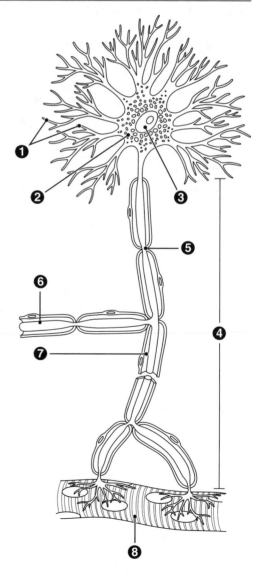

3. EXPLAIN THE CONSTRUCTION AND FUNCTION OF THE MYELIN SHEATH.

EXERCISE 9-4

Write the appropriate term in each blank from the list below.

dendrite   neurilemma   axon

node   white matter   gray matter

1. A neuron fiber that carries impulses away from the cell body   _____

2. The part of a neuron that receives a stimulus   _____

3. The sheath around some neuron fibers that aids in regeneration   _____

4. A gap in the neuron sheath   _____

5. The portion of the spinal cord made up of myelinated axons   _____

6. Neural tissue composed of cell bodies and unmyelinated axons   _____

EXERCISE 9-5: Formation of a Myelin Sheath (Text Fig. 9-4)

1. Write the name of each labeled part on the numbered lines in different colors. Structures 5, 7, 8, and 9 will not be colored, so write their names in black.
2. Color the different structures on the diagram with the corresponding colors. Make sure you color the structure in all parts of the diagram. For instance, structure 3 is visible in three locations.

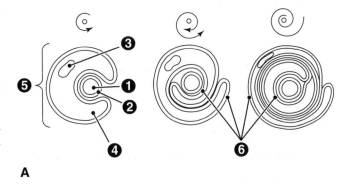

A

1. _____

2. _____

3. _____

4. _____

5. _____

6. _____

7. _____

8. _____

9. _____

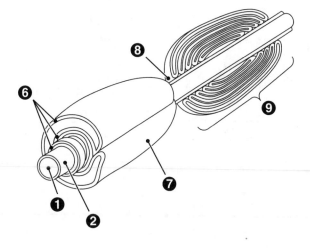

B

4. DESCRIBE HOW NEURON FIBERS ARE BUILT INTO A NERVE.

EXERCISE 9-6

Write the appropriate term in each blank from the list below.

neuron   tract   endoneurium   perineurium

epineurium   nerve   fascicle

1. The scientific name for a nerve cell   _____

2. A group of axons within a nerve or tract   _____

3. A bundle of neuron fibers located outside the central nervous
   system   _____

4. A bundle of neuron fibers located within the central nervous
   system   _____

5. The coating of an individual neuron   _____

6. The coating of an entire nerve   _____

7. The coating of a small group of axons within a nerve   _____

5. LIST FOUR TYPES OF NEUROGLIA IN THE CENTRAL NERVOUS SYSTEM, AND CITE THE FUNCTIONS OF EACH.

EXERCISE 9-7

Complete the following table. Use Figure 9-6 and information in the narrative.

Name	Appearance	Function
		Forms the myelin sheath of CNS neurons
Microglia		
	Cuboidal cells lining ventricles	
	Star-shaped cells	

6. DIAGRAM AND DESCRIBE THE STEPS IN AN ACTION POTENTIAL.

EXERCISE 9-8

1. In the space below, draw an action potential tracing as illustrated in Figure 9-7. On your diagram, indicate which event is occurring (rising phase, falling phase, or resting) and which ion is moving (sodium [Na^+] or potassium [K^+]), if any.

EXERCISE 9-9

Write the appropriate term in each blank from the list below.

Na^+ K^+ hyperpolarization   depolarization

repolarization   rising phase   falling phase

1. Any change that brings membrane potential closer to rest   _____

2. Any change that makes the membrane potential less negative
   than at rest   _____

3. Any change that makes the membrane potential more negative
   than at rest   _____

4. The phase of the action potential in which the membrane
   potential changes from about –70 mV to +55 mV   _____

5. The phase of the action potential in which the membrane
   potential changes from +55 mV to –70 mV   _____

6. The ion that enters the cell during the rising phase of the action
   potential   _____

7. The ion that leaves the cell during the falling phase of the action
   potential   _____

EXERCISE 9-10

Place the following events in order by writing the appropriate numbers (1 through 4) in the blanks.

_____ a. The action potential opens sodium channels in adjacent portions of the membrane.

_____ b. Sodium entry into the cell causes another action potential in the adjacent portion of the
        membrane.

_____ c. A stimulus initiates an action potential.

_____ d. Open sodium channels let sodium enter the cell.

EXERCISE 9-11

Label each of the following statements as true (T) or false (F).

1. Action potentials occur in axon regions surrounded by myelin. _____

2. Action potential transmission is faster in myelinated neurons. _____

3. Action potentials occur at nodes in myelinated axons. _____

7. EXPLAIN THE ROLE OF NEUROTRANSMITTERS IN IMPULSE TRANSMISSION AT A SYNAPSE.

EXERCISE 9-12: A Synapse (Text Fig. 9-10)

Label the parts of the synapse shown below.

1. _____

2. _____

3. _____

4. _____

5. _____

6. _____

7. _____

8. _____

9. _____

10. _____

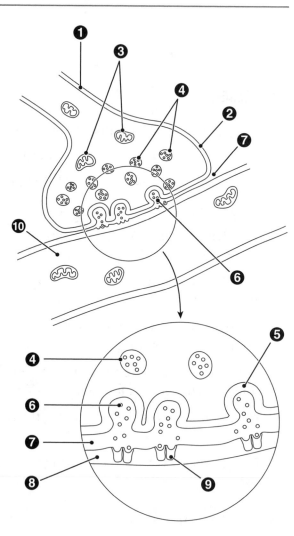

EXERCISE 9-13

Place the following events in order by writing the appropriate numbers (1 through 5) in the blanks.

_____ a. Neurotransmitter molecules bind to receptors in the postsynaptic membrane.

_____ b. Neurotransmitter molecules are released into the synaptic cleft.

_____ c. The nerve impulse arrives at the end of the presynaptic neuron.

_____ d. Vesicles containing neurotransmitter fuse with the axonal plasma membrane.

_____ e. The activity of the postsynaptic cell is altered.

EXERCISE 9-14

List three examples of neurotransmitters in the spaces below.

1. _____

2. _____

3. _____

8. DESCRIBE THE DISTRIBUTION OF GRAY AND WHITE MATTER IN THE SPINAL CORD.

EXERCISE 9-15: The Spinal Cord (Text Fig. 9-11B)

1. Write the name of each labeled part on the numbered lines. Use the following color scheme:
   - 1 and 2: red
   - 3, 4, 9: different dark colors
   - 5: pink
   - 6: any light color
   - 7: light blue
   - 10: medium blue
   - 11: purple
2. Color or outline the different structures on the diagram with the corresponding colors. Color each structure on both sides of the spinal cord, not just on the side that is labeled.

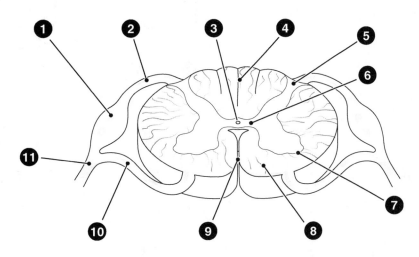

1. _____   7. _____

2. _____   8. _____

3. _____   9. _____

4. _____   10. _____

5. _____   11. _____

6. _____

9. DESCRIBE AND NAME THE SPINAL NERVES AND THREE OF THEIR MAIN PLEXUSES.

EXERCISE 9-16: The Spinal Cord (Text Fig. 9-11A)

Write the name of each structure on the appropriate line.

1. _____

2. _____

3. _____

4. _____

5. _____

6. _____

7. _____

8. _____

9. _____

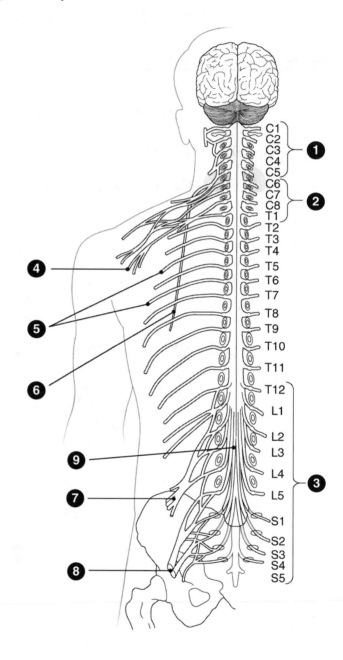

10. LIST THE COMPONENTS OF A REFLEX ARC.

EXERCISE 9-17: The Reflex Arc (Text Fig. 9-13)

1. Write the names of the five components of a reflex arc on the numbered lines 1 to 5 in different colors, and color the components with the appropriate color. Follow the color scheme provided below.

2. Write the names of the parts of the spinal cord on numbered lines 6 to 12 in different colors, and color the structures with the corresponding colors. Use the following color scheme:
   - 1, 2, 6, 7, 8: red
   - 3: green
   - 4, 10: medium blue
   - 5: purple
   - 9: do not color (write name in black)
   - 11: pink
   - 12: light blue

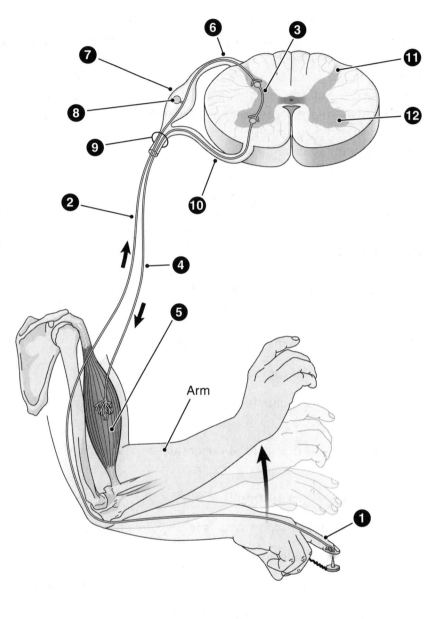

1. _____

2. _____

3. _____

4. _____

5. _____

6. _____

7. _____

8. _____

9. _____

10. _____

11. _____

12. _____

11. DEFINE A SIMPLE REFLEX, AND GIVE SEVERAL EXAMPLES OF REFLEXES.

EXERCISE 9-18

Use the terms below to complete the paragraph.

patellar reflex   simple reflex   somatic reflex   autonomic reflex

A(n) (1) _____ describes any rapid automatic response involving very few neurons. Reflexes involving skeletal muscles are called (2) _____; reflexes involving smooth muscle or glands are (3) _____. An example of the former type is the (4) _____, a stretch reflex that involves striking a tendon in the knee region.

12. COMPARE THE LOCATIONS AND FUNCTIONS OF THE SYMPATHETIC AND PARASYMPATHETIC NERVOUS SYSTEMS.

EXERCISE 9-19

Fill in the blank after each statement—does it apply to the sympathetic nervous system (S) or the parasympathetic nervous system (P)?

1. Also described as the adrenergic system   _____

2. Also described as the cholinergic system   _____

3. Motor neurons originate in the thoracolumbar region of the spinal cord   _____

4. Motor neurons originate in the craniosacral region of the spinal cord   _____

5. Activation causes the pupils to dilate   _____

6. Activation causes blood vessels in digestive organs to dilate   _____

7. Terminal ganglia located in or near the effector   _____

8. Ganglia located near the spinal cord or in collateral ganglia   _____

9. Activation decreases kidney activity   _____

10. Activation stimulates the sweat glands   _____

13. EXPLAIN THE ROLE OF CELLULAR RECEPTORS IN THE ACTION OF NEUROTRANSMITTERS IN THE AUTONOMIC NERVOUS SYSTEM.

EXERCISE 9-20

Write the appropriate term in each blank from the list below.

muscarinic receptor   nicotinic receptor   adrenergic receptor

1. Binds norepinephrine   _____

2. Binds acetylcholine and induces muscle contraction   _____

3. Acetylcholine receptor found on effector organs of the parasympathetic system   _____

14. DESCRIBE EIGHT DISORDERS OF THE SPINAL CORD AND SPINAL NERVES.

EXERCISE 9-21

Write the appropriate term in each blank from the list below. Not all terms will be used.

paraplegia   shingles   mononeuropathy   lumbar puncture
epidural   poliomyelitis   monoplegia   polyneuropathy
amyotrophic lateral sclerosis

1. Carpal tunnel syndrome and sciatica are examples of this disorder   _____

2. Technique used to extract CSF for diagnostic procedures   _____

3. Paralysis of one arm   _____

4. A viral disease resulting in paralysis   _____

5. A viral disease caused by herpes zoster   _____

6. Loss of sensation and motion in the lower part of the body   _____

7. A degenerative disorder associated with selective destruction of motor neurons   _____

8. Damage to multiple nerves, such as that resulting from diabetes mellitus   _____

15. USING THE CASE STUDY, DESCRIBE THE EFFECTS OF DEMYELINATION ON MOTOR AND SENSORY FUNCTION.

EXERCISE 9-22

Use the terms below to complete the paragraph. Not all terms will be used.

sensory	motor	descending	ascending	dorsal
ventral	myelin	multiple sclerosis	white matter	gray matter

Sue suffered from (1) _____, a disease associated with demyelination. The (2) _____ coating of neuronal axons was destroyed by an auto-immune reaction. Lesions were apparent in the type of spinal cord tissue containing myelinated axons, known as (3) _____. Sue's sense of touch was impaired because of damage to neurons in the (4) _____ tracts of her spinal cord. These neurons are known as (5) _____, or afferent neurons, and they receive input from neurons entering the spinal cord through the (6) _____ root. Sue's muscle control was impaired because of damage to (7) _____ neurons in the (8) _____ tracts of her spinal cord. Motor impulses from this pathway leave the spinal cord through the (9) _____ root.

16. SHOW HOW WORD PARTS ARE USED TO BUILD WORDS RELATED TO THE NERVOUS SYSTEM.

EXERCISE 9-23

Complete the following table by writing the correct word part or meaning in the space provided. Write a word that contains each word part in the "Example" column.

Word Part	Meaning	Example
1. _____	sheath	_____
2. para-	_____	_____
3. _____	four	_____
4. soma-	_____	_____
5. hemi-	_____	_____
6. _____	paralysis	_____
7. _____	nerve, nervous tissue	_____
8. _____	remove	_____
9. aut/o	_____	_____
10. post-_____	_____	_____

Making the Connections

The following concept map deals with the organization of the nervous system. Each pair of terms is linked together by a connecting phrase into a sentence. The sentence should be read in the direction of the arrow. Complete the concept map by filling in the appropriate term or phrase. There is one right answer for each term. However, there are many correct answers for the connecting phrases (6, 7).

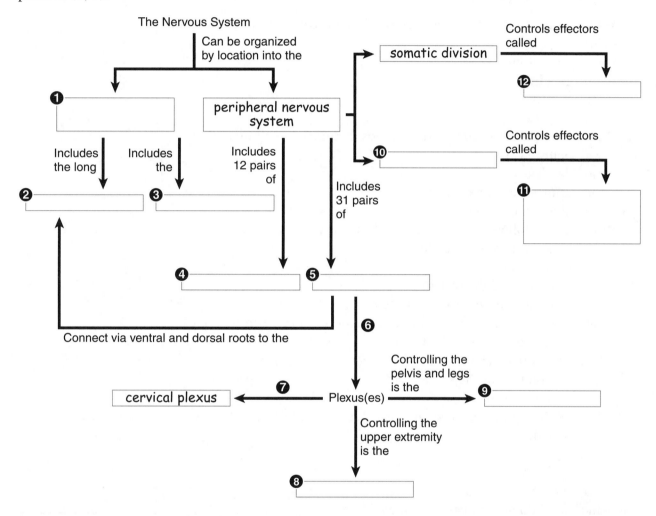

Optional Exercise: Make your own concept map based on the structures of the spinal cord and the components of a reflex loop. Use the following terms and any others you would like to include: dorsal root ganglion, gray matter, white matter, ventral root ganglion, sensory neuron, motor neuron, receptor, muscle, gland, and effector.

Testing Your Knowledge

BUILDING UNDERSTANDING

I. MULTIPLE CHOICE

Select the best answer.

1. Which of the following terms describes a skin region supplied by a single spinal nerve?   1. _____
   a. dermatome
   b. ganglion
   c. plexus
   d. synapse
2. Which of these structures would be most abundant in white matter?   2. _____
   a. neuron cell bodies
   b. dendrites
   c. unmyelinated axons
   d. myelinated axons
3. Which of the following are effectors of the nervous system?   3. _____
   a. sensory neurons and ganglia
   b. muscles and glands
   c. synapses and dendrites
   d. receptors and neurotransmitters
4. Where would you find cell bodies of somatic sensory neurons?   4. _____
   a. dorsal root of the spinal cord
   b. sympathetic chain
   c. ventral root of the spinal cord
   d. effector organ
5. Which of the following substances is a neurotransmitter?   5. _____
   a. myelin
   b. actin
   c. epinephrine
   d. sebum
6. Which fluid does a lumbar tap remove?   6. _____
   a. blood
   b. cerebrospinal fluid
   c. intracellular fluid
   d. interstitial fluid
7. Which neuron type conveys impulses *within* the spinal cord?   7. _____
   a. sensory neurons
   b. motor neurons
   c. interneurons
   d. mixed nerves

II. COMPLETION EXERCISE

Write the word or phrase that correctly completes each sentence.

1. Fibers that carry impulses toward the neuron cell body are called _____.

2. The portion of the spinal cord made up of cell bodies and unmyelinated axons is called the

   _____.

3. The fatty material that covers some axons is called _____.

4. Dilation of the bronchial tubes is increased by the part of the autonomic nervous system called

   the _____.

5. The tiny space separating the presynaptic and postsynaptic neurons is

   the _____.

6. The network of spinal nerves that supplies the pelvis and legs is

   the _____.

7. The brain and spinal cord together are referred to as the _____.

8. The neurotransmitter used at cholinergic synapses is _____.

9. The small channel in the center of the spinal cord that contains cerebrospinal fluid is

   the _____.

10. The ion responsible for the rising phase of the action potential is _____.

11. The bridge of gray matter connecting the right and left horns of the spinal cord is

    the _____.

UNDERSTANDING CONCEPTS

I. TRUE/FALSE

For each statement, write T for true or F for false in the blank to the left of each number. If a statement is false, correct it by replacing the underlined term, and write the correct statement in the blanks below the statement.

_____   1.   Motor impulses leave the <u>dorsal</u> horn of the spinal cord.

_____

_____   2.   The delivery of pain medication into the fat-filled area surrounding the spinal cord is called <u>spinal anesthesia</u>.

_____

_____   3.   A <u>tract</u> is a bundle of neuron fibers within the central nervous system.

_____

_____ 4. The <u>parasympathetic</u> system has terminal ganglia.

_____

_____ 5. The <u>brachial plexus</u> controls the shoulder and arm.

_____

_____ 6. The parasympathetic system is <u>adrenergic</u>.

_____

_____ 7. The cranial nerves are part of the <u>central</u> nervous system.

_____

_____ 8. Neurotransmitters bind to specific proteins on the postsynaptic cell called <u>transporters</u>.

_____

_____ 9. At a synapse, a neurotransmitter is released from the <u>postsynaptic</u> cell.

_____

_____ 10. A reflex arc that passes through the spinal cord but not the brain is called a <u>spinal</u> reflex.

_____

II. PRACTICAL APPLICATIONS

Study each discussion. Then write the appropriate word or phrase in the space provided.

1. Mr. W, a patient with diabetes mellitus for 10 years, complained of pain and numbness of his feet. In observing Mr. W walk, the physician noted there was weakness in the muscles responsible for dorsiflexion of the foot. These symptoms are caused by a degenerative disorder of multiple nerves supplying the extremities. This disorder is known as _____.

2. The nerves that supply the foot are found in a plexus called the _____.

3. The physician pricked Mr. W's foot with a needle. Mr. W did not feel the needle prick, suggesting that there was a problem with the nerves that carry impulses to the brain. These nerves are called _____.

4. The physician tapped below Mr. W's knee to elicit a knee-jerk response. The tendon she struck was the _____.

5. When the tendon was stretched, it activated a receptor. The type of neuron that conveyed the signal from the receptor to the spinal cord is a(n) _____.

6. Mr. W knew that his tendon had been tapped, because signals passed to his brain through a nerve tract in the spinal cord called the _____.

7. The effector in this reflex arc is the _____.

III. SHORT ESSAYS

1. List the events that occur in an action potential.

_____

_____

_____

2. Compare and contrast:
   a. poliomyelitis and amyotrophic lateral sclerosis; name one similarity and one difference.
   b. multiple sclerosis and Guillain-Barré syndrome; name two similarities and two differences.

_____

_____

_____

CONCEPTUAL THINKING

1. Ms. J is teaching English in Japan. She dines on a local delicacy called pufferfish, and shortly thereafter her lips go numb. She later discovers that pufferfish contain a toxin that blocks sodium channels. Explain why her lips are numb.

_____

_____

_____

2. Dopamine is a neurotransmitter involved in feelings of pleasure. Cocaine blocks the reuptake of dopamine. Use this information to discuss how cocaine affects mood.

_____

_____

_____

Expanding Your Horizons

Christopher Reeve of *Superman* fame is probably the best known victim of a spinal cord injury. An equestrian injury pulverized his first and second cervical vertebrae, resulting in complete paralysis of his limbs and semiparalysis of his respiratory muscles. He became a strong advocate for spinal cord research, establishing the *Christopher and Dana Reeve Foundation* for spinal cord research and patient support. Check out the Foundation's website for patient stories and breaking research news, or read the article listed below about advances.

- Griffith A. Healing broken nerves. Sci Am 2007;297:28–30.
- Christopher and Dana Reeve Foundation. Available at: http://www.christopherreeve.org

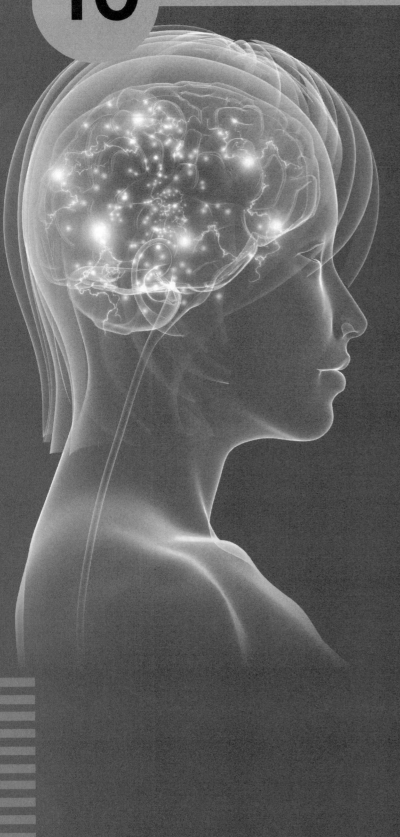

10 The Nervous System: The Brain and Cranial Nerves

▶ Overview

The brain consists of the two cerebral hemispheres, the diencephalon, the brain stem, and the cerebellum. Each cerebral hemisphere is covered by a layer of gray matter, the **cerebral cortex**, which is further divided into four lobes (the frontal, parietal, temporal, and occipital lobes). Specific functions have been localized to the different lobes. For instance, the interpretation of visual images is performed by an area of the occipital lobe. The **diencephalon** consists of the **thalamus**, an important relay station for sensory impulses, and the **hypothalamus**, which plays an important role in homeostasis. The **brain stem** links the spinal cord to the brain and regulates many involuntary functions necessary for life, whereas the **cerebellum** is involved in coordination and balance. The **limbic system** is a widespread neuronal network involved in emotion, learning, and memory.

The brain and spinal cord are covered by three layers of fibrous membranes called the **meninges**. The **cerebrospinal fluid** (CSF) also protects the brain and spinal cord by providing support and cushioning. The CSF is produced by the choroid plexuses (capillary networks) in four ventricles (spaces) within the brain.

Connected with the brain are 12 pairs of **cranial nerves**, most of which supply structures in the head. Most of these, like all the spinal nerves, are mixed nerves containing both sensory and motor fibers. A few of the cranial nerves contain only sensory fibers, whereas others are motor in function.

Addressing the Learning Objectives

1. GIVE THE LOCATIONS OF THE FOUR MAIN DIVISIONS OF THE BRAIN.

EXERCISE 10-1: Brain, Sagittal Section (Text Fig. 10-1)

1. Write the names of the four labeled brain divisions in lines 1 to 4, using four different colors. Use red for #2 and blue for #3. Do not color the diagram yet.
2. Write the name of each labeled structure on the appropriate numbered line in different colors. Use different shades of red for structures 5 and 6 and different shades of blue for structures 8 to 10.
3. Color each structure on the diagram with the corresponding colors.

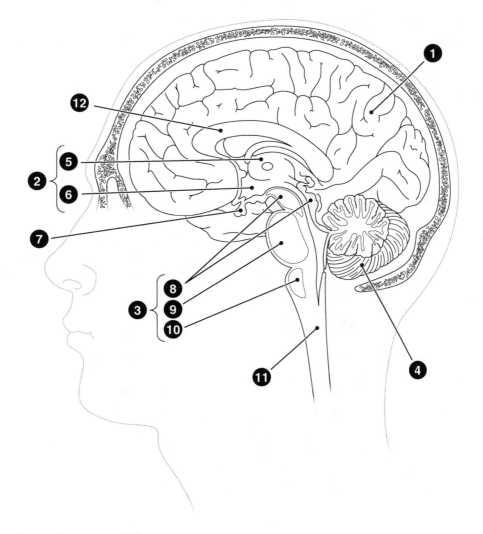

1. _____

2. _____

3. _____

4. _____

5. _____

6. _____

7. _____

8. _____

9. _____

10. _____

11. _____

12. _____

EXERCISE 10-2

Write the appropriate term in each blank from the list below.

lobe   hemisphere   cerebrum
diencephalon   brain stem   cerebellum

1. Each half of the cerebrum   _____

2. The "little" brain that coordinates voluntary muscle movements   _____

3. An individual subdivision of the cerebrum that regulates
   specific functions   _____

4. The portion of the brain that contains the thalamus and
   hypothalamus   _____

5. Connects the spinal cord with the brain   _____

6. The largest part of the brain   _____

2. NAME AND DESCRIBE THE THREE MENINGES.

(Also see Exercise 10-6.)

EXERCISE 10-3: Meninges and Related Parts (Text Fig. 10-2)

1. Write the name of each labeled part on the numbered lines in different colors. Use the same
   color for structures 7 and 8. Write the names of structures 4, 9, 10, and 12 in black.
2. Color the structures on the diagram with the corresponding colors. Do not color structures 4,
   9, 10, and 12.

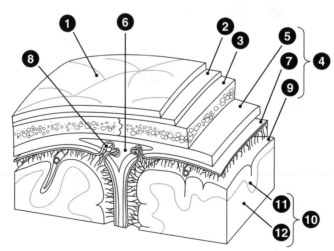

1. _____   7. _____

2. _____   8. _____

3. _____   9. _____

4. _____   10. _____

5. _____   11. _____

6. _____   12. _____

3. CITE THE FUNCTION OF CEREBROSPINAL FLUID, AND DESCRIBE WHERE AND HOW THIS FLUID IS FORMED.

EXERCISE 10-4: Ventricles of the Brain (Text Fig. 10-3A)

1. Write the names of each labeled part on the numbered lines in different colors.
2. Color the structures on the diagram with the corresponding colors. The boundaries between structures are not always well defined. For instance, structure 4 is continuous with structure 5. You can overlap your colors to signify this fact.

1. _____

2. _____

3. _____

4. _____

5. _____

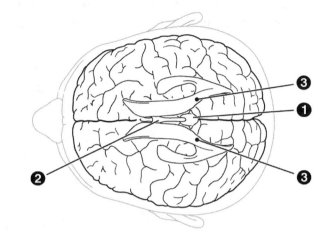

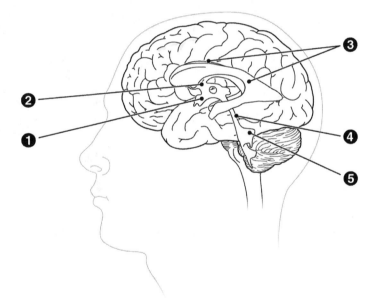

EXERCISE 10-5: Flow of Cerebrospinal Fluid (Text Fig. 10-3B)

1. Write the name of each labeled part on the numbered lines in different colors. Use light colors for structures 5 to 12.
2. Color the structures on the diagram with the corresponding colors. The boundaries between structures 5 to 12 (inclusive) are not always well defined. For instance, structure 6 is continuous with structure 7. You can overlap your colors to signify this fact.
3. Draw arrows to indicate the direction of CSF flow.

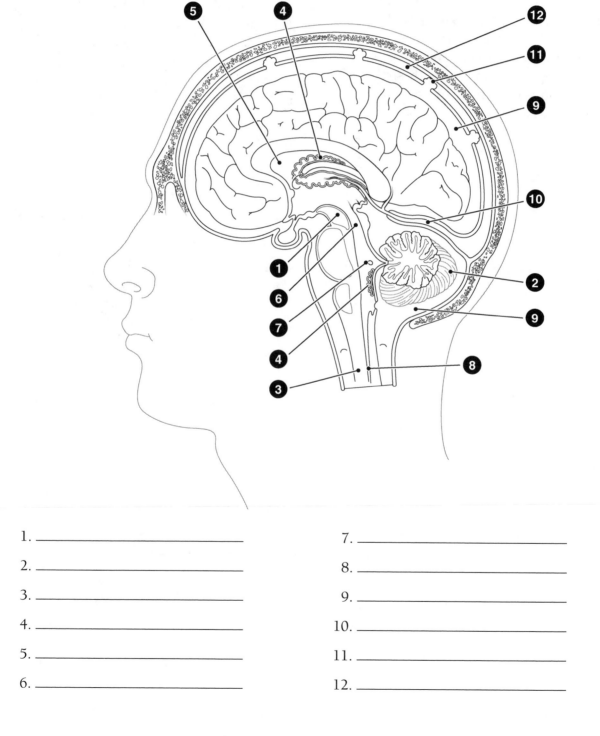

1. _____

2. _____

3. _____

4. _____

5. _____

6. _____

7. _____

8. _____

9. _____

10. _____

11. _____

12. _____

EXERCISE 10-6

Write the appropriate term in each blank from the list below. Not all terms will be used.

dura mater   pia mater   arachnoid   choroid plexus

subarachnoid space   arachnoid villi   ventricle   dural sinus

cerebral aqueduct

1. The weblike middle meningeal layer _____

2. Venous channel between the two outermost meninges _____

3. The innermost layer of the meninges, the delicate membrane in which there are many blood vessels _____

4. The area in which cerebrospinal fluid collects before its return to the blood _____

5. The vascular network in a ventricle that forms cerebrospinal fluid _____

6. The projections in the dural sinuses through which CSF is returned to the blood _____

7. The outermost layer of the meninges, which is the thickest and the toughest _____

4. NAME AND LOCATE THE LOBES OF THE CEREBRAL HEMISPHERES.

(Also see Exercise 10-8)

EXERCISE 10-7

Write the appropriate term in each blank from the list below.

gyrus   central sulcus   lateral sulcus   insula

corpus callosum   longitudinal fissure   cortex

1. A shallow groove that separates the temporal lobe from the frontal and parietal lobes _____

2. A small cerebral lobe located deep within each hemisphere _____

3. The deep groove separating the two cerebral hemispheres _____

4. An elevated portion of the cerebral cortex _____

5. The thin layer of gray matter on the surface of the cerebrum _____

6. A band of myelinated fibers that bridges the two cerebral hemispheres _____

7. A shallow groove separating the frontal and parietal lobes _____

5. CITE ONE FUNCTION OF THE CEREBRAL CORTEX IN EACH LOBE OF THE CEREBRUM.

EXERCISE 10-8: Functional Areas of the Cerebral Cortex (Text Fig. 10-5)

1. Color the boxes next to the cerebral lobe names as follows: frontal lobe, pink; parietal lobe, light purple; temporal lobe, light blue; occipital lobe, light green.
2. Lightly color the four cerebral lobes on the diagram with the appropriate colors.
3. Write the names of structures 1 to 12 on the appropriate lines in different colors. For all structures except for 3, use a darker color than the one used for the corresponding cerebral lobe. For instance, a structure found in the frontal lobe could be colored red. Use the same color for structures 6 to 8. Use a dark color for structure 3. Use black for 11, because it involves multiple cerebral lobes.
4. Color or outline the structures on the diagram with the corresponding colors.

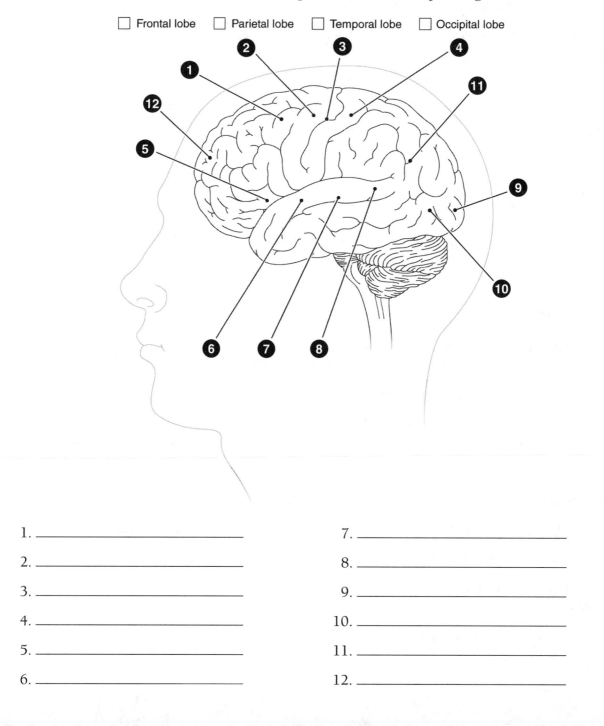

☐ Frontal lobe   ☐ Parietal lobe   ☐ Temporal lobe   ☐ Occipital lobe

1. _____   7. _____

2. _____   8. _____

3. _____   9. _____

4. _____   10. _____

5. _____   11. _____

6. _____   12. _____

EXERCISE 10-9

Write the appropriate term in each blank from the list below. If you like, use different colored pens to write each term, following this color scheme (note that one structure will need three colors!): Frontal lobe regions, orange; temporal lobe regions, blue; parietal lobe regions, purple; occipital lobe regions, green.

premotor cortex   prefrontal cortex   Broca area

primary motor area   primary somatosensory area   somatosensory association area

posterior association area   Wernicke area   visual receiving area

visual association area

1. The portion of the cerebral cortex that assembles mental "pictures" of visual stimuli   _____

2. The first brain region to receive input from the retina (organ of sight)   _____

3. Also known as the speech comprehension area   _____

4. Also known as the motor speech area   _____

5. Sends commands to specific muscles   _____

6. Responsible for conscious thought and problem solving   _____

7. Plans complex movements   _____

8. The first area of the cortex to receive input from skin receptors regarding pressure, pain, touch   _____

9. Integrates information from the primary somatosensory area with memories to identify physical sensations   _____

10. Uses input from many brain areas to construct an integrated view of the world   _____

6. NAME TWO DIVISIONS OF THE DIENCEPHALON, AND CITE THE FUNCTIONS OF EACH.

See Exercises 10-11 and 10-12.

7. LOCATE THE THREE SUBDIVISIONS OF THE BRAIN STEM, AND GIVE THE FUNCTIONS OF EACH.

See Exercises 10-11 and 10-12.

8. DESCRIBE THE CEREBELLUM, AND IDENTIFY ITS FUNCTIONS.

Also see Exercise 10-11.

EXERCISE 10-10

List three functions of the cerebellum in the spaces below.

1. _____

2. _____

3. _____

9. NAME THREE NEURONAL NETWORKS THAT INVOLVE MULTIPLE REGIONS OF THE BRAIN, AND DESCRIBE THE FUNCTION OF EACH.

EXERCISE 10-11: DIENCEPHALON, BRAIN STEM, AND NEURAL NETWORKS (TEXT FIGS. 10-6 AND 10-8)

1. Use black to write the names of structures 1, 3, 5, and 10 in the appropriate blanks.
2. Write the names of the remaining structures on the appropriate lines in different colors. Use shades of red for structures of the limbic system, shades of blue for the parts of the brain stem, shades of gray for the parts of the cerebellum, and shades of green for the parts of the diencephalon. Some structures may need multiple colors.
3. Color or outline the structures on both diagrams with the corresponding colors.

1. _____
2. _____
3. _____
4. _____
5. _____
6. _____
7. _____
8. _____
9. _____
10. _____
11. _____
12. _____
13. _____
14. _____
15. _____

EXERCISE 10-12

Write the appropriate term in each blank from the list below. Not all terms will be used.

thalamus   pons   midbrain   hypothalamus

reticular formation   vasomotor center   limbic system   medulla oblongata

basal nuclei   cardiac center

1. The portion of the brain stem composed of myelinated nerve fibers that connects to the cerebellum   _____

2. The superior portion of the brain stem   _____

3. The part of the brain between the pons and the spinal cord   _____

4. The region of the diencephalon that acts as a relay center for sensory stimuli   _____

5. A diffuse neuronal network involved in emotional states and behavior   _____

6. Nuclei that regulate the contraction of smooth muscle in blood vessel walls   _____

7. The portion of the brain controlling the autonomic nervous system   _____

8. Interconnected collections of neuronal cell bodies that modulate motor outputs and facilitate routine motor tasks   _____

9. A network of brain stem neurons involved in wakefulness and screening out routine sensory input   _____

10. DESCRIBE FOUR TECHNIQUES USED TO STUDY THE BRAIN.

EXERCISE 10-13

Write the appropriate term in each blank from the list below.

MRI   CT   PET   EEG

1. Technique that produces a picture of brain activity levels in the different parts of the brain   _____

2. Technique that measures electric currents in the brain   _____

3. X-ray technique that provides photos of bone, cavities, and lesions   _____

4. Technique used to visualize soft tissue, such as scar tissue, hemorrhages, and tumors that does not use x-rays   _____

11. DESCRIBE AT LEAST SIX DISORDERS THAT AFFECT THE BRAIN.

EXERCISE 10-14

Write the appropriate term in each blank from the list below. Not all terms will be used.

Alzheimer disease	glioma	hydrocephalus	subdural hematoma
aphasia	encephalitis	cerebrovascular accident	epidural hematoma
epilepsy	Parkinson disease		

1. A brain tumor derived from neuroglia   _____

2. A chronic brain disorder that usually can be diagnosed by electroencephalography   _____

3. Damage to brain tissue caused by a blood clot, ruptured vessel, or embolism   _____

4. Loss of the power of expression by speech or writing   _____

5. A degenerative brain disorder associated with the development of amyloid   _____

6. A condition that may result from obstruction of the normal flow of CSF   _____

7. Bleeding between the dura mater and the skull   _____

8. The general term for inflammation of the brain   _____

9. A degenerative brain disorder resulting from the loss of substantia nigra neurons   _____

12. LIST THE NAMES AND FUNCTIONS OF THE 12 CRANIAL NERVES.

Also see Exercise 10-16.

EXERCISE 10-15: Cranial Nerves (Text Fig. 10-15)

1. Write the number and name of each labeled cranial nerve on the numbered lines in different colors. Use the same color for structures 1 and 2.
2. Color the nerves on the diagram with the corresponding colors.

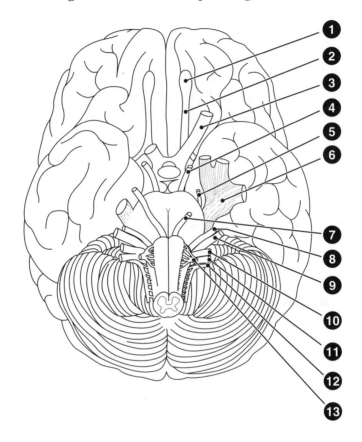

1. _____

2. _____

3. _____

4. _____

5. _____

6. _____

7. _____

8. _____

9. _____

10. _____

11. _____

12. _____

13. _____

13. DISCUSS FIVE DISORDERS THAT INVOLVE THE CRANIAL NERVES.

EXERCISE 10-16

Write the appropriate term in each blank from the list below. Not all terms will be used.

optic nerve	glossopharyngeal nerve	trochlear nerve	abducens nerve
vagus nerve	vestibulocochlear nerve	facial nerve	trigeminal nerve
accessory nerve	oculomotor nerve		

1. A motor nerve controlling the trapezius, sternocleidomastoid, and larynx muscles   _____

2. The nerve that controls contraction of a single eye muscle   _____

3. Damage to this nerve impairs vision   _____

4. This nerve carries sensory information from the face and motor impulses to the jaw   _____

5. The nerve that supplies most of the organs in the thoracic and the abdominal cavities   _____

6. Damage to one branch of this nerve causes Bell palsy   _____

7. The nerve that carries sensory impulses for hearing and equilibrium   _____

8. Damage to this nerve can cause a drooping eyelid   _____

9. Damage to this nerve would interfere with swallowing   _____

14. USING INFORMATION IN THE CASE STUDY, LIST THE POSSIBLE EFFECTS OF MILD TRAUMATIC BRAIN INJURY.

EXERCISE 10-17

Rewrite each sentence to make it true by replacing the underlined term.

1. Mild traumatic brain injury is also known as underlined cerebral palsy.

   _____

2. Individuals who experience one concussion are underlined less likely to suffer a second concussion.

   _____

3. A few weeks after her concussion, Natalie was still forgetting where she left her tea. She is suffering from underlined meningitis.

   _____

4. During the week following her concussion, Natalie underlined should stay in bed and watch television.

   _____

5. Natalie's symptoms reflect swelling of her brain, known as underlined encephalitis.

   _____

15. SHOW HOW WORD PARTS ARE USED TO BUILD WORDS RELATED TO THE NERVOUS SYSTEM.

EXERCISE 10-18

Complete the following table by writing the correct word part or meaning in the space provided. Write a word that contains each word part in the "Example" column.

Word Part	Meaning	Example
1. _____	cut	_____
2. chori/o	_____	_____
3. _____	tongue	_____
4. encephal/o	_____	_____
5. cerebr/o	_____	_____
6. _____	head	_____
7. _____	speech, ability to talk	_____
8. _____	lateral, side	_____
9. gyr/o	_____	_____
10. -rhage	_____	_____

Making the Connections

The following concept map deals with brain anatomy. Complete the concept map by filling in the appropriate term or phrase that describes the indicated structure.

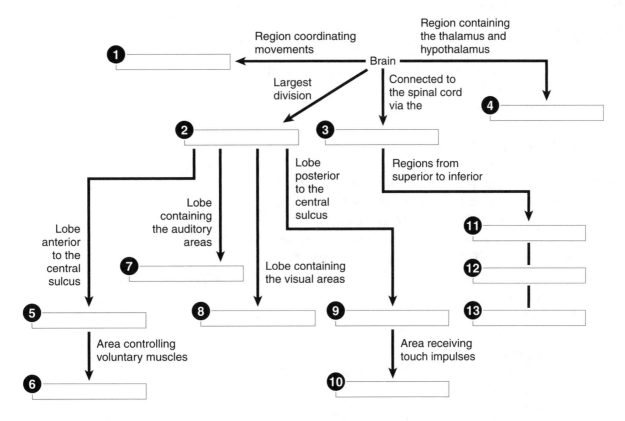

Optional Exercise: Make your own concept map based on structures involved in the synthesis and movement of CSF. Use the following terms and any others you would like to include: ventricle, choroid plexus, lateral ventricles, third ventricle, fourth ventricle, foramina, cerebral aqueduct, spinal cord, hydrocephalus, CSF, dural sinuses, subarachnoid space, and arachnoid villi.

Testing Your Knowledge

BUILDING UNDERSTANDING

I. MULTIPLE CHOICE

Select the best answer.

1. What is the name of the shallow groove separating the frontal
   and parietal lobes?   1. _____
   a. lateral sulcus
   b. central sulcus
   c. longitudinal fissure
   d. basal nuclei

2. Which of these phrases describes the dura mater?   2. _____
   a. the innermost layer of the meninges
   b. the outermost layer of the meninges
   c. the network of vessels that produces cerebrospinal fluid
   d. the part of the brain that connects with the spinal cord

3. Which brain region receives visual impulses?   3. _____
   a. parietal lobe
   b. temporal lobe
   c. hippocampus
   d. occipital lobe

4. Which of these structures *forms* cerebrospinal fluid?   4. _____
   a. cerebral aqueduct
   b. central sulcus
   c. choroid plexus
   d. internal capsule

5. Which of these regions does the abducens nerve innervate?   5. _____
   a. eye
   b. ear and pharynx
   c. face and salivary gland
   d. tongue and pharynx

6. What is the cause of multi-infarct dementia?   6. _____
   a. accumulation of an abnormal protein
   b. ischemia (lack of blood supply)
   c. obstruction of the flow of CSF
   d. infection of the brain coverings

7. What is the reticular formation?   7. _____
   a. a region of the limbic system that controls wakefulness and sleep
   b. a deep groove that divides the cerebral hemispheres
   c. the part of the temporal lobe concerned with the sense of smell
   d. the fifth lobe of the cerebrum

8. Which of these functional brain regions is found in the frontal lobe?   8. _____
   a. premotor area
   b. posterior association area
   c. speech comprehension area
   d. visual receiving area

II. COMPLETION EXERCISE

Write the word or the phrase that correctly completes each sentence.

1. A severe pain disorder affecting the fifth cranial nerve is known as tic douloureux or

   _____.

2. The four chambers within the brain where cerebrospinal fluid is produced are the

   _____.

3. Sounds are interpreted in the area of the temporal lobe called the _____.

4. The region of the diencephalon that helps maintain homeostasis (e.g., water balance, appetite, and body temperature) and controls the autonomic nervous system is the

   _____.

5. The hypothalamic nucleus controlling the muscle in blood vessel walls is called the

   _____.

6. Records of the electrical activity of the brain can be made with an instrument called a(n)

   _____.

7. Inflammation of the brain is called _____.

8. The brain region involved in conscious thought and problem solving is the

   _____.

9. Except for the first two pairs, all the cranial nerves arise from the _____.

10. The three layers of membranes that surround the brain and spinal cord are called the

    _____.

11. The clear liquid that helps support and protect the brain and spinal cord is

    _____.

12. The number of pairs of cranial nerves is _____.

13. The widespread collection of brain tissue involved in emotion and memories is called the

    _____.

14. The middle portion of the cerebellum is called the _____.

15. A cerebral concussion is also known as a(n) _____.

UNDERSTANDING CONCEPTS

I. TRUE/FALSE

For each statement, write T for true or F for false in the blank to the left of each number. If a statement is false, correct it by replacing the underlined term, and write the correct statement in the blanks below the statement.

_____ 1. The <u>pia mater</u> is the middle layer of the meninges.

_____

_____ 2. The interventricular foramina form channels between the lateral ventricles and the <u>fourth ventricle</u>.

_____

_____ 3. The primary motor cortex is found in the <u>parietal lobe</u>.

_____

_____ 4. The amygdala is part of a diffuse brain network known as the <u>basal nuclei</u>.

_____

_____ 5. The <u>reticular activating system</u> governs alertness and sleep.

_____

_____ 6. The raised areas on the surface of the cerebrum are called <u>gyri</u>.

_____

_____ 7. <u>Alzheimer disease</u> results from a lack of dopamine in the substantia nigra.

_____

_____ 8. The <u>hypoglossal</u> nerve transmits sensory information from the tongue.

_____

_____ 9. The <u>anterior</u> association area is responsible for integrating all sensory information with memories to produce an integrated view of the world.

_____

II. PRACTICAL APPLICATIONS

Study each discussion. Then write the appropriate word or phrase in the space provided.

1. Mr. B, aged 87, drove his car into a ditch. When the paramedics arrived, he was unable to speak, a disorder called _____.

2. Mr. B also experienced paralysis on his right side, indicating a problem within the largest division of the brain, the _____.

3. The damaged brain region controlling skeletal muscles is called the _____.

4. The right side of Mr. B's face droops, and he cannot control his facial expressions. The facial muscles are controlled by the cranial nerve numbered _____.

5. A CT scan at the hospital revealed the rupture of a cerebral blood vessel, resulting in bleeding, or _____.

6. The rupture occurred in the left side of the brain. The anatomical name for the left side of the brain is the left _____.

7. The physician diagnosed Mr. B with a type of brain injury called _____.

8. Mr. B's speech disorder indicates that the bleed affected his motor speech area, also called _____.

III. SHORT ESSAYS

1. Describe the structures that protect the brain and spinal cord.

_____

_____

_____

2. List some functions of the structures in the diencephalon.

_____

_____

_____

3. List the name, number, and sensory information conveyed for each of the purely sensory cranial nerves.

_____

_____

_____

_____

_____

CONCEPTUAL THINKING

1. Mr. S, age 42, has difficulty initiating movements. Other symptoms include hand tremors, frequent falls, and limb and joint rigidity. What disorder does Mr. S have? How would you explain this disorder to his family? What are some possible treatments?

_____

_____

_____

_____

_____

2. Describe the journey of CSF, beginning with its synthesis and ending with its entry into the circulatory system.

_____

_____

_____

_____

_____

Expanding Your Horizons

What do you use to help you study? Gingko biloba? Other medications? As discussed in the first _Scientific American_ article listed below, gingko biloba does boost memory—to the same extent as a candy bar! There's a free (and enjoyable) alternative to candy bars and dietary supplements—napping. See the second website for more information, or do a website search for "naps memory."

- Gold PE, Cahill L, Wenk GL. The lowdown on ginkgo biloba. Sci Am 2003;288:86–91.
- University of California-Berkeley. Midday nap markedly boosts the brain's learning capacity. ScienceDaily. 2010. Available at: www.sciencedaily.com/releases/2010/02/100221110338.htm

The Sensory System

▶ Overview

The sensory system enables us to detect changes taking place both internally and externally. These changes are detected by specialized structures called **sensory receptors**. Any change that acts on a receptor to induce a nerve impulse is termed a stimulus. The **special senses**, so called because the receptors are limited to a few specialized sense organs in the head, include the senses of vision, hearing, equilibrium, taste, and smell. The receptors of the eye are the rods and cones located in the retina. The receptors for both hearing (the spiral organ) and equilibrium (the vestibule and semicircular canals) are located within the inner ear. Receptors for the chemical senses of taste and smell are located on the tongue and in the upper part of the nose, respectively.

The **general senses** are scattered throughout the body; they respond to touch, pressure, temperature, pain, and position. Receptors for the sense of position, known as proprioceptors, are found in muscles, tendons, and joints.

The nerve impulses generated by a receptor in response to a stimulus must be carried to the central nervous system by way of a sensory (afferent) neuron. Here, the information is processed and a suitable response is made.

Disorders of the eye and ear are common. They are associated with aging, infection, environmental factors, inherited malfunctions, and injury.

This chapter is quite challenging, because it contains both difficult concepts and large amounts of detail. You can use concept maps to assemble all of the details into easy-to-remember frameworks.

Addressing the Learning Objectives

1. DESCRIBE THE FUNCTIONS OF THE SENSORY SYSTEM.

EXERCISE 11-1

Use the terms below to complete the paragraph.

central nervous system   **homeostasis**   **sensory neuron**   **sensory receptor**

The sensory system protects people by detecting changes in the internal and external environments that threaten to disrupt (1) _____, which is the maintenance of a constant internal environment. The change is detected by a (2) _____ which sends an impulse through a (3) _____ to the (4) _____.

2. DIFFERENTIATE BETWEEN THE DIFFERENT TYPES OF SENSORY RECEPTORS, AND GIVE EXAMPLES OF EACH.

EXERCISE 11-2

Classify each of the following senses as general senses (G) or special senses (S).

1. Sense of position   _____

2. Smell   _____

3. Vision   _____

4. Touch   _____

5. Temperature   _____

6. Equilibrium   _____

3. DESCRIBE SENSORY ADAPTATION, AND EXPLAIN ITS VALUE.

EXERCISE 11-3

Define *sensory adaptation* in the space below

_____

_____

4. LIST AND DESCRIBE THE STRUCTURES THAT PROTECT THE EYE.

EXERCISE 11-4: Protective Structures (Text Figs. 11-1 and 11-2)

Label the indicated parts. The three empty bullets in the lower figure label structures you identified in the upper diagram; identify these structures by writing the correct number in each bullet.

1. _____
2. _____
3. _____
4. _____
5. _____
6. _____
7. _____
8. _____
9. _____
10. _____
11. _____
12. _____
13. _____
14. _____
15. _____
16. _____

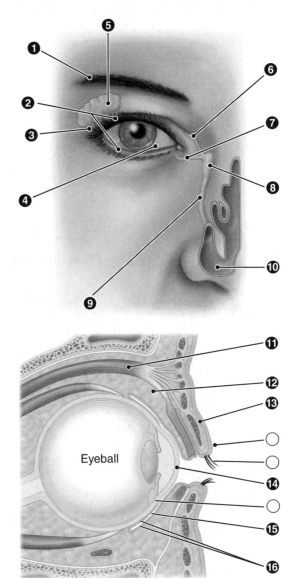

Eyeball

5. CITE THE LOCATION AND THE PURPOSE OF THE EXTRINSIC EYE MUSCLES.

EXERCISE 11-5: Extrinsic Muscles of the Eye (Text Fig. 11-3)

1. Write the name of each labeled muscle on the numbered lines in different colors.
2. Color the different muscles on the diagram with the corresponding colors.

1. _____

2. _____

3. _____

4. _____

5. _____

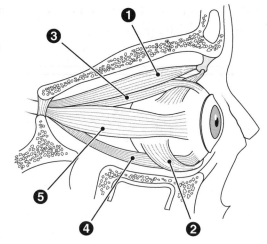

6. IDENTIFY THE THREE TUNICS OF THE EYE.

EXERCISE 11-6: The Eye (Text Fig. 11-4)

1. Color the small boxes yellow (nervous tunic), red (vascular tunic), and blue (fibrous tunic).
2. Write the name of each labeled part on the numbered lines in different colors. Use the following color scheme:
   - Yellow: 6, 7, 8, and 9
   - Red: 5, 10, 11, 13
   - Blue: 3, 4
   - Choose your own colors for the other structures.
3. Color the different structures on the diagram with the corresponding colors. Some structures are present in more than one location on the diagram. Try to color all of a particular structure in the appropriate color. For instance, only one of the suspensory ligaments is labeled, but color both suspensory ligaments.

☐ Nervous tunic   ☐ Vascular tunic   ☐ Fibrous tunic

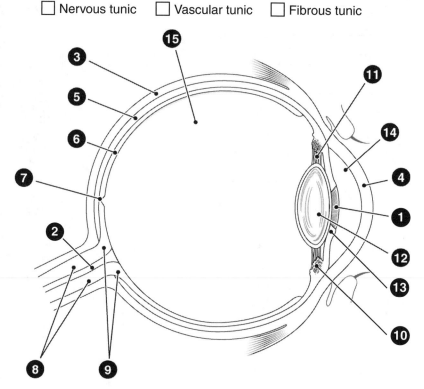

1. _____   9. _____

2. _____   10. _____

3. _____   11. _____

4. _____   12. _____

5. _____   13. _____

6. _____   14. _____

7. _____   15. _____

8. _____

7. DESCRIBE THE PROCESSES INVOLVED IN VISION.

EXERCISE 11-7

List four eye structures that bend (refract) light in the spaces below.

1. _____

2. _____

3. _____

4. _____

EXERCISE 11-8

Complete the chart below, which summarizes the process of accommodation. A few boxes have been completed to get you started.

	Distant Vision	Near Vision
Contraction state of ciliary muscle		Contracted
Shape of ciliary muscle	Thin	
Size of the central opening (ciliary muscle)		
Tension in suspensory ligaments	Strong	
Lens shape		
Degree of refraction		

EXERCISE 11-9

Complete the chart below, which summarizes how the iris muscles modify pupil size. A few boxes have been completed to get you started.

	Bright light	Dim light
Pupil size	Constricted (narrow)	
Contraction state of radial muscle fibers		
Contraction state of circular muscle fibers	Contracted	

8. DIFFERENTIATE BETWEEN THE RODS AND THE CONES OF THE EYE.

EXERCISE 11-10

Write the appropriate term in each blank from the list below. Not all terms will be used.

choroid   cone   cornea   rhodopsin   sclera

optic disk   retina   rod   fovea centralis

1. A vision receptor that is sensitive to color   _____

2. The part of the eye that light rays pass through first as they enter the eye   _____

3. Another name for the blind spot, the region where the optic nerve connects with the eye   _____

4. The innermost coat of the eyeball, the nervous tissue layer that includes the receptors for the sense of vision   _____

5. A vision receptor that functions well in dim light   _____

6. A pigment needed for vision   _____

7. The depressed area in the retina that is the point of clearest vision   _____

8. The vascular, pigmented middle tunic of the eyeball   _____

EXERCISE 11-11

Write the appropriate term in each blank from the list below. Not all terms will be used.

optic nerve   ophthalmic nerve   oculomotor nerve   aqueous humor

vitreous body   lens   ciliary muscle   conjunctiva

pupil   iris

1. The structure that alters the shape of the lens for accommodation   _____

2. The watery fluid that fills much of the eyeball in front of the crystalline lens   _____

3. Structure with two sets of muscle fibers that regulate the amount of light entering the eye   _____

4. The jelly-like material located behind the crystalline lens that maintains the spherical shape of the eyeball   _____

5. The central opening of the iris   _____

6. The nerve supplying most of the extrinsic eye muscles   _____

7. The nerve that carries visual impulses from the photoreceptors to the brain   _____

8. The nerve that carries somatosensory impulses from the eyeball   _____

9. LIST SEVEN DISORDERS OF THE EYE AND VISION.

EXERCISE 11-12

Write the appropriate term in each blank from the list below. Not all terms will be used.

strabismus   glaucoma   myopia   hyperopia   presbyopia

macular degeneration   neonatal conjunctivitis   cataract   trachoma   astigmatism

1. A serious eye infection of the newborn that can be prevented with a suitable antiseptic
   _____

2. The scientific name for nearsightedness, in which the focal point is in front of the retina and distant objects appear blurred
   _____

3. The visual defect caused by irregularity in the curvature of the lens or cornea
   _____

4. Condition in which the eyes do not move in a coordinated fashion
   _____

5. Condition caused by continued high pressure of the aqueous humor, which may result in destruction of the optic nerve fibers
   _____

6. The refraction disorder resulting from an abnormally short eyeball
   _____

7. A chronic eye infection for which antibiotics and proper hygiene have reduced the incidence of reinfection and blindness
   _____

8. The age-related decline in lens elasticity that impairs close vision
   _____

9. Condition in which the lens becomes too cloudy to transmit light properly
   _____

10. DESCRIBE THE THREE DIVISIONS OF THE EAR.

EXERCISE 11-13: The Ear (Text Fig. 11-12)

1. Write the names of the three ear divisions on the appropriate lines (1 to 3).
2. Write the names of the labeled parts on the numbered lines in different colors. Use black for structures 12 to 14 because they will not be colored.
3. Color each part with the corresponding colors (except for parts 12 to 14).

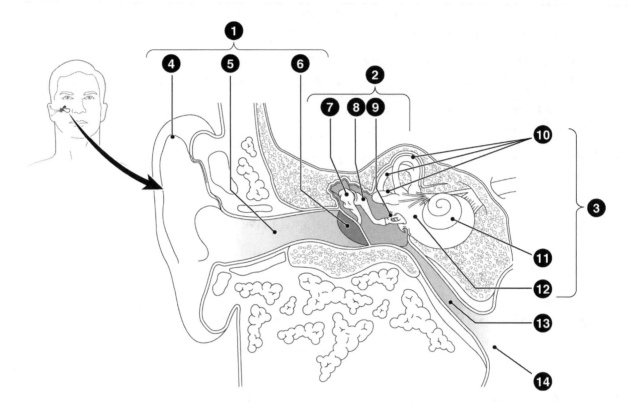

1. _____   9. _____

2. _____   10. _____

3. _____   11. _____

4. _____   12. _____

5. _____   13. _____

6. _____   14. _____

7. _____

8. _____

EXERCISE 11-14: The Inner Ear (Text Fig. 11-13)

1. Label the bony labyrinth and the membranous labyrinth in the upper diagram using yellow and green (respectively). Color the upper diagram using these two colors.
2. Label the structures in the upper diagram by writing the correct term in each blank.
3. Look at the lower diagram. The three empty bullets label structures you identified in the upper diagram—identify these structures by writing the correct numbers in the bullets, and color them yellow if they contain perilymph and green if they contain endolymph.
4. Label structures 13 to 16.

1. _____

2. _____

3. _____

4. _____

5. _____

6. _____

7. _____

8. _____

9. _____

10. _____

11. _____

12. _____

13. _____

14. _____

15. _____

16. _____

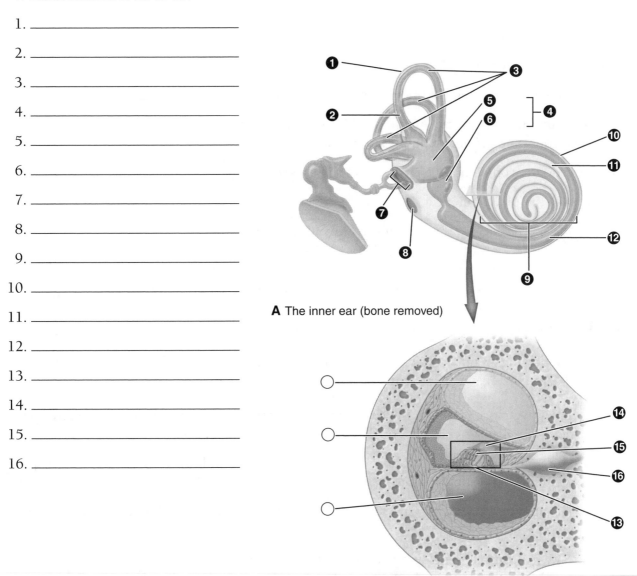

A The inner ear (bone removed)

B Cross-section of the cochlea

EXERCISE 11-15

Write the appropriate term in each blank from the list below. Not all terms will be used.

spiral organ	bony labyrinth	malleus
eustachian tube	cochlear duct	endolymph
perilymph	incus	pinna

1. The fluid contained within the membranous labyrinth of the inner ear   _____

2. The bone in contact with the tympanic membrane   _____

3. Another name for the projecting part, or auricle, of the ear   _____

4. The channel connecting the middle ear cavity with the pharynx   _____

5. The fluid of the inner ear contained within the bony labyrinth and surrounding the membranous labyrinth   _____

6. The structure containing the receptor cells for hearing   _____

7. The skeleton of the inner ear   _____

8. The duct in the cochlea that contains endolymph   _____

11. DESCRIBE THE RECEPTOR FOR HEARING, AND EXPLAIN HOW IT FUNCTIONS.

EXERCISE 11-16

Place the following events involved in hearing in order by writing the appropriate numbers (1 through 9) in the blanks. Note that step 5 occurs at the same time as steps 6–9.

_____a. The wave passes through the cochlear membranes. The differences in their relative positions flex the hair cell tips embedded in the tectorial membrane.

_____b. Vibrations of the stapes result in vibrations of the oval window, creating a wave in the perilymph of the vestibular duct.

_____c. The amount of neurotransmitter released by the hair cell changes, altering the firing rate of the sensory neuron.

_____d. A wave begins in the membranes of the cochlear duct.

_____e. The tympanic membrane and ossicles vibrate at the same frequency as the original sound wave.

_____f. The sound wave travels through the ear canal.

_____g. The bending of the cilia changes their membrane potential.

_____h. Action potentials travel through the vestibulocochlear nerve to the brain.

_____i. The cochlear movements initiate a fluid wave in the tympanic duct, which is dissipated when it hits the round window. The movement of the round window allows clear hearing by preventing the wave from bouncing back and creating an echo.

12. COMPARE THE LOCATIONS AND FUNCTIONS OF THE EQUILIBRIUM RECEPTORS.

EXERCISE 11-17

Fill in the blank after each characteristic—does it apply to to the vestibule (V), the semicircular canals (S), or both (B)?

1. Contain maculae _____

2. Activated by rotational acceleration _____

3. Hair cell tips embedded in gelatinous membrane floating
   in endolymph _____

4. Send signals through the vestibular nerve _____

5. Activated by linear acceleration _____

6. There are three of them _____

7. Hair cell tips embedded in gelatinous membrane studded
   with tiny crystals _____

8. Receptors are called cristae _____

9. Sense gravity _____

13. DESCRIBE THREE DISORDERS THAT INVOLVE THE EAR.

EXERCISE 11-18

Write the appropriate term in each blank from the list below. Not all terms will be used.

otitis media	otitis externa	conductive hearing loss
otosclerosis	presbycusis	sensorineural hearing loss

1. The scientific name for swimmer's ear _____

2. A hereditary bone disorder that prevents normal vibration of
   the stapes _____

3. Slow, progressive hearing loss associated with aging _____

4. Hearing loss resulting from damage to the cochlea or to nerves
   associated with hearing _____

5. Infection and inflammation of the middle ear cavity _____

14. DISCUSS THE LOCATIONS AND FUNCTIONS OF THE SENSE ORGANS FOR TASTE AND SMELL.

EXERCISE 11-19

Replace the underlined terms in each of the following statements to make each statement true.

1. The sense of smell is <u>gustation</u>, and the sense of taste is <u>olfaction</u>.

   _____ ; _____

2. Substances to be smelled and tasted must always be <u>free in the air</u> surrounding the receptor cell.

   _____

3. We have hundreds of different types of <u>taste receptors</u> but less than 10 types of <u>odor receptors</u>.

   _____ ; _____

4. Organic compounds activate <u>sour receptors</u>, and hydrogen ions activate <u>salty receptors</u>.

   _____ ; _____

15. DESCRIBE FIVE GENERAL SENSES.

EXERCISE 11-20

Complete the following table, summarizing the five general senses and their receptors.

Sense	Receptor	Receptor location
	Tactile corpuscle	
Pressure		
Temperature		
	Proprioceptors	
Pain		

16. LIST FIVE APPROACHES FOR TREATMENT OF PAIN.

EXERCISE 11-21

Write the appropriate term in each blank from the list below. Not all terms will be used.

NSAID   narcotic   analgesic

anesthetic   endorphin

1. Term describing any drug that relieves pain   _____

2. A substance produced by the brain that relieves pain   _____

3. Drug that acts on the CNS to alter pain perception, such as
   morphine   _____

4. Drug that acts to reduce inflammation   _____

17. REFERRING TO THE CASE STUDY, DISCUSS THE PURPOSE AND MECHANISM OF A COCHLEAR IMPLANT.

EXERCISE 11-22

Imagine that you were Evan's parent. How would you explain Evan's cochlear implant to other parents?

_____

_____

_____

_____

18. SHOW HOW WORD PARTS ARE USED TO BUILD WORDS RELATED TO THE SENSORY SYSTEM.

EXERCISE 11-23

Complete the following table by writing the correct word part or meaning in the space provided. Write a word that contains each word part in the "Example" column.

Word Part	Meaning	Example
1. presby-	_____	_____
2. _____	stone	_____
3. -opia	_____	_____
4. -stomy	_____	_____
5. _____	drum	_____
6. _____	yellow	_____
7. propri/o	_____	_____
8. _____	pain	_____
9. -esthesia	_____	_____
10. _____	hearing	_____

Making the Connections

The following concept map deals with the structure and function of the eye. Each pair of terms is linked together by a connecting phrase into a sentence. The sentence should be read in the direction of the arrow. Complete the concept map by filling in the appropriate term or phrase. There is one right answer for each term. However, there are many correct answers for the connecting phrases (2, 9).

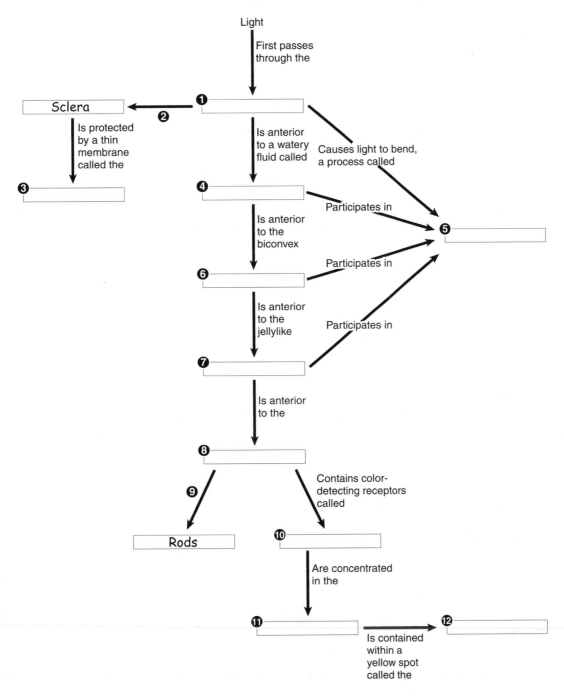

Optional Exercise: Make your own concept map of terms relating to the ear. Use the following terms and any others you would like to include: tympanic membrane, stapes, malleus, incus, pinna, bony labyrinth, spiral organ, oval window, round window, cochlear duct, tectorial membrane, and cochlear nerve. You may also want to construct concept maps relating to the other special senses (equilibrium, taste, and smell) and the general senses (touch, pressure, temperature, and proprioception).

Testing Your Knowledge

BUILDING UNDERSTANDING

I. MULTIPLE CHOICE

Select the best answer.

1. Which of the following receptor types responds to light?   1. _____
   a. photoreceptor
   b. mechanoreceptor
   c. chemoreceptor
   d. thermoreceptor
2. Which of the following terms relates to the sense of touch?   2. _____
   a. tactile
   b. gustatory
   c. proprioceptive
   d. thermal
3. What process alters the lens's shape to allow for near or far vision?   3. _____
   a. accommodation
   b. convergence
   c. divergence
   d. dark adaptation
4. Which secretion is involved in the process of *lacrimation*?   4. _____
   a. mucus
   b. wax
   c. tears
   d. aqueous humor
5. Which of the following chemicals are painkillers released by certain
   brain regions?   5. _____
   a. narcotics
   b. endorphins
   c. anesthetics
   d. nonsteroidal antiinflammatory drugs
6. What disorder results from a deficiency in retinal cones?   6. _____
   a. blindness
   b. color blindness
   c. glaucoma
   d. trachoma
7. Which of the following participates in hearing?   7. _____
   a. spiral organ
   b. macula
   c. olfactory bulb
   d. retina
8. What is a cataract?   8. _____
   a. an irregularity in the cornea's shape
   b. an infection of the conjunctiva
   c. an abnormally short eyeball
   d. loss of lens transparency

9. Which of the following is a component of the eye's vascular tunic?   9. _____
    a. retina
    b. cornea
    c. choroid
    d. lens
10. Which of the following structures is involved in sensing gravity?   10. _____
    a. cupula
    b. vestibular duct
    c. otoliths
    d. crista

II. COMPLETION EXERCISE

1. The transparent portion of the sclera is the _____.

2. The glands that secrete earwax are called _____.

3. The receptors that aid in judging position and changes in location of body parts are the

    _____.

4. The sense of position is partially governed by equilibrium receptors in the internal ear, including two small chambers in the vestibule and the three _____.

5. The ciliated cells that function in hearing and equilibrium are called _____.

6. Any drug that relieves pain is called a(n) _____.

7. When you enter a darkened room, it takes a while for the rods to begin to function. This interval is known as the period of _____.

8. The nervous tunic of the eye is the _____.

9. The muscles that adjust the shape of the lens are called _____.

UNDERSTANDING CONCEPTS

I. TRUE/FALSE

For each statement, write T for true or F for false in the blank to the left of each number. If a statement is false, correct it by replacing the underlined term, and write the correct statement in the blanks below the statement.

_____ 1. Extrinsic eye muscles control the diameter of the pupil.

_____

_____ 2. A crista contains receptor (hair) cells and a gelatinous otolith membrane.

_____

_____ 3. The incus is in contact with the oval window of the inner ear.

_____

_____ 4. The cochlear duct contains <u>endolymph</u>.

_____

_____ 5. The <u>rods</u> of the eye function in bright light and detect color.

_____

_____ 6. When the eyes are exposed to a bright light, the pupils <u>constrict.</u>

_____

_____ 7. The scientific name for nearsightedness is <u>hyperopia</u>.

_____

_____ 8. The ciliary muscle <u>contracts</u> to thicken the lens.

_____

_____ 9. The sense of smell is also called <u>olfaction</u>.

_____

II. PRACTICAL APPLICATIONS

Study each discussion. Then write the appropriate word or phrase in the space provided.

➤ Group A

Baby L was brought in by his mother because he awakened crying and holding the right side of his head. He had been suffering from a cold, but now he seemed to be in pain. Complete the following descriptions relating to his evaluation and treatment.

1. Examination revealed a bulging red eardrum. The eardrum is also called the
   _____.

2. The cause of Baby L's painful bulging eardrum was an infection of the middle ear, a condition called _____.

3. Antibiotic treatment of Baby L's middle ear infection was begun because this early treatment usually prevents complications. In this case, however, it was necessary to cut the eardrum to prevent its rupture. Another name for this surgical procedure is _____.

4. The mother was warned that Baby L may be particularly susceptible to middle ear infections. To prevent further damage to his eardrum, a special tube was inserted. This tube is called a(n)
   _____.

5. Baby L will have to be careful in the future because repeated middle ear infections can lead to a type of hearing loss called _____.

6. Baby L returned to the emergency room the next day because he was falling down repeatedly. The physician suspected a problem with his sense of balance, or _____.

7. Baby L's mother asked how an ear infection could affect balance. The physician explained that two structures were located within the inner ear that are involved with balance, named the semicircular canals and the _____.

8. In particular, the physician feared that the middle ear infection had spread to the fluid within the membranous labyrinth. This fluid is called _____.

➤ Group B

Mr. S, age 60, rode his scooter over some broken glass. A fragment of glass bounced up and flew into one eye. Complete the following descriptions relating to his evaluation and treatment.

1. Examination by the eye specialist showed that there was a cut in the transparent window of the eye, the _____.

2. On further examination of Mr. S, the colored part of the eye was seen to protrude from the wound. This part of the eye is the _____.

3. Mr. S's treatment included antiseptics, anesthetics, and suturing of the wound. Medication was instilled in the saclike structure at the anterior of the eyeball. This sac is lined with a thin epithelial membrane, the _____.

4. The eye specialist evaluated Mr. S's vision in his uninjured eye. Like virtually all older adults, Mr. S was shown to have difficulties with near vision. This condition in the aging adults is called _____.

5. The eye specialist also observed that the pressure in his aqueous humor was abnormally high. This finding signifies that Mr. S suffers from _____.

6. Mr. S returned to the emergency room a week later with a severe infection in the injured eye. Despite proper wound care and several changes of antibiotics, the damaging infection persisted. The eye specialist reluctantly decided to remove the eyeball, a procedure called _____.

➤ Group C

You are conducting hearing tests at a senior citizens' home. During the course of the afternoon, you encounter the following patients. Complete the following descriptions relating to the evaluation and treatment of hearing loss.

1. Mrs. B complained of some hearing loss and a sense of fullness in her outer ear. Examination revealed that her ear canal was plugged with hardened ear wax, which is scientifically called

    _____.

2. Mr. J, age 72, complained of gradually worsening hearing loss, although he had no symptoms of pain or other ear problems. Examination revealed that his hearing loss was due to nerve damage. The cranial nerve that carries hearing impulses to the brain is called the _____.

3. In particular, the endings of this nerve were damaged. These nerve endings are located in the spiral-shaped part of the inner ear, a part of the ear that is known as the _____.

4. Because it reflects nerve damage, Mr. J's hearing loss is known as _____.

5. Mrs. C requested surgical treatment to correct her hereditary form of conductive hearing loss. The surgery would restore normal movement of her stapes. Mrs. C suffers from a disorder known as _____.

III. SHORT ESSAYS

1. Describe several different structural forms of sensory receptors, and give examples of each.

_____

_____

_____

2. Describe some changes that occur in the sensory receptors with age.

_____

_____

_____

3. List three methods to relieve pain that do not involve administration of drugs.

_____

_____

_____

CONCEPTUAL THINKING

1. You have probably been sitting in a chair for quite a while, yet you have not been constantly aware of your legs contacting the chair. Why not?

_____

_____

_____

2. Write your name at the bottom of this sheet of paper. Explain the contributions of different sensory receptors that were required to successfully complete that simple task. For instance, proprioceptors are required to indicate where the fingers are at each moment of time.

_____

_____

_____

3. Hair cells function as receptor cells in three different sensory organs. Compare and contrast these three sensory organs by completing the following table. Some boxes have been filled in for you.

Sensory Organ Name	Membrane Overlying Hair Cell	Stimulus Sensed by the Sensory Organ
		Gravity and linear acceleration
	Tectorial membrane	
Crista		

Expanding Your Horizons

1. Imagine if you could taste a triangle or hear blue. This is reality for individuals with a disorder called *synesthesia*. Read about some exceptional artists that suffer from this disorder, and how synesthesia has helped us understand how the brain processes sensory information in the article below. You can get a taste of how synesthetics view the world in the exercises at the website listed below.

   ▪ Ramachandran VS, Hubbard EM. Hearing colors, tasting shapes. Sci Am 2003;288:52–59.
   ▪ Synesthesia and the Synesthetic Experience. Available at: http://web.mit.edu/synesthesia/www/

2. Here is an exercise you can do to find your own blind spot. Draw a cross (on the left) and a circle (on the right) on a piece of paper that are separated by a hand width. Focus on the cross, and notice (but do not focus on) the circle. Move the paper closer and further away until the circle disappears. Weird activities to investigate your blind spot can be found at this website:

   ▪ Find Your Blind spot! Available at: http://www.moillusions.com/find-your-blind-spot-trick/

The Endocrine System: Glands and Hormones

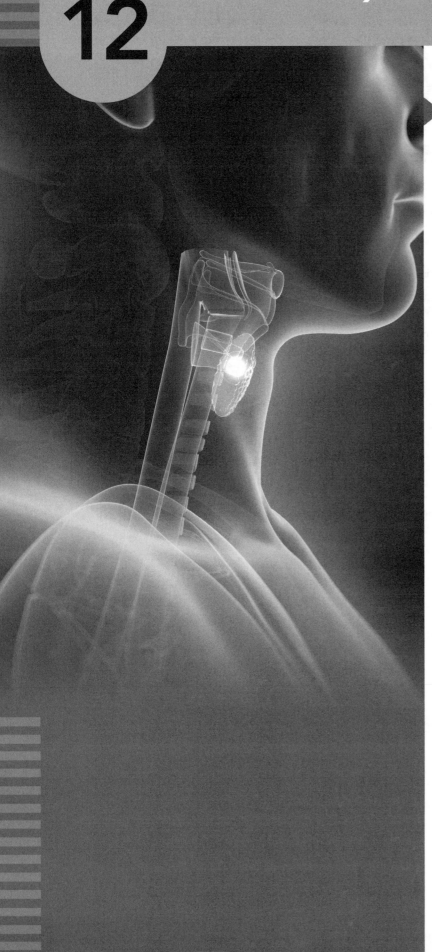

▶ Overview

The endocrine system and the nervous system are the main coordinating and controlling systems of the body. Chapters 9 and 10 discuss how the nervous system uses chemical and electrical stimuli to control very rapid, short-term responses. This chapter discusses the endocrine system, which uses specific chemicals called **hormones** to induce short-term or long-term changes in the reproduction, development, and function of specific cells. Although some hormones act locally, most travel in the blood to distant sites and exert their effects on any cell (the target cell) that contains a specific receptor for the hormone. Although hormones are produced by many tissues, certain glands, called **endocrine glands**, specialize in hormone production. These endocrine glands include the pituitary (hypophysis), thyroid, parathyroids, adrenals, pancreas, gonads, and pineal. Together, these glands constitute the endocrine system.

The activity of endocrine glands is regulated by negative feedback, other hormones, nervous stimulation, and/ or biological rhythms. One of the most important endocrine glands is the **pituitary gland**, which comprises the anterior pituitary and the posterior pituitary. The **posterior pituitary** gland is made of nervous tissue—it contains the axons and terminals of neurons that have their cell bodies in a part of the brain called the hypothalamus. Hormones are synthesized in the hypothalamus and released from the posterior pituitary. The **anterior**

pituitary secretes a number of hormones that act on other endocrine glands. The cells of the anterior pituitary are controlled in part by releasing hormones made in the hypothalamus. These releasing hormones pass through the blood vessels of a portal circulation to reach the anterior pituitary. Hormones are also made outside the endocrine glands. Other structures that secrete hormones include the stomach, small intestine, kidney, heart, skin, and placenta. Hormones are extremely potent, and small variations in hormone concentrations can have significant effects on the body.

This chapter contains a lot of details for you to learn. Try to summarize the material using concept maps and summary tables. You should also understand negative feedback (Chapter 1) before you tackle the concepts in this chapter. Objectives 6 through 8 ask you to integrate information from different chapter topics. So, make sure you read about all of the different hormones before you attempt the relevant activities.

Addressing the Learning Objectives

1. COMPARE THE EFFECTS OF THE NERVOUS SYSTEM AND THE ENDOCRINE SYSTEM IN CONTROLLING THE BODY.

EXERCISE 12-1

Fill in the blank after each characteristic—does it apply to the nervous system (N) or the endocrine system (E)?

1. Controls rapid responses   _____

2. Used to regulate growth   _____

3. Uses chemical stimuli only   _____

4. Uses both chemical and electrical stimuli   _____

2. DESCRIBE THE FUNCTIONS OF HORMONES.

EXERCISE 12-2

Define the following terms:

1. Receptor _____

2. Target tissue _____

3. Hormone _____

3. IDENTIFY THE GLANDS OF THE ENDOCRINE SYSTEM ON A DIAGRAM.

EXERCISE 12-3: The Endocrine Glands (Text Fig. 12-1)

1. Write the names of each labeled part on the numbered lines in different colors.
2. Color the different structures on the diagram with the corresponding colors. Make sure you color all four parathyroid glands.

1. _____

2. _____

3. _____

4. _____

5. _____

6. _____

7. _____

8. _____

9. _____

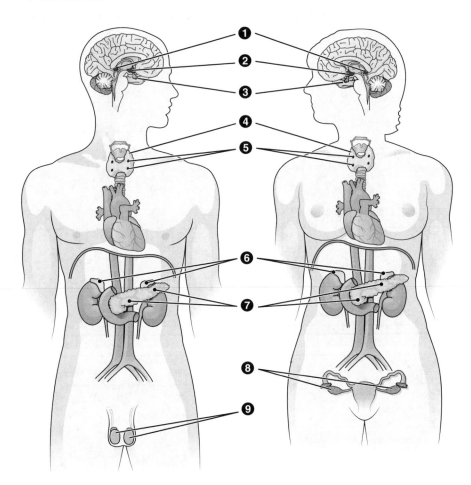

EXERCISE 12-4

Write the appropriate term in each blank from the list below. Not all terms will be used.

parathyroid   pineal   hypothalamus

thyroid   adrenal   pancreas

1. One of the tiny glands located behind the thyroid gland   _____

2. The largest of the endocrine glands, located in the neck   _____

3. The gland in the brain that is regulated by light   _____

4. An organ that contains islets   _____

5. The endocrine gland composed of a cortex and medulla, each with specific functions   _____

4. DISCUSS THE CHEMICAL COMPOSITION OF HORMONES.

EXERCISE 12-5

Fill in the blank after each statement—does it apply to amino acid compounds (A) or lipid-derived hormones (L)?

1. Protein hormones   _____

2. Can be formed by modifying cholesterol   _____

3. Hormones of the sex glands and the adrenal cortex   _____

4. Hormones not produced by the sex glands or the adrenal cortex   _____

5. Often contain the word part *sterone* in their name   _____

5. EXPLAIN HOW HORMONES ARE REGULATED.

EXERCISE 12-6

1. Define *negative feedback* in the blank lines below.

_____

_____

2. Which of the situations below are examples of negative feedback? Circle all that apply.

    a. When the concentration of thyroid hormone in blood increases, the secretion of thyroid-stimulating hormone from the pituitary gland decreases.

    b. When the concentration of growth hormone in blood increases, the production of growth hormone from the pituitary gland decreases.

    c. When the house heats up, the furnace produces less heat.

    d. When an astute investor makes money on a stock deal, he can make further investments to make even more money.

6. LIST THE HORMONES PRODUCED BY EACH ENDOCRINE GLAND, AND DESCRIBE THE EFFECTS OF EACH ON THE BODY.

EXERCISE 12-7

Fill in the missing information in the chart on the next page. This chart does not cover all of the hormones and their actions discussed in the text. Do not fill in the three columns at the far right yet. These columns will be discussed under Objectives 8 and 11.

Hormone	Gland	Effects	Hyposecretion	Hypersecretion	Medical Uses
ADH (antidiuretic hormone)					N/A
	Adrenal cortex	Increases blood glucose concentration in response to stress			
		Uterine contraction, milk ejection	N/A	N/A	
	Anterior pituitary	Promotes growth of all body tissues			
Parathyroid hormone					N/A
	Pancreas	Reduces blood glucose concentrations by promoting glucose uptake into cells and glucose storage; promotes fat and protein synthesis			
Thyroid hormones: thyroxine (t4) and triiodothyronine (t3)					
	Adrenal cortex	Promotes salt (and thus water) retention and potassium excretion			N/A
		Stimulates glucose release from the liver, thereby increasing blood glucose levels	N/A	N/A	N/A
Melatonin			N/A	N/A	N/A
Estrogens and progesterone			N/A	N/A	
Androgens (testosterone)			N/A	N/A	

EXERCISE 12-8

Write the appropriate term in each blank from the list below.

epinephrine	antidiuretic hormone	ACTH
follicle-stimulating hormone	estrogen	
aldosterone	prolactin	

1. The anterior pituitary hormone that stimulates milk synthesis _____

2. The main hormone of the adrenal medulla that, among other actions, raises blood pressure and increases the heart rate _____

3. The anterior pituitary hormone that stimulates the adrenal cortex _____

4. A hormone produced by the ovaries _____

5. The hormone from the adrenal cortex that regulates sodium and potassium reabsorption in the kidney tubules _____

6. A gonadotropic hormone _____

7. A hormone synthesized in the hypothalamus _____

EXERCISE 12-9

Write the appropriate term in each blank from the list below.

insulin	glucagon	parathyroid hormone
testosterone	cortisol	

1. A hormone that raises the blood calcium level _____

2. A hormone that lowers the blood glucose level _____

3. A pancreatic hormone that raises the blood glucose level _____

4. An adrenal hormone that raises the blood glucose level _____

5. A hormone that promotes development of secondary sex characteristics _____

7. DESCRIBE HOW THE HYPOTHALAMUS CONTROLS THE POSTERIOR AND ANTERIOR PITUITARY.

EXERCISE 12-10: The Pituitary Gland (Text Fig. 12-2)

1. Label the parts of the hypothalamo–pituitary system (structures 1 to 5).
2. Write the names of each pituitary hormone in the appropriate blanks (6 to 13). Try to remember the full names as well as the abbreviations. If you would like, use the same color to write the name and color the target tissue.
3. Write the name of each hormone produced by the target tissue in the correct box (14 to 18).

6. _____   10. _____

7. _____   11. _____

8. _____   12. _____

9. _____   13. _____

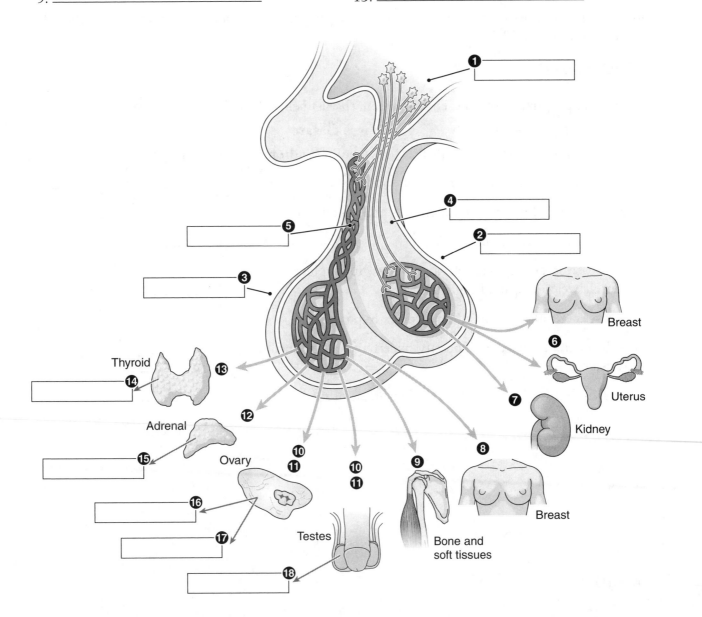

EXERCISE 12-11

Fill in the blank after each statement—does the characteristic apply to the anterior pituitary (AP) or the posterior pituitary (PP)?

1. Secretes hormones synthesized in the hypothalamus

   _____

2. Consists of neural tissue

   _____

3. Releases hormones under the regulation of hypothalamic releasing hormones

   _____

8. DESCRIBE THE EFFECTS OF HYPOSECRETION AND HYPERSECRETION OF THE VARIOUS HORMONES.

EXERCISE 12-12

Add the disorders associated with each hormone's hyposecretion and hypersecretion to the table prepared in Exercise 12-7.

EXERCISE 12-13

Write the appropriate term in each blank from the list below. Not all terms will be used.

goiter   acromegaly   Graves disease   gigantism

Addison disease   diabetes mellitus   metabolic syndrome   Cushing syndrome

1. Any enlargement of the thyroid gland

   _____

2. A disease in which insulin function is abnormally low

   _____

3. A form of hyperthyroidism

   _____

4. A disease associated with elevated blood glucose levels, muscle weakness, and fragile bones

   _____

5. A disease resulting from adrenal cortex hypoactivity

   _____

6. A disorder associated with obesity, hypertension, and insulin resistance

   _____

7. Growth hormone excess that develops in adulthood

   _____

9. LIST SEVEN TISSUES OTHER THAN THE ENDOCRINE GLANDS THAT PRODUCE HORMONES.

EXERCISE 12-14

Fill in the missing information in the chart below.

Hormone	Site of Synthesis	Effects
Atrial natriuretic peptide		
Thymosin		
	Kidney	Stimulates erythrocyte production
	Fat	Controls appetite
Osteocalcin		

10. EXPLAIN THE ORIGIN AND FUNCTION OF PROSTAGLANDINS.

EXERCISE 12-15

Rewrite each of these statements to make them true by replacing the underlined text.

1. Prostaglandins act on cells <u>far away</u> from the site of synthesis. _____

2. Prostaglandins are produced by <u>specific</u> body cells. _____

3. A common prostaglandin action is <u>inhibiting</u> inflammation. _____

4. Prostaglandins are derived from <u>amino acids</u>. _____

11. LIST EIGHT MEDICAL USES OF HORMONES.

EXERCISE 12-16

Fill in the column "Medical Uses" in the table prepared for Exercise 12-7.

12. EXPLAIN HOW THE ENDOCRINE SYSTEM RESPONDS TO STRESS.

EXERCISE 12-17

Explain how the actions of cortisol help an individual cope with a physical stressor such as a predator (or more realistically, a 5-km road race).

_____

_____

_____

_____

13. REFERRING TO THE CASE STUDY, DISCUSS THE SIGNS, SYMPTOMS, AND TREATMENT OF DIABETES MELLITUS TYPE 1.

EXERCISE 12-18

In the list below, circle all of the signs and symptoms of diabetes mellitus type 1. You should circle one option from each pair of signs/symptoms.

a. hunger or lack of appetite

b. weight gain or weight loss

c. lethargy or nervousness

d. glucose in the urine or protein in the urine

e. hypoglycemia or hyperglycemia

14. SHOW HOW WORD PARTS ARE USED TO BUILD WORDS RELATED TO THE ENDOCRINE SYSTEM.

EXERCISE 12-19

Complete the following table by writing the correct word part or meaning in the space provided. Write a word that contains each word part in the "Example" column.

Word Part	Meaning	Example
1. trop/o	_____	_____
2. _____	cortex	_____
3. -poiesis	_____	_____
4. natri	_____	_____
5. _____	male	_____
6. _____	enlargement	_____
7. ren/o	_____	_____
8. acro	_____	_____
9. oxy	_____	_____
10. nephr/o	_____	_____

Making the Connections

The following concept map deals with the relationship among the hypothalamus, pituitary gland, and some target organs. Each pair of terms is linked together by a connecting phrase into a sentence. The sentence should be read in the direction of the arrow. Complete the concept map by filling in the appropriate term or phrase. There is one right answer for each term. However, there are many correct answers for the connecting phrases (1, 4, 5, 9, 12).

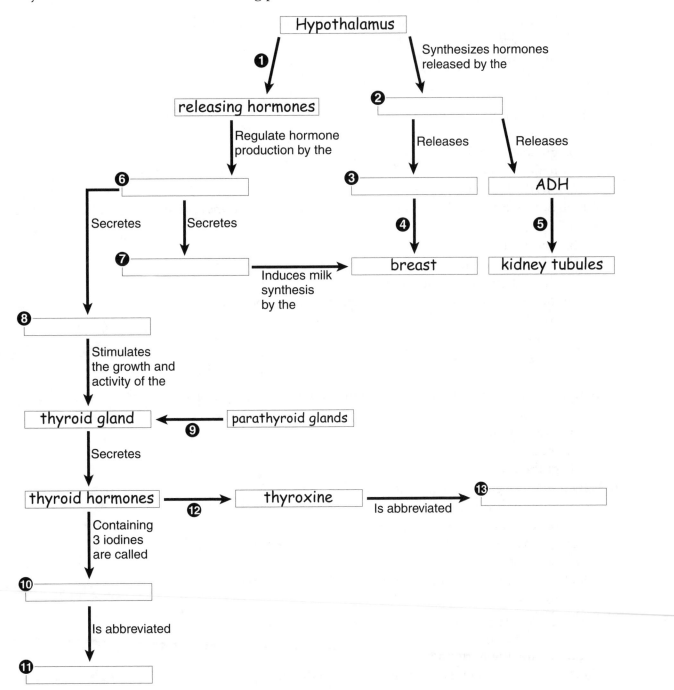

Optional Exercise: Make your own concept map involving the endocrine system. Use the following terms and any others you would like to include (e.g., target sites, hormone actions): pancreas, insulin, glucagon, adrenal gland, medulla, cortex, epinephrine, cortisol, aldosterone, raises blood glucose, lowers blood glucose, Cushing syndrome, diabetes mellitus, and Addison disease.

Testing Your Knowledge

BUILDING UNDERSTANDING

I. MULTIPLE CHOICE

Select the best answer.

1. What disease results from an excess of growth hormone in an adult?   1. _____
   a. acromegaly
   b. dwarfism
   c. tetany
   d. goiter

2. What is an androgen?   2. _____
   a. female sex hormone
   b. glucocorticoid
   c. male sex hormone
   d. atrial hormone

3. Which of the following hormones is NOT produced by the thyroid gland?   3. _____
   a. T_4
   b. thyroxine
   c. triiodothyronine
   d. thyroid-stimulating hormone

4. Which of these hormones causes milk ejection from the breasts?   4. _____
   a. oxytocin
   b. prolactin
   c. progesterone
   d. estrogen

5. Which of these hormones decreases blood glucose concentrations?   5. _____
   a. cortisol
   b. aldosterone
   c. insulin
   d. glucagon

6. Which of these glands releases antidiuretic hormone?   6. _____
   a. anterior pituitary
   b. posterior pituitary
   c. adrenal
   d. pancreas

7. Which of the following hormones is derived from cholesterol?   7. _____
   a. progesterone
   b. thyroid hormone
   c. growth hormone
   d. luteinizing hormone

8. Underactivity of which gland causes Addison disease?   8. _____
   a. adrenal medulla
   b. anterior pituitary
   c. parathyroid
   d. adrenal cortex

9. Which pituitary hormone regulates the activity of the thyroid gland?   9. _____
   a. TSH
   b. GH
   c. ACTH
   d. MSH
10. Which of these organs synthesizes erythropoietin?   10. _____
    a. kidneys
    b. skin
    c. heart
    d. placenta

II. COMPLETION EXERCISE

1. An abnormal increase in production of the hormone epinephrine may result from a tumor of the _____.

2. Releasing hormones are sent from the hypothalamus to the anterior pituitary by way of a special circulatory pathway called a(n) _____.

3. When the blood glucose level decreases below average, the islet cells of the pancreas release less insulin. The result is an increase in blood glucose. This is an example of the regulatory mechanism called _____.

4. The hypothalamus stimulates the anterior pituitary to produce ACTH, which, in turn, stimulates hormone production by the _____.

5. The element needed for the production of thyroxine is _____.

6. Local hormones that have a variety of effects, including the promotion of inflammation and the production of uterine contractions, are the _____.

7. A hormone secreted from the posterior pituitary that is involved in water balance is _____.

8. The primary target tissue for prolactin is the _____.

UNDERSTANDING CONCEPTS

I. TRUE/FALSE

For each statement, write T for true or F for false in the blank to the left of each number. If a statement is false, correct it by replacing the underlined term, and write the correct statement in the blanks below the statement.

_____ 1. Diabetics who still secrete detectable amounts of insulin suffer from type 2 diabetes.

_____

_____ 2. Cortisol and the pancreatic hormone insulin both raise blood glucose levels.

_____

_____ 3. <u>Diabetes insipidus</u> results from hyposecretion of antidiuretic hormone.

_____

_____ 4. The ovaries and testes produce <u>steroid</u> hormones.

_____

_____ 5. Cortisol is produced by the <u>adrenal cortex</u>.

_____

_____ 6. Islet cells are found in the <u>adrenal gland</u>.

_____

_____ 7. ADH and oxytocin are secreted by the <u>anterior</u> lobe of the pituitary.

_____

_____ 8. Atrial natriuretic peptide (ANP) is produced by the <u>kidneys</u>.

_____

II. PRACTICAL APPLICATIONS

Study each discussion. Then write the appropriate word or phrase in the space provided.

1. Ms. H, age 8, was brought to the clinic because she had been losing weight, was very thirsty, and urinated frequently. A blood chemistry panel taken when she had not eaten for eight hours showed an extremely high blood glucose reading of 186 mg/dL as well as other abnormal readings. A pancreatic disorder was suspected, leading to a tentative diagnosis of _____.

2. Mr. G, age 38, had been taking high doses of oral steroid drugs for asthma for six years. He noted fat deposits between his shoulders and around his abdomen, muscle loss in his arms and legs, and thin skin that was prone to bruising. These symptoms are common with excess of a hormone produced normally by the _____.

3. Mr. G was told that he was manifesting the symptoms of an endocrine disease. This disease is called _____.

4. Mr. L, age 42, reported to the hospital emergency room with complaints of shortness of breath and heart palpitations. The initial assessment by the nurse included the following findings: rapid heart rate, nervousness with tremor of the hands, skin warm and flushed, sweating, rapid respiration, and protruding eyes. Laboratory tests confirmed the diagnosis of overactivity of the _____.

5. After surgery for his endocrine problem, Mr. L had tetany, or contractions of the muscles of the hands and face. This was caused by the incidental surgical removal of the glands that control the release of calcium into the blood. The glands that maintain adequate blood calcium levels are the _____.

6. Ms. M has just been stung by a bee. She is extremely allergic to bees. Her sister gives her a lifesaving injection of an adrenal hormone. This hormone is called _____.

7. Ms. Q was supposed to have her baby 10 days ago. Her relatives are anxiously awaiting the birth of her child, so she asks her obstetrician if he can do something to hasten the birth of her child. The obstetrician agrees to induce labor using a hormone called _____.

8. Mr. S is preparing for his final law exams and is feeling very stressed. His partner is studying for a physiology exam and mentions that he probably has elevated levels of an anterior pituitary hormone that acts on the adrenal cortex. This hormone is called _____.

9. Ms. J, age 35, is preparing for a weight-lifting competition. She wants to build muscle tissue very rapidly, so a friend recommends that she use injections of a male steroid known to stimulate tissue building. This steroid is normally produced by the _____.

III. SHORT ESSAYS

1. Explain why hormones, although they circulate throughout the body, exercise their effects only on specific target cells.

_____

_____

_____

2. List two differences between the endocrine system and the nervous system.

_____

_____

_____

3. List three chronic complications associated with untreated diabetes mellitus.

_____

_____

_____

4. Name three organs other than endocrine glands that produce hormones, and name a hormone produced in each organ.

_____

_____

_____

5. Compare the anterior and the posterior lobes of the pituitary.

_____

_____

_____

CONCEPTUAL THINKING

1. Both Ms. J and Ms. K have a goiter. However, Ms. J has hyperthyroidism, and Ms. K has hypothyroidism. List the specific diseases suffered by the two women, and explain why each patient has a goiter.

_____

_____

_____

2. Young Ms. K suffers from asthma. She uses an inhaler containing epinephrine to treat her attacks. However, lately she has been suffering from a very rapid heartbeat. Her physician advises her to use her inhaler less frequently. Why?

_____

_____

_____

Expanding Your Horizons

In the past, hormones were not available for therapeutic use or were obtained only from cadavers. As you can imagine, hormone supplies were limited and reserved for cases of hormone deficiency. But many hormones now can be produced in unlimited quantities in the laboratory. This widespread availability has led to abuse by athletes and others trying to gain a competitive advantage. Read about all of the hormones—and other substances—that are banned by the World Anti-Doping Agency (WADA) at the website listed below. You can also access the online articles listed below to read more about the "arms race" between some unscrupulous athletes and the authorities.

- Catlin DH, Fitch KD, Ljungqvist A. Medicine and science in the fight against doping in sport. J Intern Med 2008;264(2): 99–114. PMID 18702750. Available at: http://onlinelibrary.wiley.com/doi/10.1111/j.1365-2796.2008.01993.x/pdf
- Saugy M, Robinson N, Saudan C, et al. Human growth hormone doping in sport. Br J Sports Med 2006;40(Suppl 1): i35–i39. PMCID PMC2657499. Available at: http://www.ncbi.nlm.nih.gov/pmc/articles/PMC2657499/
- World Anti Doping Agency Web site. Available at: http://www.wada-ama.org/

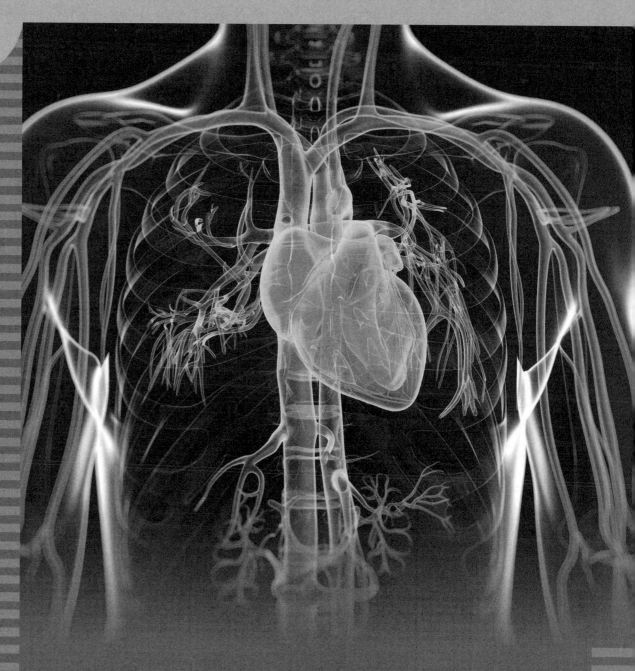

UNIT V

Circulation and Body Defense

The Blood

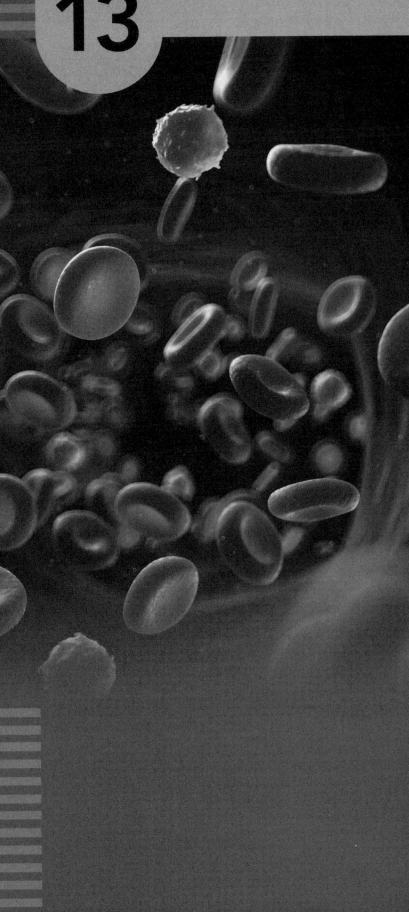

Overview

The blood maintains the constancy of the internal environment through its functions of transportation, regulation, and protection. Blood is composed of two portions: the liquid portion, or **plasma**, and the formed elements, consisting of cells and cell fragments. The plasma is 91% water and 9% solutes, including proteins, carbohydrates, lipids, electrolytes, and waste products. The formed elements are composed of the **erythrocytes**, which carry oxygen to the tissues; the **leukocytes**, which defend the body against invaders; and the **platelets**, which help prevent blood loss. The forerunners of the blood cells are called **hematopoietic stem cells**. These are formed in the red bone marrow, where they then develop into the various types of blood cells.

Hemostasis (the prevention of blood loss) is a series of protective mechanisms that occur when a blood vessel is ruptured by an injury. The steps in hemostasis include constriction of the blood vessels, formation of a platelet plug, and formation of a clot (coagulation), a complex series of reactions involving many different factors.

If the quantity of blood in the body is severely reduced because of hemorrhage or disease, the cells suffer from lack of oxygen and nutrients. In such instances, a **transfusion** may be given after typing and matching the blood of the donor with that of the recipient. Donor red cells with different surface antigens (proteins or glycoproteins) than the recipient's red cells will react with antibodies in the recipient's blood plasma, causing harmful agglutination reactions and destruction of the donated cells. Blood is most commonly tested for the **ABO system** involving antigens A and B. Blood can be packaged and stored in blood banks for use when transfusions are needed. Whenever possible, blood components such as cells, plasma, plasma fractions, or platelets are used. This practice is more efficient and can reduce the chances of incompatibility and transmission of disease.

The **Rh factor**, another red blood cell antigen, also is important in transfusions. If blood containing the Rh factor (Rh positive) is given to a person whose blood lacks that factor (Rh negative), the recipient will produce antibodies to the foreign Rh factor. If an Rh-negative mother is thus sensitized by an Rh-positive fetus, her antibodies may damage fetal red cells in a later pregnancy, resulting in **hemolytic disease of the newborn** (erythroblastosis fetalis).

Anemia is a common blood disorder. It may result from loss or destruction of red blood cells or from impaired production of red blood cells or hemoglobin. Other abnormalities are **leukemia**, a neoplastic disease of white cells, and clotting disorders.

Scientists have devised numerous studies to measure the composition of blood. These include the hematocrit, hemoglobin measurements, cell counts, blood chemistry tests, and coagulation studies. These techniques can diagnose blood diseases, some infectious diseases, and some metabolic diseases. Modern laboratories are equipped with automated counters, which rapidly and accurately count blood cells, and with automated analyzers, which measure enzymes, electrolytes, and other constituents of blood serum.

Addressing the Learning Objectives

1. LIST THE FUNCTIONS OF THE BLOOD.

EXERCISE 13-1

List three functions of blood, and provide an example for each.

1. _____
2. _____
3. _____

2. IDENTIFY THE MAIN COMPONENTS OF PLASMA.

EXERCISE 13-2

Which of the following substances are constituents of plasma? Circle all that apply.

a. erythrocytes
b. water
c. proteins
d. platelets
e. nutrients
f. electrolytes

3. DESCRIBE THE FORMATION OF BLOOD CELLS.

EXERCISE 13-3

Name the type of stem cell that can develop into all types of blood cells.

_____

4. NAME AND DESCRIBE THE THREE TYPES OF FORMED ELEMENTS IN THE BLOOD, AND GIVE THE FUNCTIONS OF EACH.

EXERCISE 13-4: Composition of Whole Blood (Text Fig. 13-1)

1. Write the names of the different blood components on the appropriate numbered lines in different colors. Use the color red for parts 1, 2, and 3.
2. Color the blood components on the diagram with the corresponding colors.

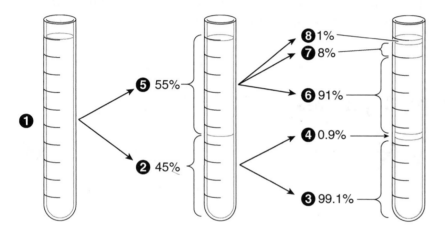

1. _____

2. _____

3. _____

4. _____

5. _____

6. _____

7. _____

8. _____

5. CHARACTERIZE THE FIVE TYPES OF LEUKOCYTES.

EXERCISE 13-5: Leukocytes (Table 13-2)

1. Write the names of the five types of leukocytes on lines 1 to 5.
2. Write the names of the labeled cell parts and other blood cells on lines 6 to 9.

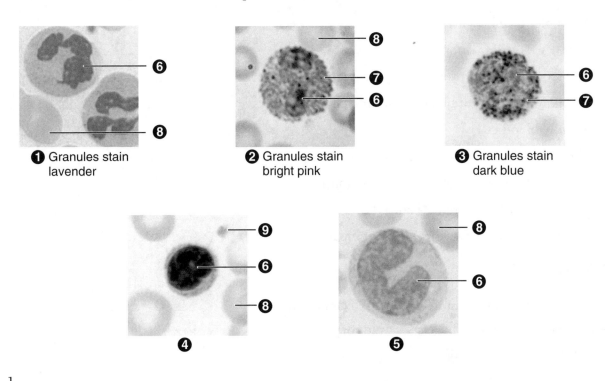

❶ Granules stain lavender

❷ Granules stain bright pink

❸ Granules stain dark blue

❹

❺

1. _____

2. _____

3. _____

4. _____

5. _____

6. _____

7. _____

8. _____

9. _____

EXERCISE 13-6

Write the appropriate term in each blank from the list below.

erythrocyte   platelet   leukocyte   plasma   hemoglobin

albumin   antibodies   complement   serum

1. A red blood cell   _____

2. Another name for thrombocyte   _____

3. The most abundant protein(s) in plasma   _____

4. The liquid portion of blood, including clotting factors   _____

5. A white blood cell   _____

6. A group of enzymes that participate in immunity   _____

7. The liquid portion of blood, without clotting factors   _____

8. The protein that fills red blood cells   _____

9. Nonenzyme blood proteins that participate in immunity   _____

EXERCISE 13-7

Write the appropriate term in each blank from the list below.

neutrophil ✓   macrophage ✓   monocyte ✓   pus ✓

eosinophil ✓   basophil ✓   plasma cell

1. The most abundant type of white blood cell in whole blood   _____

2. A tissue cell that develops from monocytes   _____

3. A lymphocyte that produces antibodies   _____

4. A leukocyte that stains with acidic dyes   _____

5. The largest blood leukocyte   _____

6. A substance that often accumulates when leukocytes are
   actively destroying bacteria   _____

7. A leukocyte that stains with basic dyes   _____

6. DEFINE *HEMOSTASIS*, AND CITE THREE STEPS IN HEMOSTASIS.

Also see Exercise 13-9.

EXERCISE 13-8

What is the difference between hemostasis and homeostasis?

_____

_____

9. EXPLAIN THE BASIS OF RH INCOMPATIBILITY AND ITS POSSIBLE CONSEQUENCES.

See Exercise 13-12.

10. LIST FOUR POSSIBLE REASONS FOR TRANSFUSIONS OF WHOLE BLOOD AND BLOOD COMPONENTS.

EXERCISE 13-12

Write the appropriate term in each blank from the list below. Not all terms will be used.

hematocrit   Rh factor   autologous   AB antigens   agglutination

hemapheresis   plasmapheresis   antibody   antigen

1. A general term describing any substance that can activate an immune response

2. The procedure for removing plasma and returning formed elements to the donor

3. The procedure for removing specific components and returning the remainder of the blood to the donor

4. Term describing blood donated by an individual for use by the same individual

5. The volume percentage of red cells in whole blood

6. These blood antigens are lacking in individuals with O-positive blood

7. The blood antigen involved in hemolytic disease of the newborn, which results from a blood incompatibility between a mother and fetus

8. The process by which cells become clumped when mixed with a specific antiserum

11. DEFINE *ANEMIA*, AND LIST SIX CAUSES OF ANEMIA.

EXERCISE 13-13

Write the appropriate term in each blank from the list below.

hemolytic anemia   pernicious anemia   aplastic anemia

thalassemia   hemorrhagic anemia

1. A disease resulting from a lack of vitamin B_{12} _____

2. A hereditary form of anemia associated with excess iron in the blood   _____

3. A disease in which red blood cells are excessively destroyed   _____

4. A disease resulting in insufficient red cell production in the bone marrow   _____

5. The type of anemia resulting from a bleeding ulcer   _____

12. DEFINE *LEUKEMIA*, AND NAME THE TWO TYPES OF LEUKEMIA.

See Exercise 13-14.

13. DESCRIBE FOUR FORMS OF CLOTTING DISORDERS.

EXERCISE 13-14

Write the appropriate term in each blank from the list below.

thrombocytopenia   disseminated intravascular coagulation   hemophilia

lymphocytic leukemia   myelogenous leukemia   von Willebrand disease

1. A clotting disorder involving excessive coagulation   _____

2. A clotting disorder that also impairs platelet adherence and is treated with an ADH-like drug   _____

3. A disease resulting from abnormal proliferation of stem cells in bone marrow   _____

4. A cancer that arises in lymphoid tissue   _____

5. A clotting disorder that can be treated with factor VIII   _____

6. Platelet deficiency   _____

14. IDENTIFY SIX TYPES OF TESTS USED TO STUDY BLOOD.

EXERCISE 13-15

Which blood test would you use to diagnose each disease? Choose the appropriate test from the list below.

red cell count ✓   Chem-7 ✓   coagulation study ✓

bone marrow biopsy ✓   blood smear ✓   platelet count ✓

1. Hemophilia   _____

2. Polycythemia vera   _____

3. Thrombocytopenia   _____

4. Myelogenous leukemia   _____

5. Malaria   _____

6. Diabetes mellitus   _____

15. REFERRING TO THE CASE STUDY, DISCUSS THE ADVERSE EFFECTS OF BONE MARROW DAMAGE.

EXERCISE 13-16

Use evidence from the case study to explain the key roles of erythrocytes, platelets, and leukocytes in body function. The first row has been completed for you.

Cell Type	Role	Disorder Resulting from Deficiency	Evidence of Disorder from Case Study
Erythrocytes	Oxygen transport	Anemia	Eleanor was tired because she lacked RBCs to transport adequate amounts of oxygen.
Platelets			
Leukocytes			

16. SHOW HOW WORD PARTS ARE USED TO BUILD WORDS RELATED TO THE BLOOD.

EXERCISE 13-17

Complete the following table by writing the correct word part or meaning in the space provided. Write a word that contains each word part in the "Example" column.

Word Part	Meaning	Example
1. erythr/o	_____	_____
2. _____	blood clot	_____
3. pro-	_____	_____
4. morph/o	_____	_____
5. _____	white, colorless	_____
6. _____	producing, originating	_____
7. hemat/o	_____	_____
8. _____	lack of	_____
9. -emia	_____	_____
10. _____	dissolving	_____

Making the Connections

The following concept map deals with the classification of blood cells. Each pair of terms is linked together by a connecting phrase into a sentence. The sentence should be read in the direction of the arrow. Complete the concept map by filling in the appropriate term or phrase. There is one right answer for each term. However, there are many correct answers for the connecting phrases (5, 6, 8, 10, 12, 14, and 16).

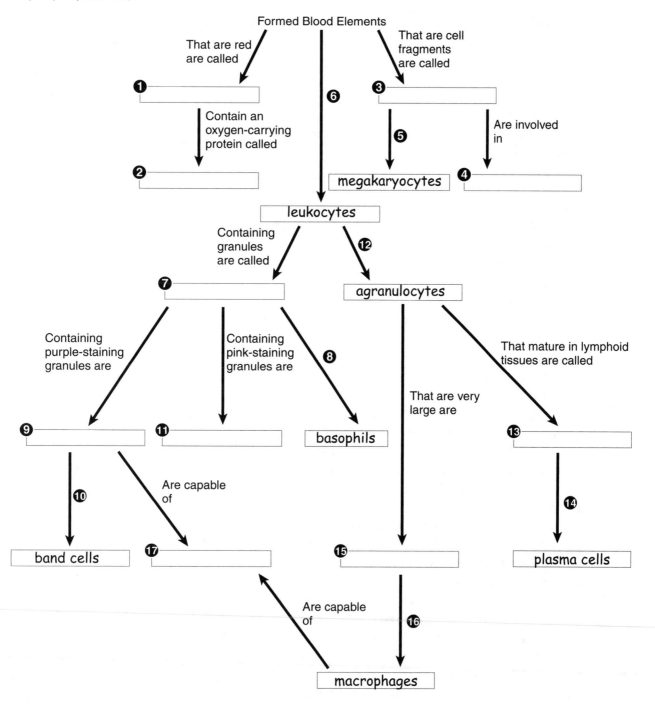

Optional Exercise: Make your own concept map involving blood. Use the following terms and any others you would like to include: procoagulants, anticoagulants, platelet plug, hemostasis, vaso-constriction, blood clot, fibrinogen, fibrin, prothrombinase, thrombin, serum, hemophilia, thrombocytopenia, and disseminated intravascular coagulation.

Testing Your Knowledge

BUILDING UNDERSTANDING

I. MULTIPLE CHOICE

Select the best answer.

1. Plasma can be given to anyone without danger of incompatibility because it lacks which of the following substances?
   a. serum
   b. red cells
   c. protein
   d. clotting factors

   1. _____

2. Which blood cell type is also called a polymorph, PMN, or seg?
   a. monocyte
   b. neutrophil
   c. basophil
   d. lymphocyte

   2. _____

3. Which of the following is NOT a type of white blood cell?
   a. thrombocyte
   b. lymphocyte
   c. eosinophil
   d. monocyte

   3. _____

4. Which of these proteins is the enzyme that activates fibrinogen?
   a. albumin
   b. thrombin
   c. thromboplastin
   d. fibrin

   4. _____

5. Which of the following might result in an Rh incompatibility problem?
   a. an Rh-positive mother and an Rh-negative fetus
   b. an Rh-negative mother and an Rh-positive fetus
   c. an Rh-negative mother and a type AB fetus
   d. an Rh-positive mother and an Rh-negative father

   5. _____

6. What is intrinsic factor?
   a. the first factor activated in blood clotting
   b. a substance needed for absorption of folic acid
   c. a type of hereditary bleeding disease
   d. a substance needed for absorption of vitamin B_{12}

   6. _____

7. Which of these problems could be diagnosed by electrophoresis?
   a. the volume proportion of red cells
   b. the presence of normal and abnormal types of hemoglobin
   c. red blood cell number
   d. white blood cell number

   7. _____

8. Which antibody classes are present in type B blood?
   a. A
   b. B
   c. both A and B
   d. neither A nor B

   8. _____

9. Which of the following cells is NOT a granulocyte?   9. _____
   a. monocyte
   b. eosinophil
   c. neutrophil
   d. polymorph
10. How do clinicians classify anemia resulting from bone
    marrow destruction?   10. _____
    a. nutritional anemia
    b. hemolytic anemia
    c. aplastic anemia
    d. pernicious anemia

II. COMPLETION EXERCISE

1. The type of leukemia that originates in lymphoid tissues is _____.

2. A deficiency in the number of circulating platelets is called _____.

3. The gas that is necessary for life and that is transported to all parts of the body by the blood
   is _____.

4. Some monocytes enter the tissues, mature, and become active phagocytes. These cells are
   called _____.

5. One waste product of body metabolism is carried to the lungs to be exhaled. This gas
   is _____.

6. The hormone that stimulates red blood cell production is _____.

7. Blood cells are formed in the _____.

8. The most important function of certain lymphocytes is to engulf disease-producing organisms
   by the process of _____.

9. The chemical element that characterizes hemoglobin is _____.

UNDERSTANDING CONCEPTS

I. TRUE/FALSE

For each statement, write T for true or F for false in the blank to the left of each number. If a statement is false, correct it by replacing the underlined term, and write the correct statement in the blanks below the statement.

_____ 1. Eosinophils and basophils are underlined granular leukocytes.

_____

_____ 2. If the formed elements are returned to the blood donor, the procedure is called
         plasmapheresis.

_____

_____ 3. Sickle cell anemia is classified as a form of <u>myelogenous</u> anemia.

_____

_____ 4. The hormone that stimulates red blood cell production is produced by the <u>kidneys</u>.

_____

_____ 5. <u>Type AB</u> blood contains antibodies to both A and B antigens.

_____

_____ 6. The disorder associated with impaired hemoglobin production and excessive iron uptake is called <u>pernicious anemia</u>.

_____

_____ 7. Substances that induce blood clotting are called <u>procoagulants</u>.

_____

_____ 8. The watery fluid that remains after a blood clot has been removed from the blood is <u>plasma</u>.

_____

II. PRACTICAL APPLICATIONS

Study each discussion. Then write the appropriate word or phrase in the space provided.

➤ **Group A**

1. A young girl named AL fell off her bike, sustaining a deep gash to her leg that bled copiously from a severed vessel. In describing this type of bleeding, the doctor in the emergency clinic used the word _____.

2. While the physician attended to the wound, the technician drew blood for typing and other studies. AL's blood did not agglutinate with either anti-A or anti-B serum. Her blood was classified as type _____.

3. Young AL required a blood transfusion. The only type of blood she could receive is type _____.

4. Further testing of AL's blood revealed that it lacked the Rh factor. She was, therefore, said to be _____.

5. If AL were to be given a transfusion of Rh-positive blood, she might become sensitized to the Rh protein. In that event, her blood would produce counteracting substances called _____.

6. The technician examined AL's blood under the microscope and noticed some red blood cells were shaped like crescent moons. AL may have the disorder called _____.

➤ Group B

1. Mr. K, age 72, had an injury riding his motorized scooter down a steep hill. He suffered only minor scrapes, but he came to the hospital because his scrapes did not stop bleeding. The process by which blood loss is prevented or minimized is called _____.

2. The physician discovered that Mr. K was the great great great grandson of Queen Victoria. Like many of her offspring, Mr. K suffers from a deficiency of a clotting factor. This hereditary disorder is called _____.

3. Mr. K must be treated with a rich source of clotting factors. The doctor gets a bag of frozen plasma, which has a white powdery substance at the bottom of the bag. The substance is called

   _____.

4. Mr. K also mentions that he has been very tired and pale lately. His cook recently retired, and he has been surviving on dill pickles and crackers. The physician suspects a deficiency in an element required to synthesize hemoglobin. This element is called _____.

5. The physician orders a test to determine the proportion of red blood cells in his blood. The result of this test is called the _____.

6. The test confirms that Mr. K does not have enough red blood cells due to a dietary deficiency. This disorder is called _____.

➤ Group C

1. Ms. J, an elite cyclist, has come to the hospital complaining of a pounding headache. The physician, Dr. L, takes a sample of her blood, and notices that it is very thick. He decides to count her red blood cells, but the automatic cell counter is broken. Dr. L calls for a technician to perform a visual count using the microscope and a special slide called a(n) _____.

2. The technician comes back with a result of 7.5 million cells/mcL. This count is abnormally high, suggesting a diagnosis of _____.

3. Dr. L is immediately suspicious. He asks Ms. J if she consumes any performance-enhancing drugs. Ms. J admits that she has been taking a hormone to increase red blood cell synthesis. This hormone is called _____.

4. Dr. L informs Ms. J that she is at risk for blood clot formation. He tells her to stop taking the hormone and to take aspirin for a few days to inhibit blood clotting. Drugs that inhibit clotting are called _____.

III. SHORT ESSAYS

1. Compare and contrast *leukopenia* and *leukocytosis*. Name one similarity and one difference.

_____

_____

_____

_____

_____

2. What kind of information can be obtained from blood chemistry tests?

_____

_____

_____

_____

_____

3. Briefly describe the final events in blood clot formation, naming the substances involved in each step.

_____

_____

_____

_____

_____

4. Name one reason to transfuse an individual with:

a. whole blood

_____

b. platelets

_____

c. plasma

_____

d. plasma protein

_____

CONCEPTUAL THINKING

1. A dehydrated individual will have an elevated hematocrit. Explain why.

   _____

   _____

   _____

2. A man named JA has a history of frequent fevers. His skin is pale and his heart rate rapid. The doctor suspects that he has cancer. His white blood cell count was revealed to be 25,000/mcL of blood.

   a. Is this white blood cell count normal? What is the normal range?

   _____

   b. Based on his white blood cell count, what is a probable diagnosis of JA's condition?

   _____

   c. JA also has a hematocrit of 30%. Is this value normal? What is the normal range?

   _____

   d. What disorder does his hematocrit value suggest?

   _____

   e. Consider the different functions of blood. Discuss some functions that might be impaired in JA's case.

   _____

   _____

   _____

3. Mr. R needs a blood transfusion. He has type AB blood. The doctor is considering a transfusion using A blood.

   a. Which antigens are present on his blood cells?

   _____

   b. Which antibodies will be present in his blood?

   _____

   c. Which antigens will be present on donor blood cells?

   _____

   d. Is this blood transfusion safe? Why or why not?

   _____

Expanding Your Horizons

Have you ever watched vampire entertainment, such as the *Twilight* movies or the *True Blood* TV series? Imagine how much easier life would be for vampires if they could buy blood in the grocery stores. And in our vampire-free world, a safe, effective artificial blood substitute would save dollars and lives. These substitutes would be free of the problems associated with normal blood transfusions: contamination with HIV or hepatitis, incompatibility reactions, short shelf life, stringent storage conditions, and limited supply. Billions of dollars have gone into the search for artificial blood, and some compounds have been used when blood supplies run short. However, none of the currently available substitutes are safe—they all appear to increase the risk of heart attack. You can read more about the search for a safe blood substitute on the websites below or do a search for "artificial blood."

- Akst J. Artificial blood is patient-ready. TheScientist April 16, 2014. Available at: http://www.the-scientist.com/?articles.view/articleNo/39718/title/Artificial-Blood-Is-Patient-Ready/
- Cooper C. Can we make a blood substitute? Available at: http://www.profchriscooper.com/Home/Artificial_Blood.html

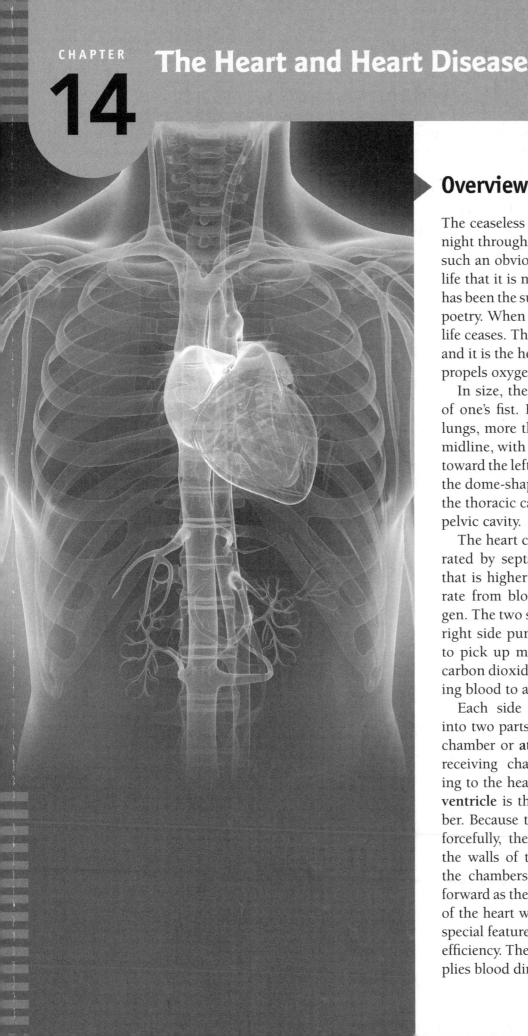

The Heart and Heart Disease

▶ Overview

The ceaseless beat of the heart day and night throughout one's entire lifetime is such an obvious key to the presence of life that it is no surprise that this organ has been the subject of wonderment and poetry. When the heart stops pumping, life ceases. The cells must have oxygen, and it is the heart's pumping action that propels oxygen-rich blood to them.

In size, the heart is roughly the size of one's fist. It is located between the lungs, more than half to the left of the midline, with the **apex** (point) directed toward the left. Below is the **diaphragm**, the dome-shaped muscle that separates the thoracic cavity from the abdomino-pelvic cavity.

The heart consists of two sides separated by septa. The septa keep blood that is higher in oxygen entirely separate from blood that is lower in oxygen. The two sides pump in unison, the right side pumping blood to the lungs to pick up more oxygen and drop off carbon dioxide, and the left side pumping blood to all other parts of the body.

Each side of the heart is divided into two parts or **chambers**. The upper chamber or **atrium** on each side is the receiving chamber for blood returning to the heart. The lower chamber or **ventricle** is the strong pumping chamber. Because the ventricles pump more forcefully, their walls are thicker than the walls of the atria. **Valves** between the chambers keep the blood flowing forward as the heart pumps. The muscle of the heart wall, the **myocardium**, has special features to enhance its pumping efficiency. The coronary circulation supplies blood directly to the myocardium.

The heartbeat originates within the heart at the **sinoatrial (SA) node**, often called the pacemaker. Electrical impulses from the pacemaker spread through special conducting fibers in the wall of the heart to induce contractions, first of the two atria and then of the two ventricles. While the atria contract, the ventricles relax and vice versa. After ventricular contraction finishes, the entire heart relaxes and fills with blood. For each heart chamber, the relaxation phase is called **diastole**, and the contraction phase is called **systole**. One **cardiac cycle** includes all of the events of a single heartbeat, that is, atrial systole, ventricular systole, and then diastole of all heart chambers. The heart rate is influenced by the nervous system and other circulating factors, such as hormones and drugs. The cardiovascular control center modifies heart rate as needed in order to maintain relatively constant blood pressure.

Heart diseases may be classified according to the cause or the area of the heart affected. Causes include congenital abnormalities, rheumatic fever, coronary artery disease, and heart failure.

Addressing the Learning Objectives

1. DESCRIBE THE THREE TISSUE LAYERS OF THE HEART WALL.

See Exercises 14-1, 14-2, and 14-3.

2. DESCRIBE THE LOCATION AND STRUCTURE OF THE PERICARDIUM, AND CITE ITS FUNCTIONS.

EXERCISE 14-1: Layers of the Heart Wall and Pericardium (Text Fig. 14-2)

1. Write the terms *heart wall* and *serous pericardium* in the appropriate boxes.
2. Write the names of structures 3 through 8 on the numbered lines in different colors. Use black for structure 6, because it will not be colored.
3. Color the structures on the diagram (except structure 6) with the corresponding colors.

3. _____

4. _____

5. _____

6. _____

7. _____

8. _____

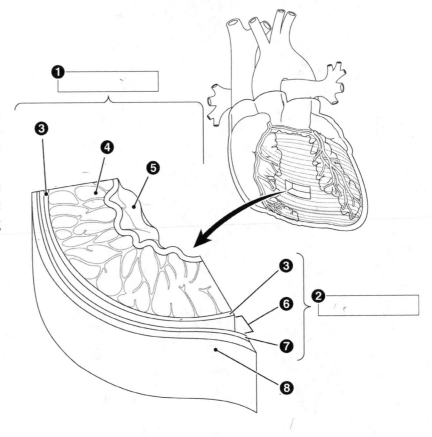

EXERCISE 14-2

Fill in the blanks of the concept map below.

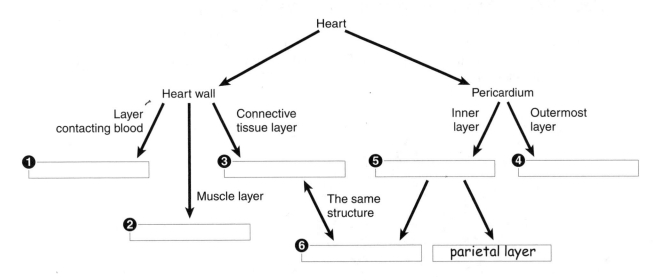

EXERCISE 14-3

Write the appropriate term in each blank from the list below. Not all terms will be used.

base   apex ✓   endocardium ✓

epicardium ✓   fibrous pericardium ✓   myocardium ✓

serous pericardium ✓   visceral layer ✓   parietal layer

1. The pointed, inferior portion of the heart   _____

2. The membrane consisting of a visceral and a parietal layer   _____

3. A layer of epithelial cells in contact with blood within
   the heart   _____

4. The heart layer containing intercalated disks   _____

5. The outermost layer of the sac enclosing the heart   _____

6. An alternate term for the epicardium   _____

7. The site where the major vessels attach to the heart   _____

8. The heart wall layer composed of connective tissue   _____

3. COMPARE THE FUNCTIONS OF THE RIGHT AND LEFT CHAMBERS OF THE HEART.

See Exercises 14-4 and 14-5.

4. NAME THE VALVES AT THE ENTRANCE AND EXIT OF EACH VENTRICLE, AND IDENTIFY THE FUNCTION OF EACH.

EXERCISE 14-4: The Heart and Great Vessels (Text Fig. 14-4)

1. Label the indicated parts. Hint: Structures 5 and 14 are chambers.
2. Use arrows to show the direction of blood flow. If you would like, use red arrows for blood high in oxygen and blue arrows for blood low in oxygen.

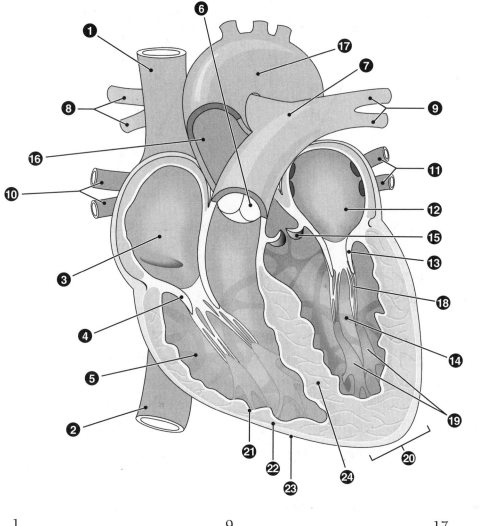

1. _____   9. _____   17. _____

2. _____   10. _____   18. _____

3. _____   11. _____   19. _____

4. _____   12. _____   20. _____

5. _____   13. _____   21. _____

6. _____   14. _____   22. _____

7. _____   15. _____   23. _____

8. _____   16. _____   24. _____

EXERCISE 14-5

Write the appropriate term in each blank from the list below.

left ventricle   right ventricle   left atrium

right atrium   atrioventricular valve   pulmonary valve

pulmonary circuit   aortic valve   systemic circuit

1. The chamber that pumps blood to the lungs   _____

2. The valve that prevents blood from returning to the right
   ventricle   _____

3. The chamber that receives blood from the lungs   _____

4. The valve that prevents blood from returning to the left
   ventricle   _____

5. The pathway that carries blood to and from the lungs   _____

6. One of two valves dividing the upper and lower chambers   _____

7. The pathway that carries blood to and from body tissues   _____

8. The chamber that pumps oxygen-rich blood to the body   _____

9. The chamber that receives oxygen-poor blood from the
   body   _____

5. BRIEFLY DESCRIBE BLOOD CIRCULATION THROUGH THE MYOCARDIUM.

EXERCISE 14-6: The Coronary Circulation (Text Fig. 14-6)

1. Label the parts of the heart (structures 1 to 10).
2. Label the vessels of the coronary circulation (structures 11 to 16).

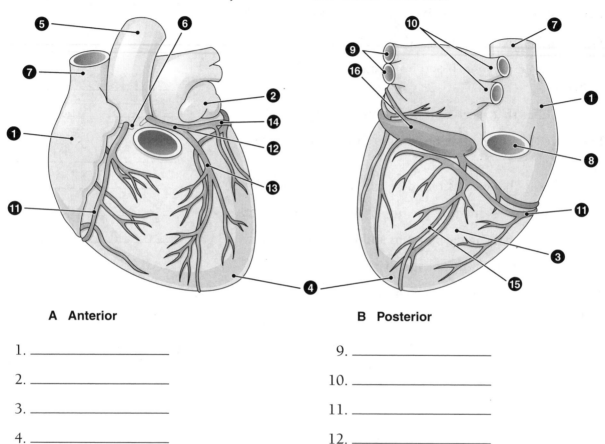

A Anterior

B Posterior

1. _____

2. _____

3. _____

4. _____

5. _____

6. _____

7. _____

8. _____

9. _____

10. _____

11. _____

12. _____

13. _____

14. _____

15. _____

16. _____

6. BRIEFLY DESCRIBE THE CARDIAC CYCLE.

EXERCISE 14-7

Fill in the blank after each event—does it occur in complete diastole (D), atrial systole (A), or ventricular systole (V)?

1. The ventricles are contracting   _____

2. The atria are contracting   _____

3. Blood is entering the aorta   _____

4. Neither the ventricles nor the atria are contracting   _____

5. The atrioventricular valves are closed   _____

EXERCISE 14-8

Write the appropriate term in each blank from the list below.

diastole   atrial systole   ventricular systole

cardiac output   stroke volume

1. The amount of blood ejected from a ventricle with each beat   _____

2. The stage of the cardiac cycle that directly follows the resting period   _____

3. The stage of the cardiac cycle that precedes the resting period   _____

4. The volume of blood pumped by each ventricle in one minute   _____

5. The resting period of the cardiac cycle   _____

7. NAME AND LOCATE THE COMPONENTS OF THE HEART'S CONDUCTION SYSTEM.

EXERCISE 14-9: The Conduction System of the Heart (Text Fig. 14-9)

1. Label the heart chambers (bullets 1 to 4).
2. Label the parts of the conducting system (bullets 5 to 10). Draw arrows to indicate the direction of impulse conduction.

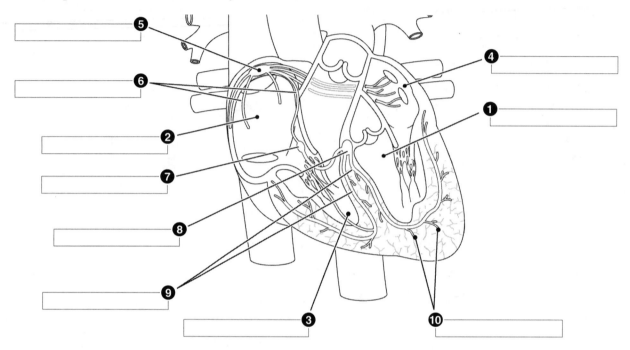

EXERCISE 14-10

Write the appropriate term in each blank from the list below. Not all terms will be used.

atrioventricular bundle ✓   Purkinje fibers ✓   sinus rhythm ✓
sinoatrial node ✓   atrioventricular node ✓   internodal pathways

1. The group of conduction fibers found in the ventricles' walls   _____

2. The mass of conduction tissue located in the septum at the bottom of the right atrium   _____

3. A normal heartbeat, originating from the normal heart pacemaker   _____

4. The group of conduction fibers carrying impulses from the AV node   _____

5. The name of the normal heart pacemaker, located in the upper wall of the right atrium   _____

8. EXPLAIN THE EFFECTS OF THE AUTONOMIC NERVOUS SYSTEM (ANS) ON THE HEART RATE.

EXERCISE 14-11

Fill in the blank after each statement—does the characteristic refer to the parasympathetic nervous system (P) or the sympathetic nervous system (S)?

1. Regulates heart activity via the vagus nerve _____

2. Uses ganglia located close to the spinal cord _____

3. Decreases the heart rate _____

4. Increases the force of each contraction _____

5. Increases the heart rate _____

9. LIST AND DEFINE SEVERAL TERMS THAT DESCRIBE VARIATIONS IN HEART RATES.

EXERCISE 14-12

Write the appropriate term in each blank from the list below.

bradycardia   tachycardia   sinus arrhythmia   extrasystole

1. A beat that comes before the normal beat _____

2. A normal variation in heart rate caused by changes in breathing rate _____

3. A heart rate of less than 60 bpm _____

4. A heart rate of greater than 100 bpm _____

10. EXPLAIN WHAT PRODUCES EACH OF THE TWO NORMAL HEART SOUNDS, AND IDENTIFY THE USUAL CAUSE OF A MURMUR.

EXERCISE 14-13

Fill in the blanks.

1. The "lub" sound is caused by the closure of the _____ valves.

2. The "dub" sound is caused by the closure of the _____ valves.

3. An abnormal sound caused by a structural problem in the heart or nearby vessels is called a(n) _____ murmur.

4. A sound heard during the workings of a healthy heart is called a(n) _____ murmur.

11. BRIEFLY DESCRIBE FIVE METHODS USED TO STUDY THE HEART.

EXERCISE 14-14

Write the appropriate term in each blank from the list below.

stethoscope   echocardiography   catheterization

coronary angiography   electrocardiography

1. Technique that uses ultrasound to study the heart as it beats   _____

2. Technique that measures the electrical activity of the heart   _____

3. Instrument used to detect heart murmurs by their sound   _____

4. Any procedure in which an extremely thin tube is inserted into a vessel   _____

5. A procedure that uses a catheter and dyes to visualize the heart's blood vessels   _____

12. DESCRIBE SIX TYPES OF HEART DISEASE.

EXERCISE 14-15

Write the appropriate term in each blank from the list below.

fibrillation   heart failure   valvular stenosis   valvular insufficiency

flutter   infarct   ischemia   plaque

1. Rapid and uncoordinated heart muscle contractions   _____

2. Condition that results from a lack of blood supply to the tissues, as from narrowing of an artery   _____

3. A hard deposit of fatty material in a blood vessel   _____

4. Term for very rapid coordinated heart contractions of up to 300 bpm   _____

5. An area of tissue damaged from a lack of blood supply   _____

6. A condition in which the heart can no longer effectively pump blood, often resulting in fluid accumulation in the lungs   _____

7. A condition in which a valve does not close properly   _____

8. A condition in which a valve does not open completely or has a narrowed opening   _____

EXERCISE 14-16

Write the appropriate term in each blank from the list below. Not all terms will be used.

occlusion   atrial septal defect   patent ductus arteriosus

atherosclerosis   thrombus   endocarditis

myocarditis   pericarditis   angina pectoris

1. Inflammation of the heart lining, often affecting the valves   _____

2. A blood clot formed within a vessel   _____

3. Persistence of an open foramen ovale after birth   _____

4. Inflammation of the heart muscle   _____

5. Complete closure, as of an artery   _____

6. Inflammation of the serous membrane on the heart surface   _____

7. Chest discomfort resulting from inadequate coronary blood flow   _____

8. A thickening and hardening of the vessels   _____

13. LIST FOUR RISK FACTORS FOR CORONARY ARTERY DISEASE THAT CANNOT BE MODIFIED.

See Exercise 14-16.

14. LIST SEVEN RISK FACTORS FOR CORONARY ARTERY DISEASE THAT CAN BE MODIFIED.

EXERCISE 14-17

Label each of the following statements as true (T) or false (F).
The risk of heart disease is increased if you:

1. smoke.   _____

2. have low blood pressure.   _____

3. have diabetes mellitus.   _____

4. exercise regularly.   _____

5. are a 30-year-old male (compared to a 30-year-old female).   _____

6. tend to deposit fat on the thighs (rather than around the abdomen).   _____

7. eat more unsaturated fat and relatively little saturated fat.   _____

8. have lower blood levels of C-reactive protein.   _____

9. suffer from sleep apnea.   _____

15. DESCRIBE THREE APPROACHES TO THE TREATMENT OF HEART DISEASE.

EXERCISE 14-18

Write the appropriate term in each blank from the list below.

digitalis   anticoagulant   angioplasty   slow calcium channel blocker

cardiac ablation   nitroglycerin   coronary atherectomy

1. Aspirin is an example of this type of drug   _____

2. A surgery that destroys abnormal heart tissue   _____

3. A drug that dilates the coronary vessels and is used to relieve angina pectoris   _____

4. A technique that uses a balloon to open a blocked blood vessel   _____

5. A drug derived from the foxglove plant that slows and strengthens heart contractions   _____

6. An agent that controls the activity of contractile and conducting cells of the heart by altering ion flux   _____

7. The use of a grinding or cutting device to remove a plaque   _____

16. LIST FOUR CHANGES THAT MAY OCCUR IN THE HEART WITH AGE.

EXERCISE 14-19

In the lines below, explain why the following changes occur in the elderly.

1. Reduced cardiac output

_____

_____

2. Increased incidence of heart murmurs

_____

_____

3. Increased arrhythmias

_____

_____

17. REFERRING TO THE CASE STUDY, LIST THE EMERGENCY AND SURGICAL PROCEDURES COMMONLY PERFORMED FOLLOWING A MYOCARDIAL INFARCTION, AND EXPLAIN WHY THEY ARE DONE.

EXERCISE 14-20

In the lines below, list three emergency measures, three administered medications, and one surgical measure that Jim, the case study patient, received. Briefly describe the purpose of each.

Emergency measures:

1. _____

2. _____

3. _____

Medications:

1. _____

2. _____

3. _____

Surgery:

1. _____

18. SHOW HOW WORD PARTS ARE USED TO BUILD WORDS RELATED TO THE HEART.

EXERCISE 14-21

Complete the following table by writing the correct word part or meaning in the space provided. Write a word that contains each word part in the "Example" column.

Word Part	Meaning	Example
1. _____	chest	_____
2. sin/o	_____	_____
3. _____	vessel	_____
4. cardi/o	_____	_____
5. scler/o	_____	_____
6. _____	lung	_____
7. _____	slow	_____
8. _____	rapid	_____
9. isch-	_____	_____
10. cyan/o	_____	_____

Making the Connections

The following flow chart deals with the passage of blood through the heart. Beginning with the right atrium, outline the structures a blood cell would pass through in the correct order by filling in the boxes. Use the following terms: left atrium, left ventricle, right ventricle, right AV valve, left AV valve, pulmonary semilunar valve, aortic semilunar valve, superior/inferior vena cava, right lung, left lung, aorta, left pulmonary artery, right pulmonary artery, left pulmonary veins, right pulmonary veins, body. You can write the names of structures that encounter blood high in oxygen in red and the names of structures that encounter blood low in oxygen in blue.

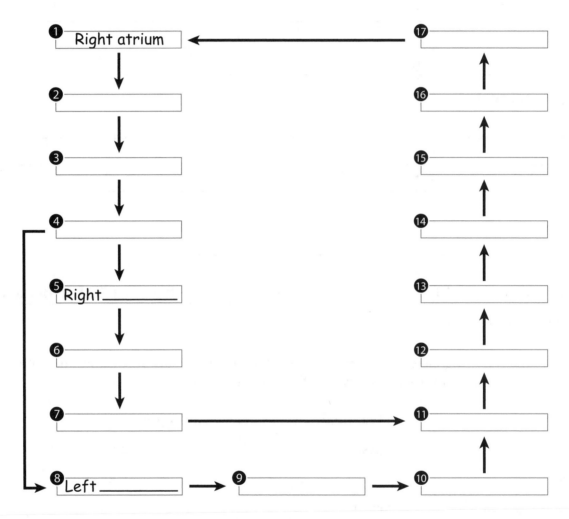

Optional Exercise: Make your own concept map based on the events of the cardiac cycle. Use the following terms and any others you would like to include: AV valves open, AV valves closed, blood flow from atria to ventricles, blood flow from ventricles to arteries, atrial diastole, atrial systole, ventricular diastole, and ventricular systole. There will be many links between the different terms.

Testing Your Knowledge

BUILDING UNDERSTANDING

I. MULTIPLE CHOICE

Select the best answer.

1. How long does it take to complete an average cardiac cycle
   in an individual at rest?   1. _____
   - a. 8 seconds
   - b. 5 seconds
   - c. 0.8 second
   - d. 30 seconds
2. Which of these terms describes the volume of blood pumped by
   each ventricle in one minute?   2. _____
   - a. stroke volume
   - b. cardiac output
   - c. heart rate
   - d. ejection rate
3. Which of the following is NOT a part of the conduction system
   of the heart?   3. _____
   - a. atrioventricular bundle
   - b. atrioventricular valve
   - c. Purkinje fibers
   - d. atrioventricular node
4. What is the function of a thrombolytic drug?   4. _____
   - a. reduce hypertension
   - b. regulate the heartbeat
   - c. dilate the blood vessels
   - d. dissolve blood clots
5. What is the effect of the parasympathetic nervous system on the heart?   5. _____
   - a. increases heart rate but not myocardial contraction strength
   - b. increases heart rate and myocardial contraction strength
   - c. decreases heart rate but not myocardial contraction strength
   - d. decreases heart rate and myocardial contraction strength
6. What can happen when clots form in the coronary arteries?   6. _____
   - a. stroke
   - b. myocardial infarction
   - c. heart failure
   - d. congenital heart disease
7. Which vein carries blood from the coronary circulation back into
   the right atrium?   7. _____
   - a. right coronary artery
   - b. interatrial septum
   - c. pulmonary vein
   - d. coronary sinus

8. What is the tetralogy of Fallot?   8. _____
    a. four specific defects associated with congenital heart disease
    b. the four cardiac valves
    c. four signs of an impending heart attack
    d. the four types of heart disease

9. Which membranes are separated by the pericardial space?   9. _____
    a. the fibrous pericardium and the serous pericardium
    b. the visceral and parietal layers of the serous pericardium
    c. the epicardium and the pericardium
    d. the myocardium and the epicardium

10. Which heart chamber receives blood from the lungs?   10. _____
    a. right ventricle
    b. left ventricle
    c. right atrium
    d. left atrium

II. COMPLETION EXERCISE

Write the word or phrase that correctly completes each sentence.

1. The autonomic nerve that slows the heartbeat is the _____.

2. The sac that surrounds the heart is the _____.

3. An abnormality in the rhythm of the heartbeat is a(n) _____.

4. The medical term for high blood pressure is _____.

5. One complete cycle of heart contraction and relaxation is called the _____.

6. The fibrous threads connecting the AV valves to muscles in the heart wall are called the
_____.

7. The right atrioventricular valve is also known as the _____.

8. Normal heart sounds heard while the heart is working are called _____.

9. The small blood vessel connecting the pulmonary artery with the aorta that is found only in the
fetus is the _____.

10. The atrioventricular valves are closed during the phase of the cardiac cycle known as ventricular _____.

UNDERSTANDING CONCEPTS

I. TRUE/FALSE

For each statement, write T for true or F for false in the blank to the left of each number. If a statement is false, correct it by replacing the <u>underlined</u> term, and write the correct statement in the blanks below the statement.

_____ 1. Discomfort felt in the region of the heart as a result of coronary artery disease is called <u>congestive heart failure</u>.

_____

_____ 2. The circulatory system of the fetus has certain adaptations for the purpose of bypassing the <u>kidneys</u>.

_____

_____ 3. A heart rate of 150 bpm is described as <u>tachycardia</u>.

_____

_____ 4. Blood in the heart chambers comes into contact with the <u>epicardium</u>.

_____

_____ 5. The aorta is part of the <u>pulmonary</u> circuit.

_____

_____ 6. Nerve impulses travel down the internodal pathways from the <u>AV node</u>.

_____

_____ 7. A heart rhythm originating at the <u>SA node</u> is termed a sinus rhythm.

_____

_____ 8. Heart disease caused by antibodies to streptococci is called <u>atherosclerosis</u>.

_____

II. PRACTICAL APPLICATIONS

Study each discussion. Then write the appropriate word or phrase in the space provided.

➤ **Group A**

1. Ms. J, age 82, was participating in a lawn bowling tournament when she suddenly collapsed with chest pain. The paramedics were preparing her for transport to the hospital when they noted a sudden onset of pale skin and unconsciousness. The heart monitor showed a rapid, uncoordinated activity of the ventricles, called _____.

2. The paramedics administered an electric shock using an automated defibrillator with the aim of restoring the normal heart rhythm, which is called a(n) _____.

3. In the hospital emergency room, Ms. J was given intravenous medications to dissolve any blood clots in her coronary circulation. These drugs are called _____.

4. The drugs were not administered in time to prevent damage to the middle layer of the heart wall, which is known as the _____.

5. The electrical activity of Ms. J's heart was analyzed using a(n) _____.

6. Blood flow through Ms. J's cardiac blood vessels was mapped using a catheter, dye, and an x-ray machine. This technique is known as _____.

7. This test revealed a narrowing in a branch of her left coronary artery that supplies the left ventricle. This vessel is known as the LAD, or the _____.

8. The cardiologist used a balloon to remove the blockage in a technique known as _____.

9. The analysis showed that the conduction system was damaged. She was scheduled to have a device inserted to supply impulses needed to stimulate heart contractions. This device is called a(n) _____.

10. Ms. J's long-term care included a daily dose of the anticoagulant drug acetylsalicylic acid, commonly known as _____.

➤ **Group B**

1. Baby L has just been born. To her parents' dismay, her skin and mucous membranes are tinged with blue. The scientific term for this coloration is _____.

2. The obstetrician listens to her heartbeat using an instrument called a(n) _____.

3. The doctor notices an abnormality in the second heart sound (or "dup"), which is largely caused by the closure of the two _____.

4. This abnormal sound probably reflects a structural problem with Baby L's heart and is thus termed a(n) _____.

5. The baby is sent for a test that uses ultrasound waves to examine her heart structure. A hole is observed between the two lower chambers of her heart. This disorder is known as _____.

6. In addition to the two defects already described, it was also observed that Baby L's aorta is not positioned properly and that her left ventricular wall was abnormally muscular. This common combination of four defects is called the _____.

7. Baby L was diagnosed with a form of heart disease present at birth, called
_____.

III. SHORT ESSAY

1. Although the heartbeat originates within the heart itself, it is influenced by factors in the internal environment. Describe some of these factors that can affect the heart.

_____

_____

_____

_____

2. What is the difference between thrombolytic drugs and anticoagulants? Describe their effects and their uses.

_____

_____

_____

_____

CONCEPTUAL THINKING

1. Atropine is a drug that inhibits activity of the parasympathetic nervous system. Discuss the effects of atropine on the heart. How does the parasympathetic nervous system affect the heart, which aspects of heart function will be affected, and which will be unaffected?

_____

_____

_____

_____

2. Mr. J is undertaking a gentle exercise program, primarily involving walking, to lose weight. His heart rate is 100 bpm, and his stroke volume is 75 mL. What is his cardiac output in liters, and what does cardiac output mean?

_____

_____

_____

_____

Expanding Your Horizons

How low can you go? You may have heard of the exploits of free divers, who dive to tremendous depths without the aid of scuba equipment. The world record for assisted free diving is held by Pipin Ferreras, who dove to 170 m (558 ft) with the aid of a sled. Free diving is not without its dangers. Pipin's wife Audrey died during a world record attempt. Free diving is facilitated by the dive reflex, which allows mammals to hold their breath for long periods of time underwater. Immersing one's face in cold water induces bradycardia and diverts blood away from the periphery. How would these modifications increase the ability to free dive? You can learn more about the underlying mechanisms, rationale, and dangers of the dive reflex by performing a website search for "dive reflex." Information is also available in the article listed below.

- Hurwitz BE, Furedy JJ. The human dive reflex: An experimental, topographical, and physiological analysis. Physiol Behav 1986;36:287–294.

CHAPTER

15

Blood Vessels and Blood Circulation

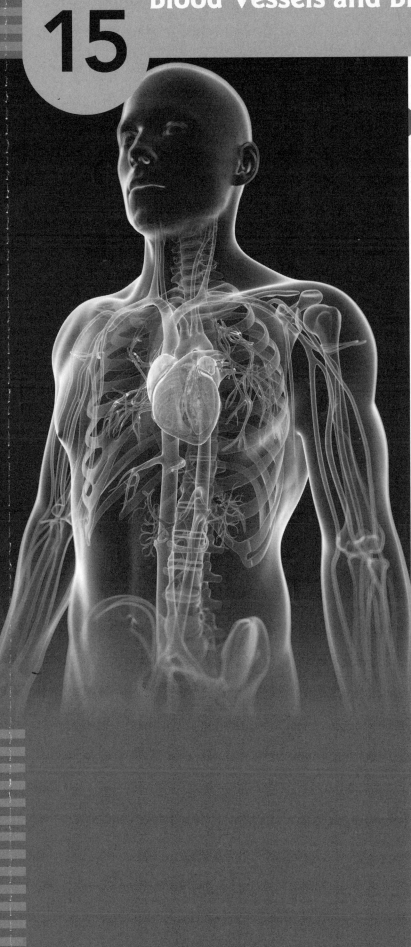

Overview

The blood vessels are classified as arteries, veins, or capillaries according to their respective functions. **Arteries** carry blood away from the heart, **veins** return blood to the heart, and **capillaries** are the site of gas, nutrient, and waste exchange between the blood and tissues. Small arteries are called **arterioles**, and small veins are called **venules**. The walls of the arteries are thicker and more elastic than are the walls of the veins in order to withstand higher pressure. All vessels are lined with a single layer of simple epithelium called **endothelium**. The smallest vessels, the capillaries, are made only of this single layer of cells. The exchange of fluid between capillaries and interstitial spaces is influenced by **blood pressure**, which pushes fluid out of the capillary, and **osmotic pressure**, which draws fluid back in. Blood pressure is regulated over the short term by the **cardiac output** and the **total peripheral resistance**. Long-term changes in overall blood volume and in the length and compliance of the blood vessels also alter blood pressure.

The vessels carry blood through two circuits. The **pulmonary circuit** transports blood between the heart and the lungs for gas exchange. The **systemic circuit** distributes blood high in oxygen to all body tissues and returns the blood low in oxygen to the heart.

The walls of the vessels, especially those of the small arteries, contain smooth muscle that is under the control of the involuntary nervous system.

Increased contraction of the muscle narrows the vessel's lumen (opening); relaxation has the opposite effect. By controlling vascular smooth muscle contraction, the autonomic nervous system can modulate blood pressure and blood distribution. Increasing the lumen diameter, a process known as **vasodilation**, reduces blood pressure if it occurs in many blood vessels at once. Vasodilation of a particular blood vessel increases the blood supply to the region supplied by the vessel. **Vasoconstriction**, or decreasing the lumen diameter, has the opposite effects.

A negative feedback loop maintains blood pressure at relatively constant levels. **Baroreceptors** act as sensors, the **cardiac control center** of the brain stem acts as the integrator, and changes in heart rate and contraction strength and in blood vessel diameter function as effectors.

Several forces work together to drive blood back to the heart in the venous system. Contraction of skeletal muscles compresses the veins and pushes blood forward, **valves** in the veins keep blood from flowing backward, and changes in intrathoracic pressure that occur during breathing help drive blood back to the heart. The **pulse rate** and **blood pressure** can provide information about an individual's cardiovascular health. Disorders of the circulatory system include hypertension, degenerative changes or obstructions that diminish blood flow in the vessels, and hemorrhage.

Addressing the Learning Objectives

1. DIFFERENTIATE AMONG THE FIVE TYPES OF BLOOD VESSELS WITH REGARD TO STRUCTURE AND FUNCTION.

EXERCISE 15-1: Sections of Small Blood Vessels (Text Fig. 15-2)

1. Write the names of the different vessel types on lines 1 to 5.
2. Write the names of the different vascular layers on the appropriate numbered lines (6 to 9) in different colors. Write the name of structure 10 in black.
3. Color the structures on the diagram with the corresponding colors (except for structure 10).
4. Draw an arrow in the box to indicate the direction of blood flow.

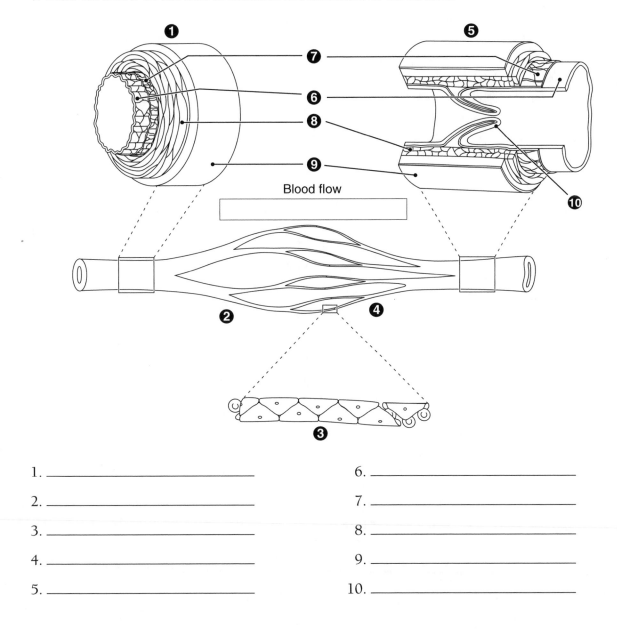

1. _____   6. _____

2. _____   7. _____

3. _____   8. _____

4. _____   9. _____

5. _____   10. _____

EXERCISE 15-2

Write the appropriate term in each blank from the list below.

artery   capillary   vein   venule   arteriole

1. A small vessel through which exchanges between the blood and the cells take place _____

2. A vessel that receives blood from the capillaries _____

3. A vessel that branches off the aorta _____

4. A small vessel that delivers blood to the capillaries _____

5. A vessel that receives blood from venules and delivers it to the heart _____

2. COMPARE THE PULMONARY AND SYSTEMIC CIRCUITS RELATIVE TO LOCATION AND FUNCTION.

EXERCISE 15-3: The Cardiovascular System (Text Fig. 15-1)

1. Label the indicated parts.
2. Color the blood high in oxygen red and the blood low in oxygen blue.
3. Use arrows to show the direction of blood flow.

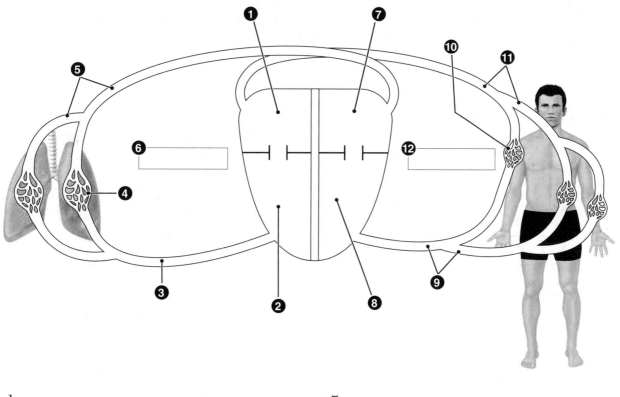

1. _____
2. _____
3. _____
4. _____
5. _____
6. _____

7. _____
8. _____
9. _____
10. _____
11. _____
12. _____

3. NAME THE FOUR SECTIONS OF THE AORTA, AND LIST THE MAIN BRANCHES OF EACH SECTION.

EXERCISE 15-4: Aorta and Its Branches (Text Fig. 15-4)

1. Write the names of the aortic sections on lines 1 to 4 in different colors, and color the appropriate structures on the diagram. Although the aorta is continuous, lines have been added to the diagram to indicate the boundaries of the different sections.
2. Write the names of the aortic branches on the appropriate lines 5 to 22 in different colors, and color the corresponding artery on the diagram. Use the same color for structures 8 and 9 and for structures 7 and 10. Use black for structure 15 because it will not be colored. Color all of the arteries, even if only one is labeled (e.g., bullet 11).

1. _____
2. _____
3. _____
4. _____
5. _____
6. _____
7. _____
8. _____
9. _____
10. _____
11. _____
12. _____
13. _____
14. _____
15. _____
16. _____
17. _____
18. _____
19. _____
20. _____
21. _____
22. _____

4. TRACE THE PATHWAY OF BLOOD THROUGH THE MAIN ARTERIES OF THE UPPER AND LOWER LIMBS.

EXERCISE 15-5: Principal Systemic Arteries (Text Fig. 15-5)

1. Write the names of the principal systemic arteries on the appropriate lines using colored felt-tip markers. Use black to list the three branches of the vessel 14 in lines 15–17 in any order. They will not be colored.
2. Outline the arteries on the diagram with the appropriate colors. If appropriate, color the left and right versions of each artery.

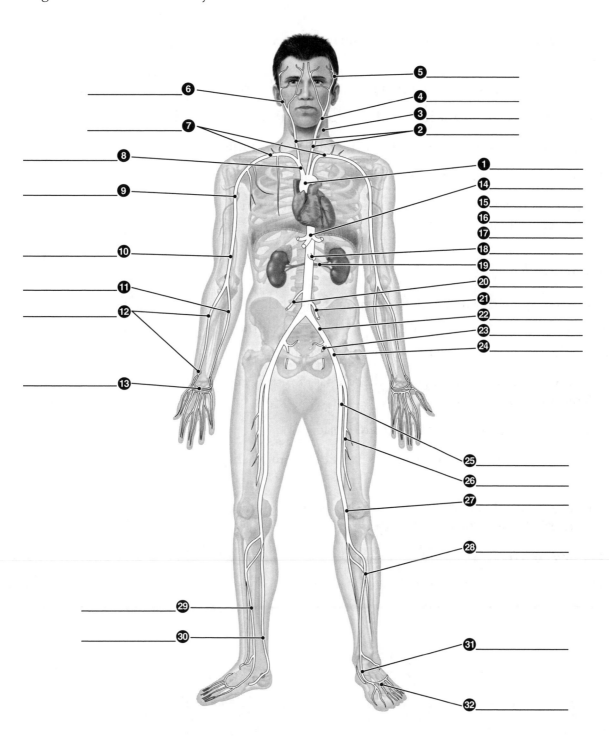

EXERCISE 15-6

Write the appropriate term in each blank from the list below. Not all terms will be used.

coronary arteries   carotid arteries   lumbar arteries   common iliac arteries   phrenic arteries

intercostal arteries   ovarian arteries   renal arteries   suprarenal arteries   brachial arteries

1. Paired branches of the abdominal aorta that supply the diaphragm   _____

2. The vessels that branch off the ascending aorta and supply the heart muscle   _____

3. The large, paired branches of the abdominal aorta that supply blood to the kidneys   _____

4. The vessels formed by final division of the abdominal aorta   _____

5. The vessels that supply the head and neck on each side   _____

6. The paired arteries that branch into the radial and ulnar arteries   _____

7. A group of paired vessels that extend between the ribs   _____

8. Paired branches of the abdominal aorta that extend into the abdominal wall musculature   _____

EXERCISE 15-7

Write the appropriate term in each blank from the list below. Not all terms will be used.

ascending aorta   aortic arch   thoracic aorta   abdominal aorta

brachiocephalic artery   celiac trunk   hepatic artery

1. The short artery that branches into the left gastric artery, the splenic artery, and the hepatic artery   _____

2. A large vessel found within the pericardial sac   _____

3. The portion of the aorta supplying the upper extremities, neck, and head   _____

4. The large vessel that branches into the right subclavian artery and the right common carotid artery   _____

5. The most inferior portion of the aorta   _____

6. The vessel supplying oxygen-rich blood to the liver   _____

5. DEFINE *ANASTOMOSIS*, CITE ITS FUNCTION, AND GIVE FOUR EXAMPLES OF ANASTOMOSES.

EXERCISE 15-8: Principal Systemic Arteries of the Head (Text Fig. 15-6)

1. Write the names of the anastomosis on line 3 in black.
2. Write the names of the arteries in the appropriate boxes in different, preferably darker, colors. Felt-tip pens would work well for this exercise. You need to write in the number for the two empty boxes in the lower diagram based on the number of the same vessel in the upper diagram.
3. Outline the arteries on the diagram with the corresponding colors. Use a colored dot for structure 12.

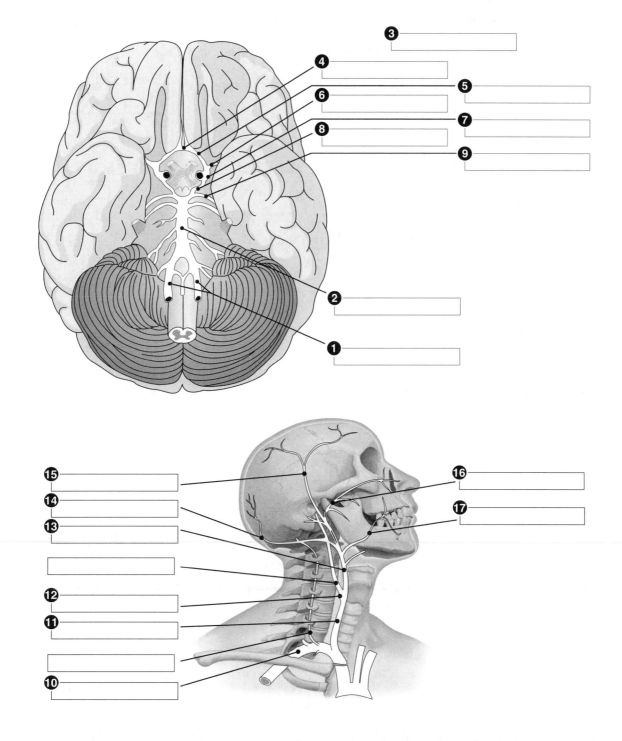

EXERCISE 15-9

Write the appropriate term in each blank from the list below.

mesenteric arch arcuate artery cerebral arterial circle

basilar artery   anastomosis superficial palmar arch

1. A general term describing a communication between two blood vessels   _____

2. An anastomosis between vessels supplying the intestines   _____

3. An anastomosis under the center of the brain formed by two internal carotid arteries and the basilar artery   _____

4. The vessel formed by the union of the two vertebral arteries   _____

5. A vessel formed by the union of the radial and ulnar arteries   _____

6. An arterial arch in the foot   _____

6. COMPARE SUPERFICIAL AND DEEP VEINS, AND GIVE EXAMPLES OF EACH TYPE.

EXERCISE 15-10

Fill in the blank after each vein—is the vein deep (D) or superficial (S)?

1. Saphenous vein   _____

2. Basilic vein   _____

3. Brachial vein   _____

4. Femoral vein   _____

5. Jugular vein   _____

7. NAME THE MAIN VESSELS THAT DRAIN INTO THE SUPERIOR AND INFERIOR VENAE CAVAE.

EXERCISE 15-11: Principal Systemic Veins (Text Fig. 15-7)

1. Write the names of the principal systemic veins on the appropriate lines using colored felt-tip markers.
2. Outline the veins on the diagram with the appropriate color.

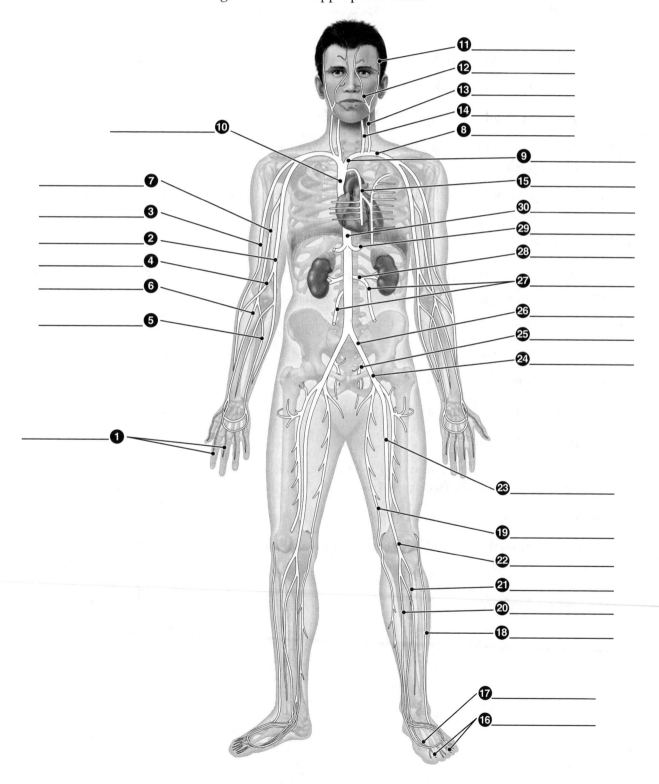

EXERCISE 15-12

Write the appropriate term in each blank from the list below. Not all terms will be used.

brachiocephalic vein   subclavian vein   internal jugular vein   external jugular vein

azygos vein   lumbar vein   superior vena cava   inferior vena cava

1. The vein that drains the area supplied by the carotid artery   _____

2. The vein that receives blood from the facial artery and the cranial venous sinuses   _____

3. Two of these veins merge to form the superior vena cava   _____

4. The two jugular veins empty directly into this vein   _____

5. An unpaired vein collecting blood from the chest wall region   _____

6. All of the blood draining from the upper body will pass through this vein   _____

7. One of four pairs of veins draining the dorsal part of the trunk   _____

EXERCISE 15-13

Write the appropriate term in each blank from the list below.

cephalic vein   saphenous vein   median cubital vein   brachial vein

gastric vein   femoral vein   common iliac vein

1. This vein drains directly into the inferior vena cava   _____

2. The longest vein   _____

3. A deep vein of the upper limb   _____

4. A deep vein of the thigh   _____

5. A vein frequently used for removing blood for testing because of its location near the surface at the front of the elbow   _____

6. The lateral superficial vein of the upper limb   _____

7. A vein that drains the stomach and empties into the hepatic portal vein   _____

8. DEFINE *VENOUS SINUS*, AND GIVE FOUR EXAMPLES OF VENOUS SINUSES.

EXERCISE 15-14: Veins of the Head and Neck and the Cranial Sinuses (Text Fig. 15-8)

1. Write the names of the cranial veins and sinuses on the appropriate lines using colored felt-tip markers.
2. Outline the vessels on the diagram with the corresponding colors.

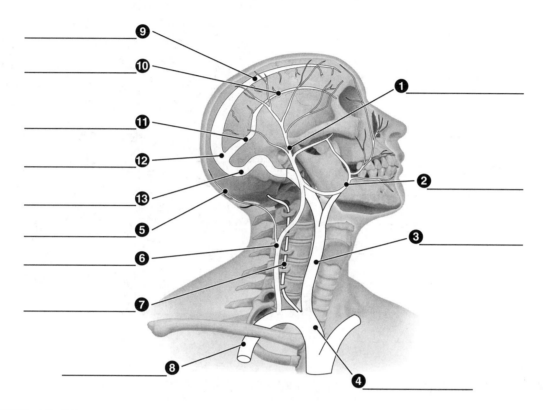

EXERCISE 15-15

Write the appropriate term in each blank from the list below. Not all terms will be used.

coronary sinus   transverse sinus   superior sagittal sinus

inferior sagittal sinus   straight sinus

1. One of two sinuses that drains directly into an internal jugular vein

   _____

2. The channel that merges with the straight sinus

   _____

3. The channel located in the fissure between the two hemispheres that ends at the confluence of sinuses

   _____

4. The channel that receives blood from most of the veins of the heart wall

   _____

9. DESCRIBE THE STRUCTURE AND FUNCTION OF THE HEPATIC PORTAL SYSTEM.

EXERCISE 15-16: Hepatic Portal Circulation (Text Fig. 15-9)

Label each of the indicated parts.

1. _____
2. _____
3. _____
4. _____
5. _____
6. _____
7. _____
8. _____
9. _____
10. _____
11. _____
12. _____
13. _____

10. EXPLAIN THE FORCES THAT AFFECT EXCHANGE ACROSS THE CAPILLARY WALL.

EXERCISE 15-17

Write the appropriate term in each blank from the list below.

diffusion   **filtration pressure**   **osmotic pressure**

1. A force that pushes water out of the capillary   _____

2. A force that moves a substance down its concentration
   gradient   _____

3. A force that draws water into the capillary   _____

11. DESCRIBE FIVE FACTORS THAT REGULATE BLOOD FLOW.

EXERCISE 15-18

Write the appropriate term in each blank from the list below.

vasomotor center   vasodilation   vasoconstriction

precapillary sphincter   valve

1. Structure that prevents blood from moving backward in the veins   _____

2. A structure that regulates blood flow into an individual capillary   _____

3. The change in a blood vessel's internal diameter caused by smooth muscle contraction   _____

4. An increase in a blood vessel's internal diameter   _____

5. A region of the medulla oblongata that controls contraction of the smooth muscle in blood vessel walls   _____

12. DEFINE *PULSE*, AND LIST SIX FACTORS THAT AFFECT PULSE RATE.

EXERCISE 15-19

Fill in the blank after each of the following situations—will the pulse rate most likely increase (I) or decrease (D)?

1. A newborn baby grows older   _____

2. An adult falls asleep   _____

3. Thyroid gland secretion increases _____

4. A teenager runs to catch a bus   _____

5. A child gets a fever   _____

13. LIST FOUR FACTORS THAT AFFECT BLOOD PRESSURE.

EXERCISE 15-20

Fill in the blank after each of the following events—will the blood pressure most likely increase (I) or decrease (D)?

1. Decreased cardiac output _____

2. Increased blood thickness _____

3. Reduced volume of blood ejected from the heart per heartbeat _____

4. Decreased vessel compliance (from atherosclerosis, for example) _____

5. Uncontrolled bleeding _____

6. Increased heart rate _____

7. Vasoconstriction _____

14. EXPLAIN THE ROLE OF BARORECEPTORS IN CONTROLLING BLOOD PRESSURE.

EXERCISE 15-21

The following list outlines the negative feedback loop controlling blood pressure. It describes how the body reacts to decreased blood pressure; however, the steps are out of sequence and are also missing some terms. Use Figure 15-13 in your textbook if needed, and fill in the missing terms in the steps by writing "increases" (or "increase") or "decreases" (or "decrease"). The first one has been filled in for you. Then place the following events in order by writing the appropriate numbers in the blanks. The first one has been done for you.

Part 1: Fill in the missing terms in the steps by writing increases (or increase) or decreases (or decrease) in the numbered blanks.

Part 2: Put the steps in the proper sequence by writing the appropriate step number from 1 to 8 in the blanks provided. Step 1 is already indicated for you.

_____ a. As a result of changes in the heart and blood vessels, peripheral resistance and cardiac output both (1) <u>increases</u>.

_____ b. In a homeostatic response, blood pressure (2) _____.

_____ c. The vasomotor center responds to the altered sensory input from the baroreceptors by altering the activity of the autonomic nervous system.

_____ d. The degree of arteriolar vasoconstriction (3) _____, the heart rate (4) _____, and the contraction strength of ventricles (5)_____.

_____ e. The firing rate of sympathetic motor nerves innervating the heart and arterioles (6) _____.

_____ f. The signal frequency from baroreceptors traveling along autonomic sensory nerves (7)

_____.

_____ g. The firing rate of baroreceptors (8) _____.

__1__ h. Bleeding (9) _____ arterial blood volume and blood pressure.

15. EXPLAIN HOW BLOOD PRESSURE IS COMMONLY MEASURED.

EXERCISE 15-22

Write the appropriate term in each blank from the list below. Not all terms will be used.

viscosity	systolic pressure	diastolic pressure	stethoscope
hypotension	sphygmomanometer	osmolarity	hypertension

1. Term for the blood pressure reading taken after ventricular relaxation   _____

2. An instrument that is used to measure blood pressure   _____

3. Term for blood pressure measured after heart muscle contraction   _____

4. A term that describes the thickness of a solution   _____

5. An abnormal increase in blood pressure   _____

6. An abnormal decrease in blood pressure   _____

7. The instrument used to hear changes in blood flow during the manual measurement of blood pressure   _____

16. DISCUSS SIX DISORDERS INVOLVING THE BLOOD VESSELS.

EXERCISE 15-23

List four complications that can result from untreated hypertension.

1. _____

2. _____

3. _____

4. _____

EXERCISE 15-24

Write the appropriate term in each blank from the list below.

thrombosis   embolus   arteriosclerosis   atherosclerosis

varicose vein   aneurysm   endarterectomy   phlebitis

1. A bulging sac in the wall of an artery that results from weakness of the vessel wall _____

2. Removal of the thickened lining of diseased blood vessels _____

3. Vein inflammation _____

4. Blood clot formation within a blood vessel _____

5. A portion of a blood clot that forms within an artery but then breaks away to travel through the circulatory system _____

6. A form of arterial wall hardening characterized by deposits of yellow, fatlike material _____

7. Any disorder associated with hardening of the arterial walls _____

8. A swollen, distorted superficial vein _____

EXERCISE 15-25

Write the appropriate pressure point for each of the following situations from the list below.

brachial artery   femoral artery

temporal artery   subclavian artery

1. A bleeding nose _____

2. A serious leg laceration _____

3. A cut to the shoulder _____

4. A lacerated finger _____

EXERCISE 15-26

Write the appropriate term in each blank from the list below.

anaphylactic shock   cardiogenic shock

septic shock   hypovolemic shock

1. Shock resulting from heart muscle damage _____

2. Shock resulting from an overwhelming bacterial infection _____

3. Shock resulting from an allergic reaction _____

4. Shock resulting from hemorrhage _____

17. BASED ON THE OPENING CASE STUDY, DISCUSS THE DANGERS OF THROMBOSIS, AND DESCRIBE ONE APPROACH TO ITS TREATMENT.

EXERCISE 15-27

In the case study, Jocelyn's doctor was worried that the blood clot in her popliteal vein might travel to her lungs. In the lines below, list all of the vessels and heart chambers that the embolus would encounter on its journey from the popliteal vein to the pulmonary artery.

_____

_____

_____

18. SHOW HOW WORD PARTS ARE USED TO BUILD WORDS RELATED TO THE BLOOD VESSELS AND CIRCULATION.

EXERCISE 15-28

Complete the following table by writing the correct word part or meaning in the space provided. Write a word that contains each word part in the "Example" column.

Word Part	Meaning	Example
1. _____	foot	_____
2. phleb/o	_____	_____
3. _____	mouth	_____
4. hepat/o	_____	_____
5. -ectomy	_____	_____
6. _____	stomach	_____
7. _____	arm	_____
8. _____	intestine	_____
9. sphygm/o	_____	_____
10. celi/o	_____	_____

Making the Connections

The following concept map deals with the measurement and regulation of blood pressure. Each pair of terms is linked together by a connecting phrase into a sentence. The sentence should be read in the direction of the arrow. Complete the concept map by filling in the appropriate term or phrase. There is one right answer for each term. However, there are many correct answers for the connecting phrases (3 and 9).

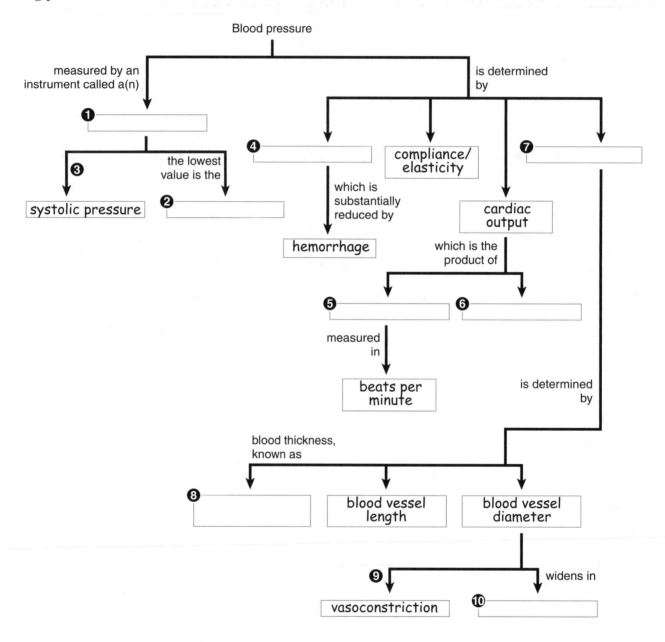

Optional Exercise: Make your own concept map based on the flow of blood from the left ventricle to the calf muscle. It is not necessary to include connecting statements between the terms. You can also make flow charts based on the flow of blood from the leg to the heart, or to and from the arm and/or head. Use the following terms and any others you would like to include: left ventricle, femoral, anterior tibial, aortic arch, abdominal aorta, descending aorta, ascending aorta, popliteal, and external iliac.

Testing Your Knowledge

BUILDING UNDERSTANDING

I. MULTIPLE CHOICE

Select the best answer.

1. Which of the following arteries is unpaired?   1. _____
   a. renal
   b. brachial
   c. brachiocephalic
   d. common carotid
2. What is a precapillary sphincter?   2. _____
   a. a ring of smooth muscle that regulates blood flow
   b. a dilated vein in the liver
   c. a tissue flap that prevents blood backflow in veins
   d. a valve at the entrance to the iliac artery
3. Which of the following arteries carries blood low in oxygen?   3. _____
   a. pulmonary artery
   b. hepatic portal artery
   c. brachiocephalic artery
   d. superior vena cava
4. Which of these terms describes a varicose vein?   4. _____
   a. stent
   b. embolus
   c. thrombus
   d. varix
5. Which of these terms describes a force drawing fluid back into the
   capillaries?   5. _____
   a. filtration pressure
   b. osmotic pressure
   c. hypertension
   d. vasoconstriction
6. Which of the following veins is found in the lower extremity?   6. _____
   a. jugular
   b. brachial
   c. basilic
   d. popliteal
7. Which of the following arteries is NOT found in the cerebral
   arterial circle?   7. _____
   a. anterior cerebral
   b. posterior communicating
   c. vertebral
   d. middle cerebral
8. Which of the following layers is found in arteries AND capillaries?   8. _____
   a. smooth muscle
   b. inner tunic
   c. outer tunic
   d. middle tunic

9. Which of these changes would increase blood pressure?   9. _____
   a. narrowing the blood vessels
   b. reducing the pulse rate
   c. increasing vasodilation
   d. decreasing blood viscosity
10. Which of the following is NOT a subdivision of the aorta?   10. _____
   a. thoracic aorta
   b. aortic arch
   c. pulmonary aorta
   d. abdominal aorta
11. Which of these pressure points would be useful to stop a hemorrhage
    of the lower extremity?   11. _____
   a. coronary artery
   b. femoral artery
   c. facial artery
   d. dorsalis pedis
12. Which two veins unite to form the inferior vena cava?   12. _____
   a. gastric veins
   b. common iliac veins
   c. jugular veins
   d. mesenteric veins

II. COMPLETION EXERCISE

Write the word or phrase that correctly completes each sentence.

1. One example of a portal system is the system that carries blood from the abdominal organs to the _____.

2. The inner, epithelial layer of blood vessels is called the _____.

3. A life-threatening condition in which there is inadequate blood flow to all body tissues is _____.

4. People whose work requires them to stand much of the time frequently suffer from varicosities of the longest vein in the body, named the _____.

5. The aorta and the venae cavae are part of the group, or circuit, of blood vessels that make up the _____.

6. A large channel that drains blood low in oxygen is called a(n) _____.

7. The smallest subdivisions of arteries have thin walls in which there is little connective tissue and relatively more muscle. These vessels are _____.

8. The sensors that detect blood pressure are called _____.

9. The cerebral arterial circle is formed by a union of the internal carotid arteries and the basilar artery. Such a union of vessels is called a(n) _____.

10. The vessel that can be compressed against the lower jaw to stop bleeding in the mouth is the _____.

11. A condition in which vein inflammation contributes to abnormal clot formation is called _____.

UNDERSTANDING CONCEPTS

I. TRUE/FALSE

For each statement, write T for true or F for false in the blank to the left of each number. If a statement is false, correct it by replacing the underlined term, and write the correct statement in the blanks below the statement.

_____ 1. The anterior and posterior communicating arteries are part of the anastomosis supplying the brain.

_____

_____ 2. Contraction of the smooth muscle in arterioles would decrease blood pressure.

_____

_____ 3. Increased blood pressure would decrease the amount of fluid leaving the capillaries.

_____

_____ 4. The external iliac artery continues in the thigh as the femoral artery.

_____

_____ 5. The transverse sinuses receive most of the blood leaving the heart.

_____

_____ 6. Sinusoids are found in the kidney.

_____

_____ 7. When blood pressure increases, the firing rate of baroreceptors increases.

_____

_____ 8. Cardiac output is equal to the pulse rate × peripheral resistance.

_____

_____ 9. Inadequate blood flow resulting from a severe allergic reaction is called anaphylactic shock.

_____

_____ 10. The brachiocephalic artery supplies the left arm.

_____

II. PRACTICAL APPLICATIONS

Study each discussion. Then write the appropriate word or phrase in the space provided.

➤ Group A

1. Mr. Q, age 87, complained of pain and swelling in the area of his saphenous vein. The term for venous inflammation is _____.

2. Further study of Mr. Q's illness indicated that, associated with the inflammation, a blood clot had formed in one vein. This serious condition is called _____.

3. Mr. Q was put on bed rest with his legs elevated and was started on anticoagulant medications. These measures were taken to reduce the serious risk that the blood clot would break loose and travel to the lungs. This potentially fatal complication is called _____.

4. Mr. Q cut the palm of his hand while eating dinner. Due to the anticoagulant medication, the cut bled profusely. Profuse bleeding is called _____.

5. To attempt to stop the bleeding, the nurse applied pressure to the artery called the _____.

➤ Group B

1. Ms. L, aged 42, had her blood pressure examined during a routine physical. Her pressure reading was 165/100. Her diastolic pressure is _____.

2. Based on this reading, the physician diagnosed Ms. L with a condition called _____.

3. The physician was very alarmed by the finding and immediately prescribed a drug to reduce the production of an enzyme produced in the kidneys that acts to increase blood pressure. This enzyme is called _____.

4. Ms. L is at greater risk for a disease associated with plaque deposits in the arteries. This disease is called _____.

5. While at the doctor's office, Ms. L suddenly collapsed, complaining of severe, crushing chest pain. Further observation yielded these signs: weak pulse of 120 bpm; blood pressure 76/40; gray, cold, clammy skin; and rapid, shallow respiration. Her symptoms were due to failure of the heart pump, a form of shock known as _____.

III. SHORT ESSAYS

1. Explain the purpose of vascular anastomoses.

2. What is the function of the hepatic portal system, and what vessels contribute to this system?

3. List the vessels a drop of blood will encounter traveling from deep within the left thigh to the right atrium.

4. Compare the structure and function of veins to those of arteries. Give at least three differences.

CONCEPTUAL THINKING

1. Mr. B, aged 19, just drank a large bottle of soda. The resulting increase in his blood volume increased his blood pressure. Describe the negative feedback loop that will bring his blood pressure back to homeostatic levels. Use the following terms in your explanation: baroreceptors, sensory nerves, vasomotor control center, sympathetic motor nerves, arterioles, heart rate, stroke volume, cardiac output, resistance, and blood pressure.

2. Ms. J has a blood pressure reading of 145/92 mm Hg. A. What is her systolic pressure? B. What is her diastolic pressure? C. Does she suffer from hypertension or hypotension? D. Should Ms. J be treated? If so, how?

Expanding Your Horizons

Eat, drink, and have happy arteries. Eat more fish. Eat less fish. Drink wine. Do not drink wine. And then there's chocolate. We receive conflicting messages about the effect of different foodstuffs on arterial health. The following articles discuss some of these claims.

- Covington MB. Omega-3 fatty acids. Am Fam Physician 2004;70:133–140.
- Klatsky AL. Drink to your health? Sci Am 2003;288:74–81.
- Ried K, Sullivan TR, Fakler P, et al. Effect of cocoa on blood pressure. Cochrane Reviews (2012), Issue 8. Summary available at: http://summaries.cochrane.org/CD008893/effect-of-cocoa-on-blood-pressure

The Lymphatic System and Lymphoid Tissue

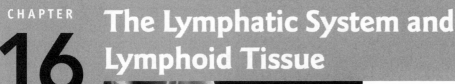

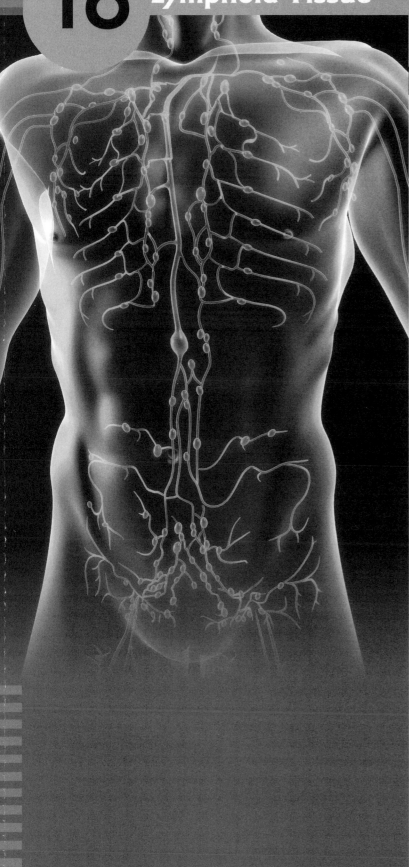

▶ Overview

Lymph is the watery fluid that flows within the lymphatic system. It originates from the tissue fluid that is found in the minute spaces around and between the body cells. The fluid moves from the **lymphatic capillaries** through the **lymphatic vessels** and then to the **right lymphatic duct** and the **thoracic duct**. These large terminal ducts drain into the subclavian veins, adding the lymph to blood that is returning to the heart. Lymphatic capillaries resemble blood capillaries, but they begin blindly, and larger gaps between the cells make them more permeable than blood capillaries. The larger lymphatic vessels are thin walled and delicate; like some veins, they have valves that prevent backflow of lymph.

The **lymph nodes**, which filter lymph, are composed of lymphoid tissue. These nodes remove impurities and house and process **lymphocytes**, cells active in immunity. Chief among them are the cervical nodes in the neck, the axillary nodes in the armpit, the tracheobronchial nodes near the trachea and bronchial tubes, the mesenteric nodes between the peritoneal layers, and the inguinal nodes in the groin.

In addition to the nodes, there are several organs of lymphoid tissue with somewhat different functions. For instance, the **thymus** is essential for development of the immune system during early life. The **spleen** has numerous functions, such as destroying worn-out red blood cells, serving as a reservoir for blood, and producing red blood cells before birth. Mucosa-associated lymphoid tissue (MALT) helps prevent microorganisms that access the respiratory and digestive tract from invading deeper tissues. MALT includes the tonsils, the appendix, and Peyer patches of the intestine.

Disorders of the lymphatic system include lymphoid tissue infections and inflammation, blockages in the lymphatic vessels, and neoplastic diseases.

Addressing the Learning Objectives

1. LIST THE FUNCTIONS OF THE LYMPHATIC SYSTEM.

EXERCISE 16-1

List three functions of the lymphatic system in the blanks below.

1. _____

2. _____

3. _____

EXERCISE 16-2: Lymphatic System in Relation to the Cardiovascular System (Text Fig. 16-1)

1. Label the indicated parts. Parts 8 and 9 refer to the two blood circuits.
2. Color the oxygen-rich blood red, the oxygen-poor blood blue, and the lymph green.

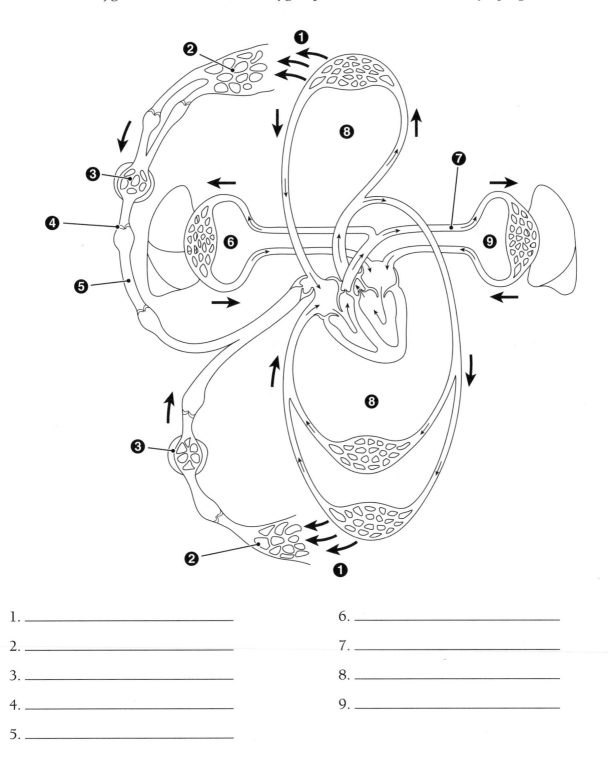

1. _____	6. _____
2. _____	7. _____
3. _____	8. _____
4. _____	9. _____
5. _____	

2. EXPLAIN HOW LYMPHATIC CAPILLARIES DIFFER FROM BLOOD CAPILLARIES.

EXERCISE 16-3

Fill in the blank after each statement—does it apply to lymphatic capillaries (L), blood capillaries (B), or both (BOTH)?

1. The vessel walls are constructed of a single layer of squamous epithelial cells. _____

2. The vessel walls are very permeable, permitting the passage of large proteins. _____

3. The cells in the vessel walls are called endothelial cells. _____

4. The gap between adjacent cells in the vessel wall is very small. _____

5. Lacteals are one example of this type of vessel. _____

6. The vessels form a bridge between two larger vessels. _____

7. The capillaries drain into vessels that have valves. _____

8. The vessel transports erythrocytes. _____

3. NAME THE TWO MAIN LYMPHATIC DUCTS, AND DESCRIBE THE AREA DRAINED BY EACH.

EXERCISE 16-4

Fill in the blank after each region—will the lymph drain into the right lymphatic duct (R) or the thoracic duct (T)?

1. Left hand _____

2. Right hand _____

3. Right breast _____

4. Left breast _____

5. Left leg _____

6. Right leg _____

4. NAME AND GIVE THE LOCATIONS OF FIVE TYPES OF LYMPHATIC TISSUE, AND LIST THE FUNCTIONS OF EACH.

EXERCISE 16-5: Lymphatic System (Text Fig. 16-3)

Label the indicated parts. Fill in the empty bullets with the correct numbers based on the labels in the other diagrams.

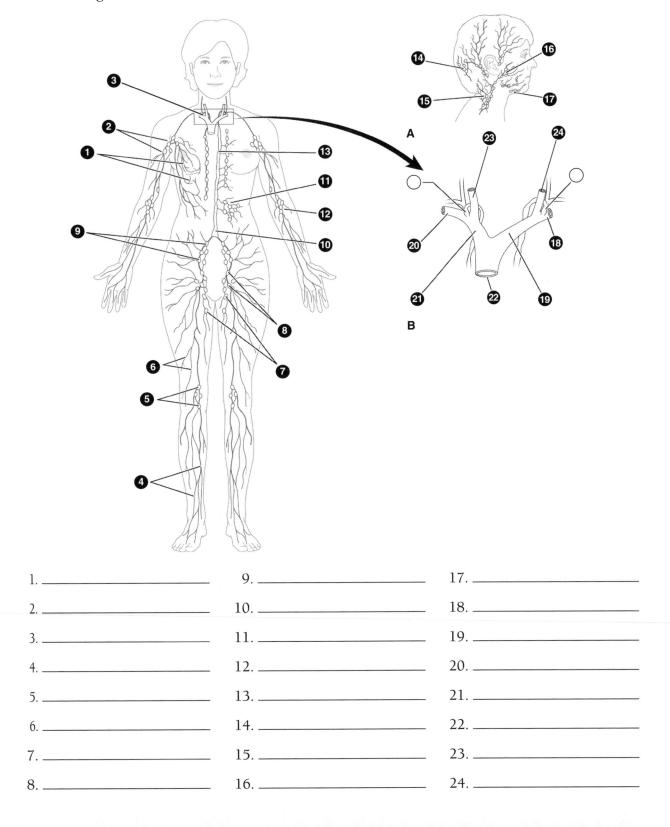

1. _____   9. _____   17. _____

2. _____   10. _____   18. _____

3. _____   11. _____   19. _____

4. _____   12. _____   20. _____

5. _____   13. _____   21. _____

6. _____   14. _____   22. _____

7. _____   15. _____   23. _____

8. _____   16. _____   24. _____

EXERCISE 16-6

Write the appropriate term in each blank from the list below. Not all terms will be used.

lacteal   superficial   mesenteric nodes   cervical nodes

deep   inguinal nodes   axillary nodes

1. The nodes that filter lymph from the lower extremities and the external genitalia _____

2. The lymph nodes located in the armpits _____

3. Term for lymphatic vessels located near the body surface _____

4. The lymph nodes found between the two peritoneal layers _____

5. A specialized vessel in the small intestine wall that absorbs digested fats _____

6. The lymph nodes located in the neck that drain certain parts of the head and neck _____

EXERCISE 16-7

Write the appropriate term in each blank from the list below. Not all terms will be used.

lymphatic capillary   right lymphatic duct   chyle   lymph

valve   subclavian vein   thoracic duct   cisterna chyli

1. The temporary storage area formed by an enlargement of the first part of the thoracic duct _____

2. The fluid formed when tissue fluid passes from the intercellular spaces into the lymphatic vessels _____

3. The large lymphatic vessel that drains lymph from below the diaphragm and from the left side above the diaphragm _____

4. The milky-appearing fluid that is a combination of fat globules and lymph _____

5. Structure that prevents backflow of fluid in lymphatic vessels _____

6. Blind-ended, thin-walled vessel that absorbs excess tissue fluid and proteins _____

EXERCISE 16-8: Lymph Node (Text Fig. 16-4A)

Label the indicated parts.

1. _____

2. _____

3. _____

4. _____

5. _____

6. _____

7. _____

8. _____

9. _____

10. _____

11. _____

12. _____

EXERCISE 16-9: Location of Lymphoid Tissue (Text Fig. 16-5)

1. Write the names of the different lymphoid organs on the numbered lines in different colors.
2. Color the structures on the diagram with the corresponding colors.

1. _____

2. _____

3. _____

4. _____

5. _____

6. _____

7. _____

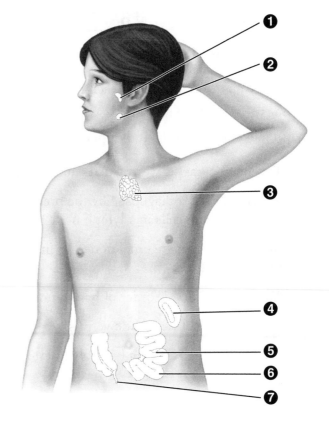

EXERCISE 16-10

Write the appropriate term in each blank from the list below. Not all terms will be used.

germinal center   trabeculae   spleen   Peyer patch   MALT

palatine tonsil   lingual tonsil   pharyngeal tonsil   thymus   hilum

1. An oval lymphoid body located at the side of the soft palate   _____

2. The organ that filters blood and is located in the upper left quadrant (left hypochondriac region) of the abdomen   _____

3. A mass of lymphoid tissue at the back of the tongue   _____

4. The mass of lymphoid tissue located in the pharynx behind the nose and commonly called the adenoids   _____

5. The organ in which T cells mature   _____

6. The indented area of a lymph node where efferent lymphatic vessels exit the node   _____

7. Lymphoid tissue associated with mucous membranes, such as the tonsils and appendix   _____

8. An area of lymphoid tissue specifically found in the small intestine wall   _____

5. DESCRIBE FOUR LYMPHATIC SYSTEM DISORDERS.

EXERCISE 16-11

Write the appropriate term in each blank from the list below. Not all terms will be used.

buboes   filariae   lymphangitis   lymphadenitis

lymphedema   Hodgkin lymphoma   non-Hodgkin lymphoma

1. The small parasitic worms that cause elephantiasis   _____

2. A form of cancer diagnosed by the presence of Reed-Sternberg cells in a lymph node biopsy   _____

3. Abnormally large inguinal nodes, as may be found in certain infections   _____

4. An inflammatory disorder of lymph nodes   _____

5. Inflammation of lymphatic vessels   _____

6. Tissue swelling resulting from obstruction in a lymphatic vessel   _____

6. CITE THE CAUSES AND SYMPTOMS OF INFECTIOUS MONONUCLEOSIS, AS DESCRIBED IN THE CASE STUDY THAT OPENS THIS CHAPTER.

EXERCISE 16-12

Fill in the blanks describing Lucas's condition.

Infectious mononucleosis is caused by a virus named the (1) _____. The virus infects (2) _____ lymphocytes. The most obvious sign is enlarged lymph nodes in the neck region, known as (3) _____ _____. Symptoms include (4) _____ (an enlarged spleen), (5)_____ (sore throat), and (6) _____ (elevated body temperature).

7. SHOW HOW WORD PARTS ARE USED TO BUILD WORDS RELATED TO THE LYMPHATIC SYSTEM.

EXERCISE 16-13

Complete the following table by writing the correct word part or meaning in the space provided. Write a word that contains each word part in the "Example" column.

Word Part	Meaning	Example
1. _____	gland	_____
2. lingu/o	_____	_____
3. _____	excessive enlargement	_____
4. -oid	_____	_____
5. -pathy	_____	_____

Making the Connections

The following concept map deals with the structure and function of the lymphatic system. Each pair of terms is linked together by a connecting phrase. Complete the concept map by filling in the appropriate term or phrase. There is one right answer for each term. However, there are many correct answers for the connecting phrases (2, 4, 8).

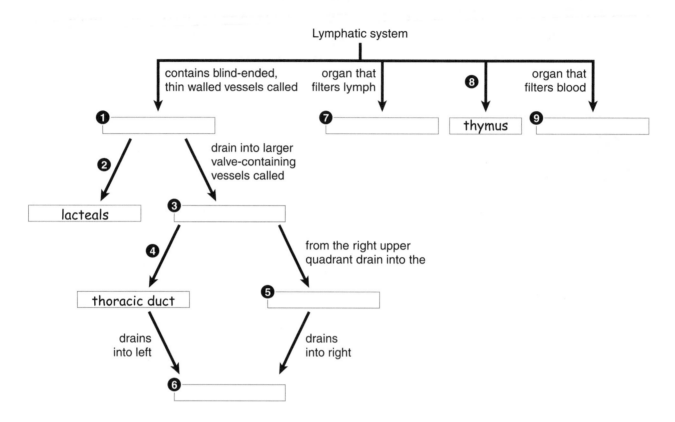

Optional Exercise: Make your own concept map based on the flow of lymph from a body part to the left atrium. It is not necessary to include connecting statements between the terms. For instance, you could map the flow of lymph from the right breast to the right atrium by using the following terms and any others you would like to include: superior vena cava, right subclavian vein, right brachiocephalic vein, mammary vessels, axillary nodes, and right lymphatic duct.

Testing Your Knowledge

BUILDING UNDERSTANDING

I. MULTIPLE CHOICE

Select the best answer.

1. Which tonsil is also known as the adenoids?   1. _____
    a. palatine
    b. lingual
    c. pharyngeal
    d. lymphatic
2. Which of the following is NOT a function of the spleen?   2. _____
    a. destruction of old red blood cells
    b. blood filtration
    c. blood storage
    d. chyle drainage
3. Where are the mesenteric nodes found?   3. _____
    a. in the groin region
    b. near the trachea
    c. between the two peritoneal layers
    d. in the armpits
4. Which of these organs shrinks the most in size after puberty?   4. _____
    a. cisterna chyli
    b. thymus
    c. spleen
    d. liver
5. Which of the following is a characteristic of lymphatic capillaries but
    not blood capillaries?   5. _____
    a. They contain a thin muscular layer.
    b. They are virtually impermeable to water and solutes.
    c. They are blind ended.
    d. They do not contain any cells.
6. Which of these terms describes the enlargement of the lymph nodes,
    as seen in Hodgkin lymphoma, AIDS, and infectious mononucleosis?   6. _____
    a. tonsillitis
    b. lymphadenopathy
    c. lymphangitis
    d. adenoidectomy
7. Which of these terms describes the enlarged portion of the thoracic duct?   7. _____
    a. cisterna chyli
    b. right lymphatic duct
    c. hilum
    d. lacteal

II. COMPLETION EXERCISE

Write the word or the phrase that correctly completes each sentence.

1. The regions within the lymph node cortex where certain lymphocytes multiply are called _____.

2. Enlargement of the spleen is called _____.

3. The milky-appearing lymph that drains from the small intestine is called _____.

4. A tumor that occurs in lymphoid tissue is called a(n) _____.

5. The fluid that moves from tissue spaces into special collecting vessels for return to the blood is called _____.

6. Lymph from the right side of the body above the diaphragm joins the bloodstream when the right lymphatic duct empties into the _____.

7. The lymph nodes surrounding the breathing passageways may become black in individuals living in highly polluted areas. The nodes involved are the _____.

8. Nearly all of the lymph from the arm, shoulder, and breast passes through the lymph nodes known as the _____.

9. The spleen contains many cells that can engulf harmful bacteria and other foreign cells by a process called _____.

UNDERSTANDING CONCEPTS

I. TRUE/FALSE

For each statement, write T for true or F for false in the blank to the left of each number. If a statement is false, correct it by replacing the underlined term(s), and write the correct statement in the blanks below the statement.

_____ 1. Lymph filtered through the mesenteric nodes will drain into the thoracic duct.

_____

_____ 2. One purpose of the lymphatic system is the absorption of protein from the small intestine.

_____

_____ 3. The spleen filters lymph, and the lymph nodes filter blood.

_____ ; _____

_____ 4. The inguinal nodes are found between the two layers of the peritoneum.

_____

_____ 5. Cancer of the breast can cause lymphadenitis of the axillary nodes.

_____

_____ 6. Lacteals are a type of blood capillary.

_____

II. PRACTICAL APPLICATIONS

Study each discussion. Then write the appropriate word or phrase in the space provided.

➤ Group A

1. Mr. W, aged 21, detected a mass in his groin region. The physician suspected that one or more of the lymph nodes that drain the lower extremities was swollen. These nodes are called the

_____.

2. When these glands are enlarged, they are often referred to as _____.

3. The physician informed Mr. W that he had a disease of the lymph nodes. The general term for these diseases is _____.

4. A malignant tumor was detected by a lymph node biopsy. This tumor, like all tumors of lymphoid tissue, is called a(n) _____.

5. Closer examination of the biopsy revealed the presence of Reed-Sternberg cells, leading to the diagnosis of _____.

➤ Group B

1. Ms. L traveled in Asia for two years before starting nursing school. During the first week of classes, she noticed swelling in her lower right leg. Fearing the worst, she tearfully told her parents that her lymph vessels were blocked by small worms called _____.

2. The physician confirmed that the swelling in Ms. L's leg was due to obstructed lymph flow. This type of swelling is called _____.

3. Next, the lymph vessels in Ms. L's swollen lower leg were palpated. These vessels are called the

_____.

4. The physician noticed red streaks along the leg near a poorly healed cut. He concluded that Ms. L was suffering from inflammation of the lymphatic vessels, or _____.

5. Ms. L was very relieved by the diagnosis. The doctor treated the infection aggressively to avoid the development of blood poisoning, or _____.

III. SHORT ESSAYS

1. Trace a lymph droplet from the interstitial fluid of the lower leg to the right atrium based on the structures shown in **Figure 16-3**.

_____

_____

_____

2. Compare and contrast lymphangitis, lymphadenitis, and lymphadenopathy. What do these three disorders have in common? How do they differ?

_____

_____

_____

CONCEPTUAL THINKING

1. Ms. Y, a healthy 24-year-old woman, is studying for her nursing finals. She has been sitting at her desk for 10 hours straight when she notices that her legs are swollen. Use your knowledge of the lymphatic system to explain the swelling, and suggest how she can prevent it in the future.

_____

_____

_____

2. What do Peter Forsberg (Colorado Avalanche hockey player) and Chris Sims (Tampa Bay Buccaneers quarterback) have in common? Each player had his spleen removed due to an injury, and both continued successful sports careers postsplenectomy. However, this operation is not without its consequences. Which of the following functions would be disrupted by a splenectomy? Circle all that apply.
   a. Destruction of red blood cells
   b. Detection of cancer cells in lymph
   c. Drainage of interstitial fluid
   d. Absorption of fats from the small intestine
   e. Red blood cell production in a 27-year-old
   f. Response to hemorrhage
   g. Detection of pathogens in blood

Expanding Your Horizons

Have you or one of your friends had the kissing disease? Infectious mononucleosis, or "mono," is an infection of lymphatic cells. It is easily transmitted between individuals living in close contact with one another (not necessarily by kissing) and is thus very common on college campuses. Learn more about this disease by reading the following articles.

- Bailey RE. Diagnosis and treatment of infectious mononucleosis. Am Fam Phys 1994;49:879–888.
- Cozad J. Infectious mononucleosis. Nurse Pract 1996;21:14–18.

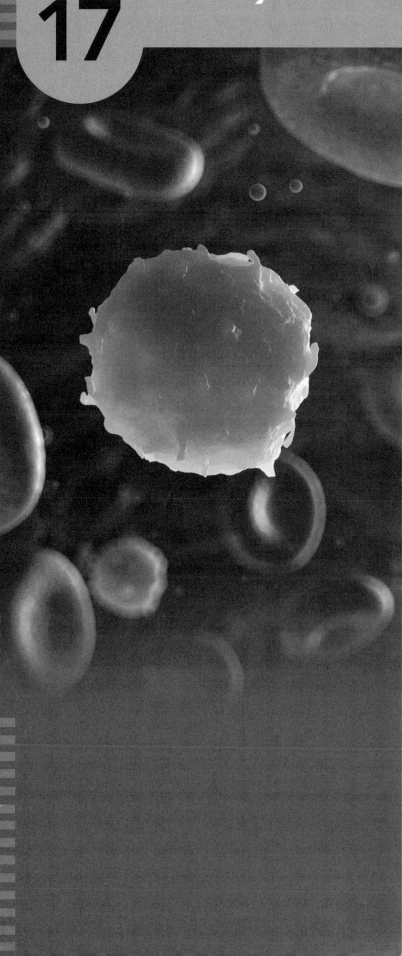

Overview

Although the body is constantly exposed to pathogenic organisms, infection develops relatively rarely. This is because the body has many "lines of defense" against pathogenic invasion. Our **innate defenses** are the first lines of defense, protecting us from all foreign substances. The intact **skin** and **mucous membranes** serve as mechanical barriers, as do certain **reflexes** such as sneezing and coughing. Body secretions wash away impurities and may kill bacteria as well. By the process of **inflammation**, the body tries to get rid of an irritant or to minimize its harmful effects. **Phagocytes** and **natural killer (NK) cells** act non-specifically to destroy invaders. **Interferon** can limit viral infections. **Fever** boosts the immune system and inhibits the growth of some organisms.

The ultimate defense against disease is **adaptive immunity**, the means by which the body resists or overcomes the effects of a particular disease or other harmful agent. It involves reactions between foreign substances or **antigens** and the white blood cells known as **lymphocytes**. The **T cells** (T lymphocytes) produce **cell-mediated immunity**. There are different types of T cells involved in immune reactions: **cytotoxic T cells** kill infected body cells; **helper T cells** magnify all aspects of the immune response; **regulatory T cells** control the response. **Antigen-presenting cells** (dendritic cells and macrophages) participate by presenting the foreign antigen to the T cell. **B cells** (B lymphocytes), when stimulated by an antigen, multiply into **plasma cells**.

These cells produce specific **antibodies**, which react with the antigen. Circulating antibodies make up the form of immunity termed **humoral immunity**. **Memory cells** for each lymphocyte type persist in the circulation long after the original antigen has been vanquished. These memory cells mount a rapid immune response upon subsequent antigen exposure and usually prevent infection.

Immunity may be **natural** (acquired by transfer of maternal antibodies or by contact with the disease) or **artificial** (provided by a vaccine or an immune serum). Immunity that involves production of antibodies and memory cells by the individual is termed **active immunity**; immunity acquired as a result of the transfer of antibodies to an individual from some outside source is described as **passive immunity**.

Harmful actions of the immune system, described as **hypersensitivity reactions**, include allergies, autoimmune diseases, and reactions against transplanted organs. The impact of hypersensitivity reactions can range from mildly uncomfortable, as in the stuffy nose of hay fever, to life-threatening allergic reactions and autoimmune diseases. Immunodeficiencies such as acquired immunodeficiency syndrome (AIDS), in contrast, leave individuals more susceptible to pathogens and cancer.

Addressing the Learning Objectives

1. LIST FOUR FACTORS THAT DETERMINE THE OCCURRENCE OF INFECTION.

EXERCISE 17-1

Write the appropriate term in each blank from the list below.

predisposition   portal of entry   virulence   toxin   dose

1. The means by which a pathogenic organism invades the body   _____

2. A poison produced by a pathogen   _____

3. The factor that includes an individual's physical and emotional health   _____

4. The power of an organism to overcome body defenses and cause disease   _____

5. The number of pathogens that invade the body   _____

2. DIFFERENTIATE BETWEEN INNATE AND ADAPTIVE IMMUNITY, AND GIVE EXAMPLES OF EACH.

EXERCISE 17-2

Fill in the blank after each characteristic—does it refer to innate immunity (I) or adaptive immunity (A)?

1. Present at birth _____

2. Develops after exposure to a pathogen _____

3. Skin and saliva are examples _____

4. Relatively nonspecific _____

5. Relatively slow _____

6. Can completely eliminate a pathogen from the body _____

7. Relatively specific _____

8. Can prevent future infections _____

3. NAME THREE TYPES OF CELLS AND THREE TYPES OF CHEMICALS ACTIVE IN THE SECOND LINE OF DEFENSE AGAINST DISEASE.

EXERCISE 17-3

Write the appropriate term in each blank from the list below.

toll-like receptor   cytokine   natural killer cell   neutrophil

interferon   macrophage   complement

1. A substance that prevents multiplication of viruses _____

2. A lymphocyte that nonspecifically destroys abnormal cells _____

3. A protein of innate immune cells that helps them recognize pathogens of different types _____

4. Interleukins and interferon are examples of this class of signaling molecules _____

5. A granular leukocyte that participates in innate defenses _____

6. A set of blood proteins that carry out many different immune activities _____

7. A phagocyte that develops from monocytes _____

4. BRIEFLY DESCRIBE THE INFLAMMATORY REACTION.

EXERCISE 17-4

Label each of the following statements as true (T) or false (F).

1. The damaged cells themselves secrete histamine.   _____

2. Mast cells are similar to macrophages.   _____

3. Histamine makes vessels constrict.   _____

4. Leukocytes actually leave the blood vessels in order to fight
   infection in injured tissues.   _____

5. The inflammatory exudate does not contain any cells.   _____

6. Cytokines result in leukocytes leaving the infected region.   _____

5. DEFINE *ANTIGEN* AND *ANTIBODY*.

See Exercise 17-5.

6. COMPARE AND CONTRAST T CELLS AND B CELLS WITH RESPECT TO DEVELOPMENT AND TYPE OF ACTIVITY.

See Exercises 17-5 through 17-9.

7. DESCRIBE THE ACTIVITIES OF FOUR TYPES OF T CELLS.

EXERCISE 17-5

Write the appropriate term in each blank from the list below.

antigen   antibody   memory cell   plasma cell

T_h cell T_{reg} cell T_c cell

1. Any substance that provokes an immune response   _____

2. An antibody-producing cell derived from a B cell   _____

3. A cell that reduces an immune response by inhibiting or destroying activated lymphocytes   _____

4. A cell that matures in the thymus and that directly destroys a foreign cell   _____

5. A circulating protein also known as an immunoglobulin   _____

6. A B or T cell that can rapidly activate an immune response when a previously encountered pathogen invades the body   _____

7. A type of T cell that produces interleukins to stimulate immune responses   _____

8. EXPLAIN THE ROLE OF ANTIGEN-PRESENTING CELLS IN ADAPTIVE IMMUNITY.

EXERCISE 17-6

Write the appropriate term in each blank from the list below. Not all terms will be used.

foreign antigen   MHC protein   lysosome   antibody

T-cell receptor   interleukin   neutrophil   dendritic cell

1. A type of antigen-presenting cell with very long processes   _____

2. The organelle in antigen-presenting cells that digests the foreign substance   _____

3. The part of the nonself substance that is inserted into the antigen-presenting cell membrane   _____

4. The self-antigen inserted into the antigen-presenting cell membrane   _____

5. The part of the helper T cell that binds to the antigen-presenting cell   _____

6. The substance released by the helper T cell after it is activated by binding to the antigen-presenting cell   _____

EXERCISE 17-7

List six ways by which the antigen–antibody reaction helps the body deal with an infection.

1. _____

2. _____

3. _____

4. _____

5. _____

6. _____

EXERCISE 17-8

Write the appropriate term in each blank from the list below.

IgG   IgM   humoral immunity

cell-mediated immunity   memory B cell   plasma cell

1. The arm of adaptive immunity that involves antibodies   _____

2. The arm of adaptive immunity that involves cytotoxic T cells   _____

3. The first type of antibody produced in an immune response   _____

4. The type of B cell that produces antibodies   _____

5. The type of B cell responsible for the secondary immune
   response   _____

6. The type of antibody produced in large amounts during
   the secondary immune response   _____

EXERCISE 17-9

The diagram below illustrates B-cell activation and antibody production.

First, describe the four steps of humoral immunity in the lines provided. Use your own words. Next, fill in the blanks, and number the bullets in the diagram.

1. Activation:

_____

2. Primary response:

_____

3. Memory B-cell formation:

_____

4. Secondary response:

_____

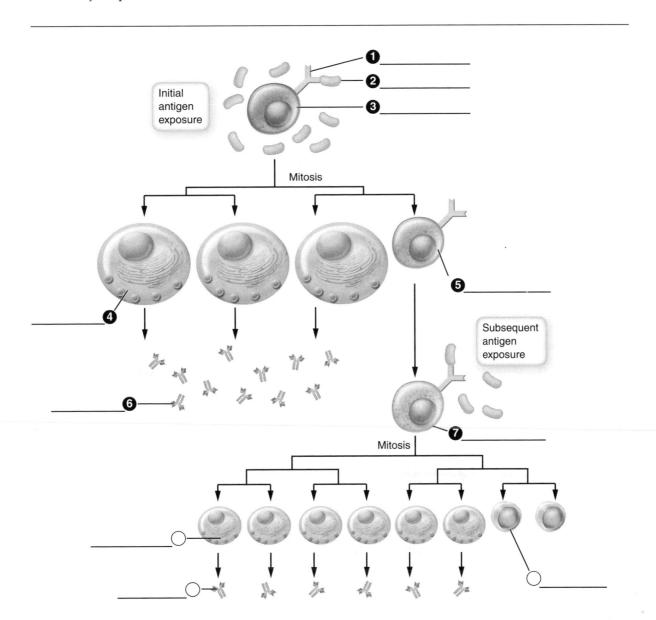

9. DIFFERENTIATE BETWEEN NATURAL AND ARTIFICIAL ADAPTIVE IMMUNITY.

See Exercise 17-10.

10. DIFFERENTIATE BETWEEN ACTIVE AND PASSIVE IMMUNITY.

EXERCISE 17-10

Fill in the blanks after each example—does it apply to active (ACT) or passive immunity (P), and natural (N) or artificial (ART) immunity?

1. Immunity resulting from exposure to a microbial toxin

   _____

2. Immunity resulting from the transfer of antibodies from mother to fetus

   _____

3. Immunity resulting from the HPV vaccination

   _____

4. Immunity resulting from the administration of antiserum

   _____

11. DEFINE THE TERM *VACCINE*, AND GIVE THREE EXAMPLES OF VACCINE TYPES.

See Exercise 17-11.

12. DEFINE THE TERM *ANTISERUM*, AND GIVE FIVE EXAMPLES OF ANTISERA.

EXERCISE 17-11

Write the appropriate term in each blank from the list below. Not all terms will be used.

toxoid   attenuation   immunization   HPV vaccine   varicella vaccine

rubella   antitoxin   gamma globulin   antivenin   MMR vaccine

1. The process of reducing the virulence of a pathogen to prepare a vaccine

   _____

2. A toxin treated with heat or chemicals to reduce its harmfulness so that it may be used as a vaccine

   _____

3. The fraction of the blood plasma that contains antibodies

   _____

4. A process designed to induce an immune response against a particular pathogen, resulting in the acquisition of artificial adaptive immunity

   _____

5. The type of immune serum used to provide passive immunity to botulism

_____

6. An immune serum administered to a snakebite victim

_____

7. A vaccine that prevents sexually transmitted genital warts

_____

8. A vaccine that prevents chickenpox

_____

9. A vaccine that prevents against three common childhood diseases

_____

13. DISCUSS FOUR TYPES OF IMMUNE DISORDERS.

EXERCISE 17-12

Write the appropriate term in each blank from the list below.

IgE	anaphylaxis	autoimmunity	allergen
Kaposi sarcoma	retrovirus	multiple myeloma	HIV

1. A protein that stimulates an allergic response, as is found in pollen and house dust

_____

2. The virus that causes AIDS

_____

3. The manufacture of antibodies to one's own tissues

_____

4. A severe, life-threatening allergic reaction

_____

5. A malignant skin cancer common in AIDS patients but rare in the general population

_____

6. Any RNA virus that uses reverse transcriptase to manufacture DNA

_____

7. A cancer of the bone marrow resulting in B-cell overproduction

_____

8. The specific class of antibodies produced by mast cells during allergic responses

_____

14. EXPLAIN THE ROLE OF THE IMMUNE SYSTEM IN PREVENTING CANCER.

See Exercise 17-13.

15. EXPLAIN THE ROLE OF IMMUNITY IN TISSUE TRANSPLANTATION.

EXERCISE 17-13

Write the appropriate term in each blank from the list below. Not all terms will be used.

immune surveillance	immunotherapy	rejection syndrome	liver
T cells	B cells	cornea	glucocorticoid

1. A cancer treatment that attempts to stimulate the patient's
immune system _____

2. The process by which immune cells continuously monitor
for cancerous cells _____

3. A common occurrence in transplant patients, in which the
patient's immune system attacks the donated organ or tissue _____

4. A drug given to transplant recipients to reduce rejection syndrome _____

5. The cells involved in rejecting transplanted organs _____

6. Transplantation of this tissue does not stimulate an immune
response _____

16. BASED ON THE CASE STUDY, DESCRIBE THE CAUSES AND SYMPTOMS OF THE AUTOIMMUNE DISORDER RHEUMATOID ARTHRITIS.

EXERCISE 17-14

Meredith's swollen and painful joints reflect adverse effects of her immune system. Consider the function of each of these chemicals, and speculate how each contributes to Meredith's symptoms in the lines provided.

1. Prostaglandins:

_____

_____

2. Histamine:

_____

_____

3. Immunoglobulins:

_____

_____

4. Complement:

_____

_____

17. SHOW HOW WORD PARTS ARE USED TO BUILD WORDS RELATED TO IMMUNITY.

EXERCISE 17-15

Complete the following table by writing the correct word part or meaning in the space provided. Write a word that contains each word part in the "Example" column.

Word Part	Meaning	Example
1. _____	marrow	_____
2. ana-	_____	_____
3. _____	poison	_____
4. erg	_____	_____

Making the Connections

The following concept map deals with some of the processes involved in specific and non-specific immunity. Each pair of terms is linked together by a connecting phrase into a sentence. The sentence should be read in the direction of the arrow. Complete the concept map by filling in the appropriate term or phrase. There is generally only one right answer for each term. However, there are many correct answers for the connecting phrases (4, 5, 8, 12, 15).

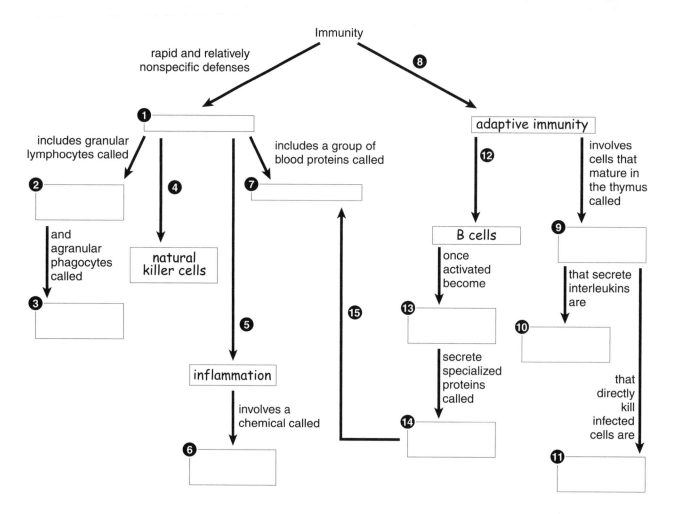

Optional Exercise: Make your own concept map based on the classification of specific immune responses. Use the following terms and any others you would like to include: immunity, natural, artificial, active, passive, vaccine, toxoid, and attenuated organism.

Testing Your Knowledge

BUILDING UNDERSTANDING

I. MULTIPLE CHOICE

Select the best answer.

1. Which type of lymphocyte is involved in innate immunity?   1. _____
   a. B cell
   b. natural killer cell
   c. memory T cell
   d. cytotoxic T cell
2. Which of the following describes an activity of B cells?   2. _____
   a. suppression of the immune response
   b. manufacture of antibodies
   c. direct destruction of foreign cells
   d. phagocytosis
3. Which of the following is an acquired defense against infection?   3. _____
   a. skin
   b. mucus
   c. antibodies
   d. cilia
4. Which of these cells acts as an antigen-presenting cell?   4. _____
   a. dendritic cells
   b. cytotoxic T cells
   c. plasma cells
   d. neutrophils
5. Which of these is an action of complement proteins?   5. _____
   a. promote inflammation
   b. attract phagocytes
   c. destroy cells
   d. all of the above
6. What are interleukins?   6. _____
   a. substances in the blood that react with antigens
   b. the antibody fraction of the blood
   c. a group of nonspecific proteins needed for agglutination
   d. substances released from T_h cells that stimulate other leukocytes
7. What are MHC antigens?   7. _____
   a. bacterial proteins
   b. foreign proteins
   c. one's own proteins
   d. antibodies
8. Which of the following will result in active immunity?   8. _____
   a. immunization
   b. antiserum administration
   c. breast-feeding
   d. none of the above

9. Which of these statements about toll-like receptors (TLRs) is false?   9. _____
    a. TLRs are found on B lymphocytes.
    b. TLRs enable cells to differentiate between bacteria and viruses
       but not between specific bacteria or specific viruses.
    c. Natural killer cells use TLRs to detect foreign antigens.
    d. Natural killer cells use TLRs to detect cancerous cells.

II. COMPLETION EXERCISE

Write the word or phrase that correctly completes each sentence.

1. Whooping cough is also called _____.

2. Transplantation of an organ or tissue may fail because the recipient's immune system will
   attempt to destroy the organ, leading to _____.

3. Circulating antibodies are responsible for the type of immunity termed

   _____.

4. Heat, redness, swelling, and pain are considered the classic symptoms of

   _____.

5. Antibodies that neutralize toxins but do not affect the organism producing the toxin are called

   _____.

6. The use of methods to stimulate the immune system in the hope of combating cancer is a form
   of treatment called _____.

7. The MMR vaccine protects against rubella, measles, and _____.

8. The site by which a pathogen enters the body is called the _____.

9. Antibodies transmitted from a mother's blood to a fetus or administered via an antiserum provide a type of short-term borrowed immunity called _____.

10. The administration of vaccine or the act of becoming infected on the other hand stimulates the
    body to produce a longer lasting type of immunity called _____.

11. The action of leukocytes in which they engulf and digest invading pathogens is known as

    _____.

UNDERSTANDING CONCEPTS

I. TRUE/FALSE

For each statement, write T for true or F for false in the blank to the left of each number. If a statement is false, correct it by replacing the underlined term, and write the correct statement in the blanks below the statement.

_____ 1. Complement is a substance that participates in nonspecific body defenses exclusively against viruses.

_____

_____ 2. Cytotoxic T cells participate in humoral immunity.

_____

_____ 3. Mast cells release histamine during an inflammatory response.

_____

_____ 4. The action of histamine in the inflammatory reaction is an example of an innate defense.

_____

_____ 5. Immunoglobulin is another name for antigen.

_____

_____ 6. Interferons are released by helper T cells in order to stimulate the activity of other leukocytes.

_____

_____ 7. Administration of an antiserum is an example of artificial passive immunity.

_____

II. PRACTICAL APPLICATIONS

1. Ms. M, age 5, was climbing trees when she was stung by a bee. Ms. M had been stung before, so her mother was shocked when Ms. M developed breathing troubles, hives, and decreased blood pressure. Ms. M reacted the second time she was exposed to bee venom because her tissues had started to produce antibodies against the venom. In other words, her tissues had been

   _____.

2. The substance in the bee venom that was inducing Ms. M's symptoms is called a(n)

   _____.

3. Ms. M was rapidly brought to the local emergency clinic. The physician immediately administered an injection of epinephrine, and her condition improved rapidly. The physician gave her prescriptions and advice on how to prevent or treat another episode of her life-threatening

   allergic reaction, which he termed _____.

4. The physician asked if there was a family history of allergies. The mother mentioned that she often suffered from a stuffy nose during pollen season, a type of allergic reaction commonly

   called _____.

5. The physician also inquired as to the status of Ms. M's immunizations. Another term for immu-

   nization is _____.

6. Ms. M's mother replied that she had not received any immunizations for several years. The doctor administered several booster shots, which help maintain high levels of circulating proteins

   called _____.

7. The doctor also warned Ms. M's mother that Ms. M was at a greater risk for diseases in which the body launches an immune attack against one's own tissues. This class of diseases is called

   _____.

III. SHORT ESSAYS

1. Discuss the role of macrophages in innate and adaptive body defenses.

   _____

   _____

   _____

2. The immune system protects against disease. Are there any instances in which this system is harmful to the individual?

   _____

   _____

   _____

CONCEPTUAL THINKING

1. Baby G was born without a thymus but was otherwise normal. Speculate as to which aspects of her immune system will be affected by this deficiency and which aspects will be unaffected.

_____

_____

_____

_____

2. Mr. A stepped on a fish while beachcombing in the South Pacific. A spine was embedded in his foot, and he rapidly became dizzy and the foot swelled up alarmingly. Mr. A was told by a local doctor that the fish spine contains an incredibly potent poison and that he will be dead in eight hours unless the poison is neutralized.

   A. What would be the best treatment for Mr. A—vaccination with fish toxoid or an injection of an antitoxin? Defend your answer.

   _____

   _____

   _____

   B. What form of immunity will be induced by this treatment?

   _____

   C. Will Mr. A be protected from future encounters with this particular fish poison? Explain why or why not.

   _____

   _____

   _____

3. In advanced stages, individuals with AIDS frequently suffer from opportunistic infections and from rare cancers. Explain why.

   _____

   _____

   _____

   _____

Expanding Your Horizons

A vaccine for AIDS? Although the treatments for AIDS have improved enormously, the disease still results in considerable suffering. The "holy grail" of AIDS research is the development of a vaccine to protect individuals from infection. However, the quest for a vaccine is not an easy one. Consult these websites and articles, or do a search for "AIDS vaccine news" for the most up-to-date information.

- International AIDS Vaccine Initiative. Available at: http://www.iavi.org/Pages/default.aspx
- World Health Organization. HIV Vaccines. Available at: http://www.who.int/hiv/topics/vaccines/Vaccines/en/
- Watkins DI. The vaccine search goes on. Sci Am 2008;299(5):69–74, 76.

UNIT

VI

Energy: Supply and Use

The Respiratory System

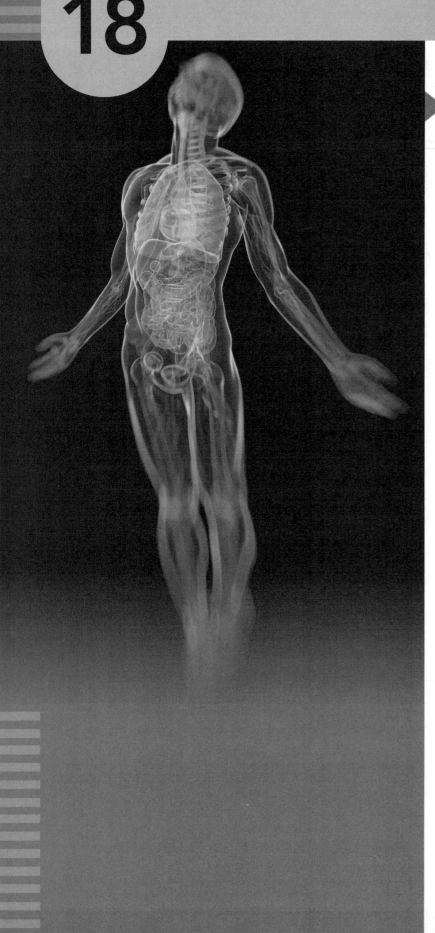

Overview

Oxygen is taken into the body, and carbon dioxide is released by means of the organs and passageways of the **respiratory system**. This system contains the **nasal cavities**, the **pharynx**, the **larynx**, the **trachea**, the **bronchi**, and the **lungs**.

The process of **respiration** transfers oxygen from the atmosphere to body cells and carbon dioxide from body cells to the atmosphere. It includes four stages. In terms of oxygen transfer, the first phase of respiration is **pulmonary ventilation**, which is normally accomplished by breathing. During normal, quiet breathing, air enters the lungs (**inhalation**) because the diaphragm and some of the intercostal muscles contract to expand the thoracic cavity. Air leaves the lungs (**exhalation**) when the muscles relax. Deeper breathing requires additional muscles during both inhalation and exhalation. **External gas exchange** transfers gases between the alveoli of the lungs and the bloodstream. The specialized mechanisms of **gas transport** enable blood to carry large amounts of oxygen and carbon dioxide. Finally, **internal gas exchange** transfers gases between the blood and the tissues. Note that the latter three stages of respiration involve the cardiovascular system. The order of the stages is reversed for carbon dioxide transfer, which begins with internal gas exchange and ends with pulmonary ventilation.

Oxygen is transported to the tissues almost entirely by the **hemoglobin** in red blood cells. Some carbon dioxide is transported in the red blood cells as well, but most is converted into **bicarbonate ions** and **hydrogen ions**. These hydrogen ions increase the acidity of the blood.

Breathing is primarily controlled by the **respiratory control centers** in the medulla and the pons of the brain stem. These centers are influenced by **chemoreceptors** located on either side of the medulla that respond to changes in the acidity of the interstitial fluid surrounding medullary neurons. The acidity reflects the concentration of carbon dioxide in arterial blood.

Disorders of the respiratory tract include infection, allergy, chronic obstructive pulmonary disease (COPD), diseases and disorders of the pleura, and cancer.

Addressing the Learning Objectives

1. DEFINE RESPIRATION, AND DESCRIBE ITS FOUR PHASES.

EXERCISE 18-1

Write the appropriate term in each blank from the list below.

external gas exchange	internal gas exchange	gas transport
cellular respiration	pulmonary ventilation	

1. The exchange of air between the atmosphere and the alveoli   _____

2. The exchange of specific gases between the alveoli and the blood   _____

3. The exchange of specific gases between the blood and the cells   _____

4. This process involves only blood   _____

5. The process by which cells use oxygen and nutrients to generate energy   _____

2. NAME AND DESCRIBE ALL THE STRUCTURES OF THE RESPIRATORY SYSTEM.

EXERCISE 18-2: Respiratory System (Text Fig. 18-2)

Label the indicated parts. Use the same color for the structure and the label if you wish.

1. _____

2. _____

3. _____

4. _____

5. _____

6. _____

7. _____

8. _____

9. _____

10. _____

11. _____

12. _____

13. _____

14. _____

EXERCISE 18-3: The Larynx (Text Fig. 18-3)

1. Write the name of each labeled part on the numbered lines in different colors.
2. Color the different parts on the diagram with the corresponding colors.

1. _____

2. _____

3. _____

4. _____

5. _____

6. _____

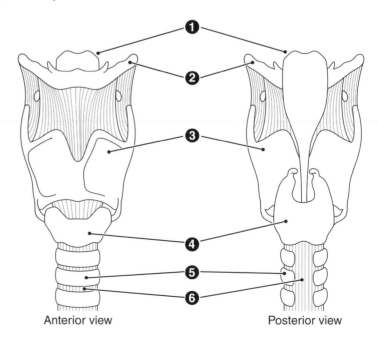

Anterior view   Posterior view

EXERCISE 18-4: The Lungs (Text Fig. 18-5A)

1. Write the name of each labeled part on the numbered lines in different colors. Use black for parts 13 and 15, because they won't be colored. Use light colors for structures 1, 3, 4, 6, and 9, and indicate whether each is part of the left lung or the right lung.

2. Color or outline the different parts on the diagram with the corresponding colors.

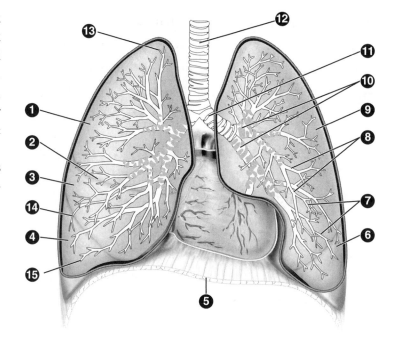

1. _____

2. _____

3. _____

4. _____

5. _____

6. _____

7. _____

8. _____

9. _____

10. _____

11. _____

12. _____

13. _____

14. _____

15. _____

EXERCISE 18-5

Write the appropriate term in each blank from the list below. Not all terms will be used.

larynx	conchae	pharynx	glottis
epiglottis	nares	parietal	mainstem bronchus
segmental bronchus	lobar bronchus	bronchiole	visceral

1. The openings of the nose   _____

2. The three projections arising from the lateral walls of each nasal cavity   _____

3. The scientific name for the voice box   _____

4. The leaf-shaped structure that helps prevent the entrance of food into the trachea   _____

5. One of the two branches formed by division of the trachea   _____

6. The pleural layer attached to the lung   _____

7. The area below the nasal cavities that is common to both the digestive and respiratory systems   _____

8. A small air-conducting tube containing a smooth muscle layer but little or no cartilage   _____

9. There are three of these in the right lung but only two in the left lung   _____

10. The space between the vocal folds   _____

3. EXPLAIN THE MECHANISM FOR PULMONARY VENTILATION.

EXERCISE 18-6: A Spirogram (Text Fig. 18-8)

Write the names of the different lung volumes and capacities in the boxes on the diagram.

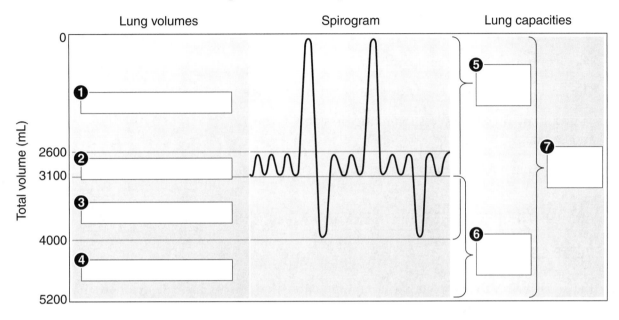

EXERCISE 18-7

Write the appropriate term in each blank from the list below. Not all terms will be used.

alveoli   surfactant   inhalation   tidal volume   intercostals

exhalation   compliance   spirometer   vital capacity

1. The substance in the fluid lining the alveoli that prevents their collapse

   _____

2. The phase of pulmonary ventilation in which air is expelled from the alveoli

   _____

3. The phase of pulmonary ventilation in which the diaphragm contracts

   _____

4. Subsets of these muscles can either increase or decrease the size of the thoracic cavity

   _____

5. The only respiratory structures involved in external gas exchange

   _____

6. The amount of air inhaled or exhaled during a relaxed breath

   _____

7. The ease with which the lungs and thorax can be expanded

   _____

8. The maximum volume of air that can be inhaled after maximum expiration

   _____

EXERCISE 18-8

Label each of the following statements as true (T) or false (F).

1. Inhalation is the active phase of quiet breathing.   _____

2. During quiet breathing, exhalation does not require any muscle contraction.   _____

3. The diaphragm rises when it contracts.   _____

4. The size of the thoracic cavity decreases during inhalation.   _____

5. Atmospheric pressure is less than lung pressure during inhalation.   _____

6. The diaphragm relaxes during exhalation.   _____

4. DISCUSS THE PROCESSES OF INTERNAL AND EXTERNAL GAS EXCHANGE.

EXERCISE 18-9

Use the terms below to complete the paragraph. Note that terms may be used more than once, and not all terms will be used.

into	out of	body cells	diffusion
active transport	partial pressure gradient	osmotic gradient	alveoli

External gas exchange transfers gases between the (1) _____ and blood. Oxygen moves (2) _____ blood, and carbon dioxide moves (3) _____ blood. Internal gas exchange transfers gases between (4) _____ and blood. Oxygen moves (5) _____ blood, and carbon dioxide moves (6) _____ blood. Gases move by the process of (7) _____. Each gas moves down its individual (8) _____.

5. LIST THE WAYS IN WHICH OXYGEN AND CARBON DIOXIDE ARE TRANSPORTED IN THE BLOOD.

EXERCISE 18-10

Write the appropriate term in each blank from the list below. Not all terms will be used.

bicarbonate ion   hemoglobin   carbonic anhydrase   carbon dioxide   15%

hydrogen ion   oxygen   10%   75%

1. The gas converted into bicarbonate and hydrogen ions   _____

2. An important blood buffer produced from carbon dioxide   _____

3. The substance that carries most of the oxygen in the blood   _____

4. The gas that is more concentrated in the blood than in metabolically active tissues   _____

5. An ion that renders blood more acidic   _____

6. The proportion of total blood carbon dioxide dissolved in plasma   _____

7. The proportion of total blood carbon dioxide transported in the form of bicarbonate   _____

8. The proportion of total blood carbon dioxide bound to plasma proteins and hemoglobin   _____

6. DESCRIBE FACTORS THAT CONTROL RESPIRATION.

EXERCISE 18-11

Write the appropriate term in each blank from the list below. Not all terms will be used.

hypercapnia   hydrogen ion   bicarbonate ion   phrenic nerve   vagus nerve

medulla   aortic arch   carbon dioxide   oxygen

1. The location of the main breathing regulatory center   _____

2. A rise in the blood carbon dioxide level   _____

3. The location of a peripheral chemoreceptor   _____

4. The substance that acts directly on the central chemoreceptors to stimulate breathing   _____

5. A rise in the arterial partial pressure of this gas stimulates breathing   _____

6. The nerve that controls the diaphragm   _____

7. DESCRIBE NORMAL AND ALTERED BREATHING PATTERNS.

See Exercises 18-12 and 18-13.

8. COMPARE HYPERVENTILATION AND HYPOVENTILATION, AND CITE THREE POSSIBLE RESULTS OF HYPOVENTILATION.

EXERCISE 18-12

Fill in the blank after each statement—does it apply to hypoventilation (HYPO) or hyperventilation (HYPER)?

1. The breathing pattern that causes hypocapnia _____

2. The breathing pattern resulting from respiratory obstruction _____

3. The breathing pattern that causes hypercapnia _____

4. The breathing pattern that causes acidosis _____

5. The breathing pattern that sometimes occurs during anxiety attacks _____

EXERCISE 18-13

Write the appropriate term in each blank from the list below.

orthopnea   dyspnea   Cheyne-Stokes respiration   hyperpnea   hypoxia

hypopnea   tachypnea   apnea   hyperventilation   hypoxemia

1. Difficult or labored breathing _____

2. An abnormal increase in the depth and rate of breathing _____

3. A temporary cessation of breathing _____

4. Difficult breathing that is relieved by sitting upright _____

5. An abnormal decrease in the depth and rate of breathing _____

6. Increased depth and rate of breathing reflecting increased activity levels _____

7. A variable respiratory rhythm observed in some critically ill patients _____

8. An abnormally low oxygen partial pressure in arterial blood _____

9. An abnormally low oxygen level in tissues _____

10. Increased rate of breathing reflecting increased activity levels _____

9. LIST NINE TYPES OF RESPIRATORY DISORDERS.

See Exercises 18-14 and 18-16.

10. LIST SIX TYPES OF RESPIRATORY INFECTIONS.

EXERCISE 18-14

Write the appropriate term in each blank from the list below. Not all terms will be used.

croup   acute coryza   influenza   lobar pneumonia   tuberculosis

bronchopneumonia   effusion   asthma   hay fever   *Pneumocystis* pneumonia

1. Technical name for the common cold, based on the discharge of fluid from the nose

    _____

2. A collection of fluid, as may occur in the pleural space

    _____

3. A bacterial or viral infection that affects an entire lobe at once

    _____

4. An infectious lung disease characterized by the presence of small lung lesions

    _____

5. A seasonal allergic reaction that affects the upper respiratory tract and eyes

    _____

6. A disorder of the bronchial tubes frequently associated with allergies

    _____

7. A condition in young children in which the airways are constricted as a result of a viral infection

    _____

8. Infection of the alveoli occurring mainly in patients with suppressed immune systems

    _____

9. A lung infection in which infected alveoli are scattered throughout the lung

    _____

11. IDENTIFY THE DISEASES INVOLVED IN CHRONIC OBSTRUCTIVE PULMONARY DISEASE (COPD).

EXERCISE 18-15

Fill in the blanks of the following description referring to your textbook as necessary.

Chronic obstructive pulmonary disease describes two disorders commonly occurring together. The first, (1) _____ _____, describes the chronic inflammation of the airways. The second, (2) _____, describes the destruction of the lung sacs, known as the (3) _____. These changes impede airflow, particularly during the phase of ventilation known as (4) _____. This change increases the amount of air remaining in the lungs after a maximal exhalation, known as the (5) _____ volume. Also, destruction of the air sacs reduces gas transfer between the lungs and blood, a process known as (6) _____ gas exchange. Eventually, carbon dioxide accumulates in arterial blood, a change known as (7) _____, and the oxygen partial pressure of blood decreases, a change known as (8) _____.

12. DESCRIBE SOME DISORDERS THAT INVOLVE THE PLEURA.

EXERCISE 18-16

Write the appropriate term in each blank from the list below. Not all terms will be used.

acute respiratory distress syndrome   SIDS   surfactant deficiency disorder

atelectasis   pleurisy   hemothorax

pneumothorax

1. A disorder common in premature babies in which lung inflation is very difficult   _____

2. Inflammation of the serous membrane covering the lungs   _____

3. The accumulation of blood in the pleural space   _____

4. The scientific term for a collapsed lung   _____

5. A sudden, potentially fatal inflammatory lung condition resulting from injury, severe allergy, or infection   _____

6. The scientific abbreviation for "crib death"   _____

13. DESCRIBE FIVE TYPES OF EQUIPMENT USED TO TREAT RESPIRATORY DISORDERS.

EXERCISE 18-17

Write the appropriate term in each blank from the list below. Not all terms will be used.

bronchoscope   tracheotomy   spirometer

suction apparatus   oxygen therapy   tracheostomy

1. An operation to insert a metal or a plastic tube into the trachea to serve as an airway for ventilation   _____

2. The incision in the trachea through which the tube is inserted   _____

3. An instrument used to inspect the bronchi and their branches   _____

4. An apparatus used to remove mucus from the respiratory tract   _____

5. An apparatus used to measure lung function   _____

14. REFERRING TO THE CASE STUDY, DISCUSS HOW ASTHMA CAN BE DIAGNOSED AND TREATED.

EXERCISE 18-18

Fill in the blanks in the following discussion referring to your textbook as needed.

In adults, the most reliable test used to diagnose asthma is (1) _____, which measures lung volumes and capacities. Emily's x-ray showed evidence of inflammation of the small airways, known as the (2) _____. Her history showed that she was sensitive to two common asthma triggers, (3) _____ and (4) _____. In order to reduce the inflammation, Emily was prescribed a low-dose (5) _____ inhaler. If this medication does not adequately control her symptoms, she will start to take an oral medication that inhibits inflammatory chemicals called (6) _____.

15. SHOW HOW WORD PARTS ARE USED TO BUILD WORDS RELATED TO RESPIRATION.

EXERCISE 18-19

Complete the following table by writing the correct word part or meaning in the space provided. Write a word that contains each word part in the "Example" column.

Word Part	Meaning	Example
1. _____	incomplete	_____
2. spir/o	_____	_____
3. _____	nose	_____
4. -centesis	_____	_____
5. -pnea	_____	_____
6. _____	carbon dioxide	_____
7. _____	lung	_____
8. _____	air, gas	_____
9. orth/o	_____	_____
10. or/o	_____	_____

Making the Connections

The following concept map deals with the four phases of respiration. Complete the concept map by filling in the appropriate term or phrase. There is one right answer for each term. However, there are many correct answers for the connecting phrase (5).

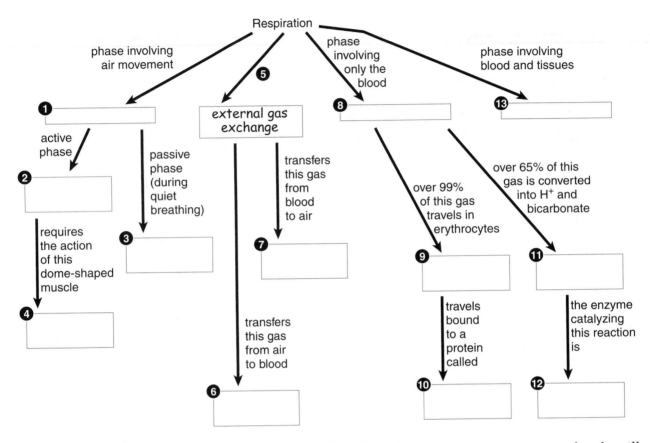

Optional Exercise: Make your own concept map based on the structures an oxygen molecule will pass through from the atmosphere to a tissue. Use the following terms and any others you would like to include: nostrils, atmosphere, nasopharynx, oropharynx, nasal cavities, laryngopharynx, trachea, larynx, bronchioles, blood cell, hemoglobin, plasma, tissue, bronchi, and alveoli.

Testing Your Knowledge

BUILDING UNDERSTANDING

I. MULTIPLE CHOICE

Select the best answer.

1. During which respiratory phase does carbon dioxide diffuse out of blood?
   a. internal gas exchange
   b. external gas exchange
   c. pulmonary ventilation
   d. none of the above

   1. _____

2. Which of these terms describes a structural defect of the partition in the nose?
   a. polyp
   b. epistaxis
   c. bifurcation
   d. deviated septum

   2. _____

3. Which of these terms describes a rhythmic abnormality in breathing that is seen in critically ill patients?
   a. hypoxemia
   b. hyperpnea
   c. Cheyne-Stokes respiration
   d. hypocapnia

   3. _____

4. Which of these is a term describing a type of allergic rhinitis?
   a. hives
   b. nosebleed
   c. hay fever
   d. asthma

   4. _____

5. Which of the following terms does NOT apply to the cells that line the conducting passages of the respiratory tract?
   a. pseudostratified
   b. connective
   c. columnar
   d. ciliated

   5. _____

6. Which of these changes would result from an increase in blood carbon dioxide partial pressure?
   a. fewer bicarbonate ions in the blood
   b. more hydrogen ions in the blood
   c. more alkaline blood
   d. hypocapnia

   6. _____

7. Which of the following is NOT an upper respiratory infection?
   a. acute coryza
   b. influenza
   c. croup
   d. tuberculosis

   7. _____

8. What chemical reduces surface tension in the alveoli?   8. _____
   a. surfactant
   b. bicarbonate
   c. exudate
   d. effusion
9. What is the residual volume?   9. _____
   a. The amount of air that is always in the lungs, even after a
      maximal expiration.
   b. The total amount of air in the lungs after a maximal inspiration.
   c. The amount of air remaining in the lungs after a normal exhalation.
   d. The amount of air that can be forced out of the lungs after a
      normal exhalation.
10. What is a pneumothorax?   10. _____
   a. presence of blood in the pleural space
   b. presence of pus in the pleural space
   c. removal of air from the pleural space
   d. presence of air in the pleural space

II. COMPLETION EXERCISE

Write the word or phrase that correctly completes each sentence.

1. The space between the vocal cords is called the _____.

2. An abnormal decrease in the depth and rate of respiration is termed _____.

3. A lower-than-normal partial pressure of oxygen in tissues is called _____.

4. Heart disease and other disorders may cause the bluish color of the skin and visible mucous

   membranes clinically known as _____.

5. The space between the lungs is called the _____.

6. The number of lobes in the left lung is _____.

7. The peripheral chemoreceptors involved in respiratory control are found in the aortic arch and

   the _____.

8. Each heme region of a hemoglobin molecule contains an inorganic element called

   _____.

9. Certain diplococci, staphylococci, chlamydias, and viruses may cause an inflammation of the

   lungs. This disease is called _____.

10. The scientific name for the organism that causes tuberculosis is _____.

11. The nerve that innervates the diaphragm is the _____.

UNDERSTANDING CONCEPTS

I. TRUE/FALSE

For each statement, write T for true or F for false in the blank to the left of each number. If a statement is false, correct it by replacing the underlined term, and write the correct statement in the blanks below the statement.

_____ 1. <u>Hypercapnia</u> results in greater blood acidity.

_____

_____ 2. Most <u>carbon dioxide</u> in the blood is carried bound to hemoglobin.

_____

_____ 3. The wall of an alveolus is made of <u>stratified</u> squamous epithelium.

_____

_____ 4. During <u>internal</u> exchange of gases, oxygen moves down its concentration gradient out of blood.

_____

_____ 5. The activity of the respiratory center in the medulla can be modified by signals from a nearby brain stem region known as the <u>midbrain.</u>

_____

_____ 6. The alveoli become filled with <u>exudate</u> in patients suffering from pneumonia.

_____

_____ 7. As a result of a chronic lung disease, Ms. L's lungs do not expand very easily. Her lungs are said to be <u>more</u> compliant than normal.

_____

_____ 8. <u>Inhalation</u> during quiet breathing involves muscle contraction.

_____

_____ 9. A tumor resulting from chronic sinusitis that obstructs air movement is called a <u>bronchogenic carcinoma.</u>

_____

_____ 10. Hyperventilation results in an <u>increase</u> of carbon dioxide in the blood.

_____

II. PRACTICAL APPLICATIONS

Study each discussion. Then write the appropriate word or phrase in the space provided.

➤ **Group A**

1. Ms. L's complaints included shortness of breath, a chronic cough productive of thick mucus, and a "chest cold" of two months' duration. She was advised to quit smoking, a major cause of lung irritation in a group of chronic lung diseases collectively known as _____.

2. Symptoms in Ms. L's case were due in part to the obstruction of groups of alveoli by mucus. The condition in which the bronchial linings are constantly inflamed and secreting mucus is called _____.

3. Ms. L was told that the mucus is not moving normally out of her airways because the toxins in cigarette smoke paralyze small cell extensions that beat to create an upward current. These extensions are called _____.

4. Ms. L's respiratory function was evaluated by quantifying different lung volumes and capacities using a machine called a(n) _____.

5. Evaluation of Ms. L's respiratory function showed a reduction in the amount of air that could be moved into and out of her lungs. The amount of air that can be expelled by maximum exhalation following maximum inhalation is termed the _____.

6. The other abnormality in Ms. L's evaluation was also characteristic of her disease. There was an increase in the amount of air remaining in her lungs after a normal expiration. This amount is the _____.

7. An x-ray of Ms. L's lungs revealed large spaces indicative of alveoli destruction. This lung change is characteristic of a disorder known as _____.

8. This destruction reduced the available surface area for gas movement between the alveoli and pulmonary blood. This phase of respiration is known as _____.

➤ **Group B**

1. Baby L was born at 32 weeks' gestation. She is obviously struggling for breath, and her skin is bluish in color. This skin coloration is called _____.

2. The obstetrician fears that one or both of her lungs are not able to inflate. The incomplete expansion of a lung is called _____.

3. Baby L is placed on pressurized oxygen to inflate the lung. The physician tells the worried parents that her lungs have not yet matured sufficiently to produce the lung substance that reduces the surface tension in alveoli. This substance is called _____.

4. A lack of this substance in a newborn results in a disorder called _____.

5. Baby L is administered the substance she is lacking and is soon resting more comfortably. The neonatologist tells her parents that Baby L has an excellent prognosis, but because of her prematurity, she will be at greater risk for "crib death," which is now known as _____.

III. SHORT ESSAYS

1. Are lungs passive or active players in pulmonary ventilation? Explain.

_____

_____

_____

2. Name some parts of the respiratory tract where gas exchange does NOT occur.

_____

_____

_____

CONCEPTUAL THINKING

1. a. Name the phase of respiration regulated by the respiratory control center.

_____

   b. Explain how the respiratory control center can alter this phase of respiration.

_____

_____

_____

   c. Name the chemical factor(s) that regulate(s) the activity of the respiratory control center.

_____

_____

_____

2. Use the equation below to answer the following questions.

$$CO_2 + H_2O \Leftrightarrow H_2CO_3 \Leftrightarrow H^+ + HCO_3$$

   a. Tom is holding his breath under water. What happens to the carbon dioxide content of his blood, and how does this change affect his blood acidity?

_____

_____

b. Notice that the arrows go in both directions, indicating the reaction can proceed in both directions. If Tom consumes a large amount of lemon juice, increasing the H^+ concentration in his blood, what do you think will happen to the amount of CO_2 in his blood? *Only consider the equation above in your answer.*

_____

_____

_____

3. Perform the following actions as you answer the questions.
   a. Inhale as deeply as possible. Which volume do your lungs contain: the vital capacity or the total lung capacity?

   _____

   b. Beginning with a maximum inhalation, exhale all of the air you can. Which volume did you exhale: the vital capacity or the total lung capacity?

   _____

   c. Breathe quietly for a few minutes, and stop after a normal exhale. Which volume remains in your lungs: the functional residual capacity or the residual volume?

   _____

   d. Breathe quietly for a few minutes, and then actively exhale all of the air you can. Which volume remains in your lungs: the functional residual capacity or the residual volume?

   _____

Expanding Your Horizons

Everyone would agree that oxygen is a very useful molecule. Commercial enterprises, working on the premise that "more is better," market water supplemented with extra oxygen. They claim that consumption of hyperoxygenated water increases alertness and exercise performance. Do these claims make sense? Do we obtain oxygen from the lungs or from the digestive tract? A logical extension of this premise is that soda pop, supplemented with carbon dioxide, would increase carbon dioxide in the blood and thus increase the breathing rate. Is this true? You can read about another oxygen gimmick, oxygen bars, on the website of the Food and Drug Administration.

- Bren L. Oxygen bars: Is a breath of fresh air worth it? 2002. Available at: http://permanent.access.gpo.gov/lps1609/www.fda.gov/fdac/features/2002/602_air.html

The Digestive System

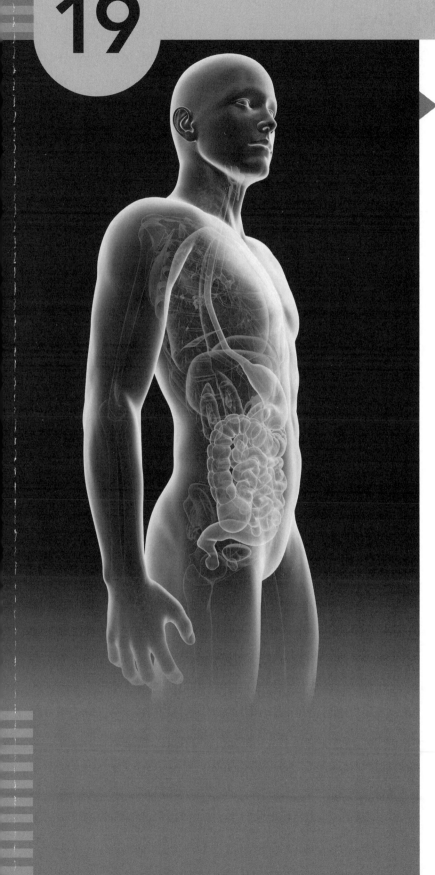

Overview

The food we eat is made available to cells throughout the body by the complex processes of **digestion** and **absorption**. These are the functions of the **digestive system**, composed of the **digestive tract** and the **accessory organs**.

The digestive tract, consisting of the **mouth**, the **pharynx**, the **esophagus**, the **stomach**, and the small and large **intestines**, forms a continuous passageway in which ingested food is prepared for use by the body and waste products are collected to be expelled from the body. The accessory organs, the **salivary glands**, **liver**, **gallbladder**, and **pancreas**, manufacture and store various enzymes and other substances needed in digestion.

Digestion begins in the mouth with the digestion of starch. It continues in the stomach, where protein digestion begins, and is completed in the small intestine. Most absorption of digested food also occurs in the small intestine through small projections of the lining called **villi**. The products of carbohydrate (monosaccharides) and protein (amino acids) digestion are absorbed into capillaries, but most products of fat digestion (glycerol and most fatty acids) are absorbed into **lacteals**. The process of digestion is controlled by both nervous and hormonal mechanisms, which regulate the secretory activity of the digestive organs and the rate at which food moves through the digestive tract.

Addressing the Learning Objectives

1. NAME THE THREE MAIN FUNCTIONS OF THE DIGESTIVE SYSTEM.

EXERCISE 19-1

List and describe the three main functions of the digestive system in the blanks below in the order in which they occur.

1. _____

2. _____

3. _____

2. NAME AND LOCATE THE TWO MAIN LAYERS AND THE SUBDIVISIONS OF THE PERITONEUM.

EXERCISE 19-2: Abdominopelvic Cavity and Peritoneum (Text Fig. 19-2)

1. Write the names of the abdominal organs on the appropriate numbered lines 1 to 6 in different colors.
2. Color the organs on the diagram with the corresponding colors.
3. Use contrasting colors to label and color the body cavities.
4. Use two contrasting dark colors to label and outline the two layers of the peritoneum (structures 8 and 9).
5. Use three shades of yellow or brown to label and color the mesenteries (structures 10 to 13).

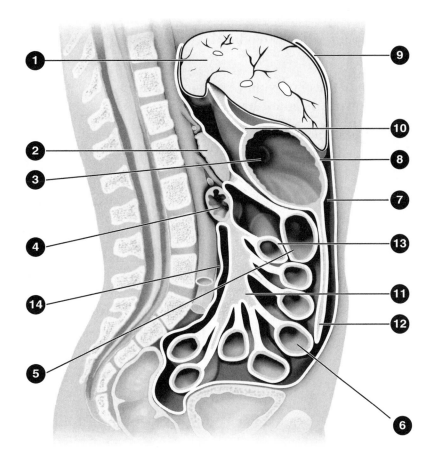

1. _____   8. _____

2. _____   9. _____

3. _____   10. _____

4. _____   11. _____

5. _____   12. _____

6. _____   13. _____

7. _____   14. _____

EXERCISE 19-3

Write the appropriate term in each blank from the list below.

| parietal peritoneum | visceral peritoneum | mesentery |
| mesocolon | greater omentum | lesser omentum |

1. The innermost layer of the serous membrane in contact with abdominal organs _____

2. The outer layer of the serous membrane lining the abdominopelvic cavity _____

3. The subdivision of the peritoneum that contains fat and hangs over the front of the intestines _____

4. The subdivision of the peritoneum extending between the stomach and liver _____

5. The fan-shaped portion of the peritoneum that contains the vessels and nerves supplying the intestine _____

6. The subdivision of the peritoneum that extends from the colon to the posterior abdominal wall _____

3. DESCRIBE THE FOUR LAYERS OF THE DIGESTIVE TRACT WALL.

Also see Exercise 19-5.

EXERCISE 19-4: The Digestive Tract Wall (Text Fig. 19-3)

1. Write the names of the four layers of the digestive tract wall in boxes 1 to 4 on the diagram, using four light colors. Lightly shade each layer with the appropriate color on the diagram.
2. Write the names of the structures in the appropriate lines.

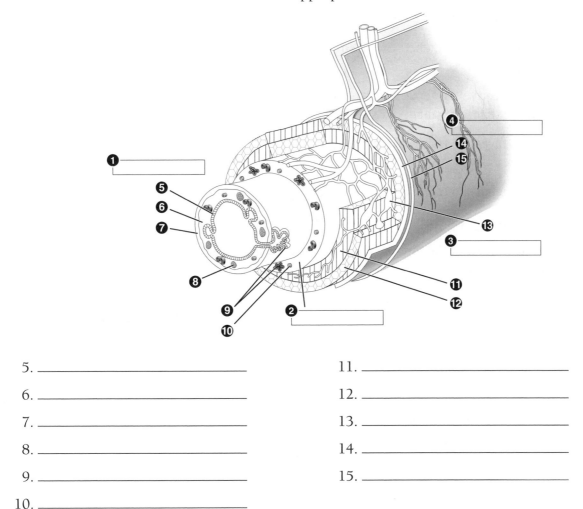

5. _____   11. _____

6. _____   12. _____

7. _____   13. _____

8. _____   14. _____

9. _____   15. _____

10. _____

4. DESCRIBE THE TWO TYPES OF MUSCULAR CONTRACTIONS IMPORTANT IN THE DIGESTIVE PROCESS.

EXERCISE 19-5

Write the appropriate term in each blank from the list below.

mucosa	muscularis externa	peristalsis	segmentation
submucosa	serosa	squamous epithelium	simple columnar epithelium

1. The digestive tract layer in contact with the intestinal contents   _____

2. The visceral peritoneum attached to the surface of a digestive organ   _____

3. The layer of connective tissue beneath the mucous membrane in the wall of the digestive tract   _____

4. The layer of the digestive tract wall that is responsible for peristalsis   _____

5. The type of epithelial tissue lining the esophagus   _____

6. The type of epithelial tissue lining the stomach   _____

7. Rhythmic contractions of the circular muscle layer that mixes food with digestive juices   _____

8. A wave of muscular contraction that propels food rapidly down the digestive tract   _____

5. NAME AND DESCRIBE THE FUNCTIONS OF THE DIGESTIVE TRACT ORGANS.

EXERCISE 19-6: Digestive System (Text Fig. 19-1)

1. Label each structure by writing the correct term in each numbered blank.
2. Draw a box around the name of each structure that is part of the gastrointestinal tract and a circle around the name of each structure that is an accessory organ.

1. _____

2. _____

3. _____

4. _____

5. _____

6. _____

7. _____

8. _____

9. _____

10. _____

11. _____

12. _____

13. _____

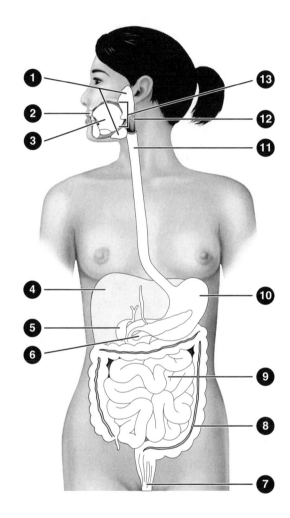

EXERCISE 19-7: Longitudinal Section of the Stomach (Text Fig. 19-6)

1. Label the parts of the stomach, esophagus, and duodenum (parts 1 to 7).
2. Label the layers of the stomach wall (parts 8 to 11). You can use colors to highlight the different muscular layers.
3. Label the two stomach curvatures (labels 12 and 13).

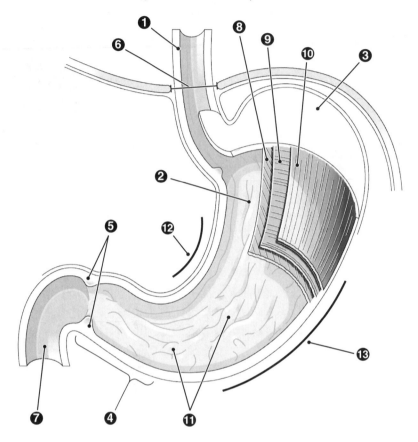

1. _____   8. _____

2. _____   9. _____

3. _____   10. _____

4. _____   11. _____

5. _____   12. _____

6. _____   13. _____

7. _____

EXERCISE 19-8: The Intestines (Text Fig. 19-7A, C, D)

Label the indicated parts. Also write the names of parts 17 and 19 in the numbered boxes.

1. _____
2. _____
3. _____
4. _____
5. _____
6. _____
7. _____
8. _____
9. _____
10. _____
11. _____
12. _____
13. _____
14. _____
15. _____
16. _____
17. _____
18. _____
19. _____
20. _____
21. _____
22. _____

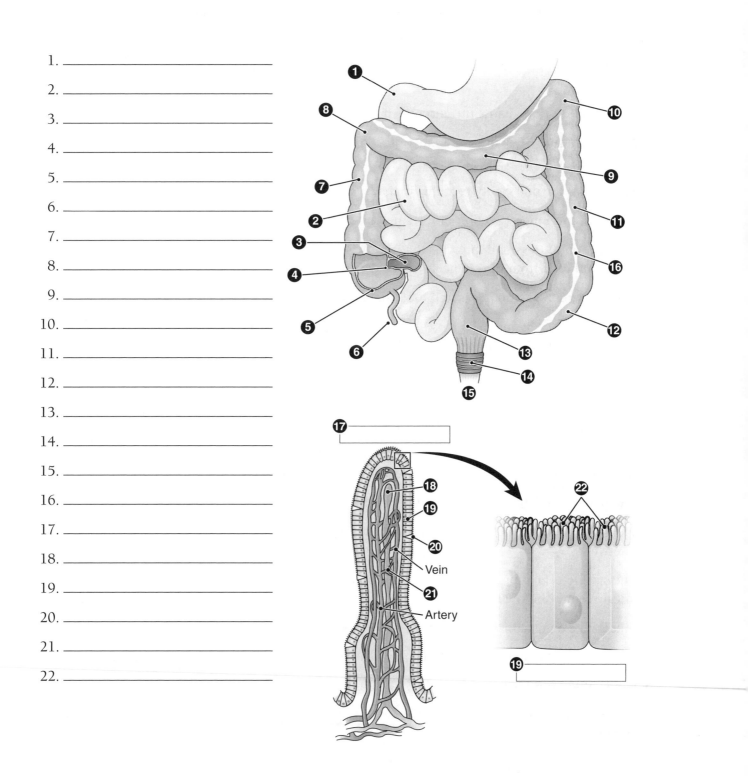

EXERCISE 19-9

Write the appropriate term in each blank from the list below. Not all terms will be used.

ileum   hard palate   soft palate   epiglottis   pyloric sphincter

LES   rugae   chyme   duodenum   jejunum

1. The valve between the distal end of the stomach and the small intestine   _____

2. The structure that guards the entrance into the stomach   _____

3. A structure that covers the opening of the larynx during swallowing   _____

4. The part of the oral cavity roof that extends to form the uvula   _____

5. The final and longest section of the small intestine   _____

6. The section of the small intestine that receives gastric juices and food from the stomach   _____

7. The mixture of gastric juices and food that enters the small intestine   _____

8. Folds in the stomach that are absent if the stomach is full   _____

EXERCISE 19-10

Write the appropriate term in each blank from the list below.

villi   teniae coli   cecum   transverse colon

vermiform appendix   lacteal   rectum   ileocecal valve

1. The part of the large intestine just proximal to the anus   _____

2. The small blind tube attached to the first part of the large intestine   _____

3. The sphincter that prevents food moving from the large intestine into the small intestine   _____

4. Finger-like extensions of the mucosa in the small intestine   _____

5. A blind-ended lymphatic vessel that absorbs fat   _____

6. Bands of longitudinal muscle in the large intestine   _____

7. The portion of the large intestine that extends across the abdomen   _____

8. The most proximal part of the large intestine   _____

6. NAME AND LOCATE THE DIFFERENT TYPES OF TEETH.

EXERCISE 19-11: The Mouth (Text Fig. 19-5A)

1. Write the names of the teeth on the appropriate lines 1 to 6 in different colors. Use the same color to label parts 5 and 6.
2. Color all of the teeth with the corresponding colors.
3. Label the other parts of the mouth.

1. _____

2. _____

3. _____

4. _____

5. _____

6. _____

7. _____

8. _____

9. _____

10. _____

11. _____

12. _____

13. _____

14. _____

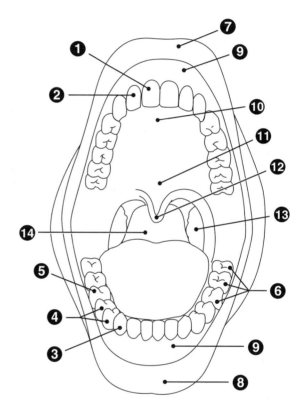

EXERCISE 19-12: A Molar Tooth (Text Fig. 19-5B)

1. Write the names of the two divisions of a tooth in the numbered boxes.
2. Write the names of the parts of the tooth and gums on the appropriate numbered lines in different colors. Use the same colors for parts 3 and 4, and for 6 and 7. Use a dark color for structure 8, because it will be outlined.

3. _____

4. _____

5. _____

6. _____

7. _____

8. _____

9. _____

10. _____

EXERCISE 19-13

Write the appropriate term in each blank from the list below. Not all terms will be used.

deglutition   mastication   incisors   enamel   periodontal ligament

deciduous   cuspids   dentin   gingiva   cementum

1. Term that describes the baby teeth, based on the fact that they are lost   _____

2. The process of chewing   _____

3. The act of swallowing   _____

4. The medical term for the gum   _____

5. The eight cutting teeth located in the front part of the oral cavity   _____

6. A calcified substance making up most of the tooth structure   _____

7. The fibrous connective tissue joining the tooth to the tooth socket   _____

8. The calcified substance coating the tooth   _____

7. NAME AND DESCRIBE THE FUNCTIONS OF THE ACCESSORY ORGANS OF DIGESTION.

EXERCISE 19-14: Accessory Organs of Digestion (Text Fig. 19-9)

1. Write the names of the labeled parts on the appropriate lines in different colors.
2. Color the structures on the diagram with the corresponding colors.

1. _____
2. _____
3. _____
4. _____
5. _____
6. _____
7. _____
8. _____
9. _____
10. _____
11. _____

EXERCISE 19-15

Write the appropriate term in each blank from the list below. Not all terms will be used.

parotid glands   submandibular glands   sublingual glands

gallbladder   liver   pancreas

1. The gland that secretes bicarbonate and digestive enzymes   _____

2. An organ that stores nutrients and releases them as needed into the bloodstream   _____

3. The accessory organ that stores bile   _____

4. The salivary glands that are inferior and anterior to the ear   _____

5. Glands found just under the tongue that secrete into the oral cavity   _____

8. DESCRIBE HOW BILE TRAVELS INTO THE DIGESTIVE TRACT AND FUNCTIONS IN DIGESTION.

EXERCISE 19-16

Write the appropriate term in each blank from the list below. Not all terms will be used.

common bile duct   urea   cystic duct   pancreatic duct

bile   bilirubin   glycogen   common hepatic duct

1. A substance that emulsifies fat   _____

2. The form in which glucose is stored in the liver   _____

3. A waste product produced from the destruction of red blood cells   _____

4. A waste product synthesized by the liver as a result of protein metabolism   _____

5. The duct connecting the hepatic duct to the gallbladder   _____

6. The duct that connects to the pancreatic duct   _____

7. The duct that carries bile from both lobes of the liver to the common bile duct   _____

9. EXPLAIN THE ROLE OF ENZYMES IN DIGESTION, AND LIST ENZYMES INVOLVED IN THE DIGESTION OF FATS, CARBOHYDRATES, PROTEINS, AND NUCLEIC ACIDS.

See Exercise 19-17.

10. DEFINE THE TERM *HYDROLYSIS*, AND EXPLAIN ITS ROLE IN DIGESTION.

EXERCISE 19-17

Write the appropriate term in each blank from the list below.

protein   hydrolysis   fat   nuclease   pepsin

sodium bicarbonate   lipase   maltose   trypsin   maltase

1. An enzyme that acts on a particular type of disaccharide

    _____

2. A substance (NOT an enzyme) released into the small
   intestine that neutralizes the acidity in chyme

    _____

3. A substance that digests DNA

    _____

4. A pancreatic enzyme that splits proteins into amino acids

    _____

5. An enzyme secreted into the stomach that splits proteins
   into amino acids

    _____

6. The splitting of food molecules by the addition of water

    _____

7. This nutrient is usually ingested in the form of triglycerides

    _____

8. The nutrient type that is partially digested by gastric juice

    _____

9. This enzyme digests fat

    _____

11. NAME THE DIGESTION PRODUCTS OF FATS, PROTEINS, AND CARBOHYDRATES.

EXERCISE 19-18

Fill in the blank after each statement—does it apply to carbohydrates (C), proteins (P), or fats (F)?

1. This nutrient type is digested into sugars.

    _____

2. This nutrient type is digested into amino acids.

    _____

3. This nutrient type is digested into glycerol and fatty acids.

    _____

4. This nutrient type is partially digested in the stomach in
   both children and adults.

    _____

5. This nutrient type can be broken down into disaccharides.

    _____

12. DEFINE *ABSORPTION*, AND STATE HOW VILLI FUNCTION IN ABSORPTION.

EXERCISE 19-19

Write a definition of *absorption* in the space below.

_____

EXERCISE 19-20

Label each of the following statements as true (T) or false (F).

1. Digested carbohydrates are absorbed into lacteals.   _____

2. Villi are small extensions of the plasma membranes
   of individual intestinal cells.   _____

3. Villi are folds in the mucosa, each composed
   of many cells.   _____

4. Most digested fats are absorbed into lacteals.   _____

13. EXPLAIN NERVOUS CONTROL OF DIGESTION AND THE ROLE OF THE ENTERIC NERVOUS SYSTEM.

EXERCISE 19-21

Fill in the blanks of the following discussion referring to your textbook as necessary.

Digestion is regulated by (1) _____ feedback, a type of feedback that maintains homeostasis. The nervous system of the digestive tract, known informally as the "gut brain," but technically called the (2) _____ nervous system, participates in homeostasis. This "gut brain" includes sensors, a(n) (3) _____, and effectors such as smooth muscles and (4) _____. The activity of the "gut brain" is modified by the involuntary branch of the nervous system, known as the (5) _____ nervous system. The (6) _____ branch usually increases digestive activity, and the (7) _____ usually decreases digestive activity.

14. LIST FOUR HORMONES INVOLVED IN REGULATING DIGESTION, AND EXPLAIN THE FUNCTION OF EACH.

EXERCISE 19-22

Write the appropriate term in each blank from the list below.

insulin   leptin   ghrelin   gastrin

gastric inhibitory peptide   cholecystokinin (CCK)   secretin

1. A hormone released from fat cells that inhibits appetite   _____

2. A duodenal hormone that stimulates insulin release   _____

3. A hormone that stimulates the secretion of gastric juice and
   increases stomach motility   _____

4. An intestinal hormone that causes the gallbladder to contract,
   releasing bile   _____

5. A duodenal hormone that controls bicarbonate production
   in the pancreas   _____

6. A gastric hormone that stimulates hunger   _____

7. A pancreatic hormone that inhibits hunger   _____

15. DESCRIBE COMMON DISORDERS OF THE DIGESTIVE TRACT AND ITS ACCESSORY ORGANS.

EXERCISE 19-23

Write the appropriate term in each blank from the list below.

gastritis   hepatitis   peritonitis   periodontitis   hiatal hernia

pyloric stenosis   cirrhosis   GERD   emesis   Vincent disease

1. Any infection of the gums and supporting bone   _____

2. Inflammation of the membrane lining the abdominal cavity   _____

3. A chronic condition in which stomach contents often
   flow into the esophagus   _____

4. A protrusion of the stomach through the diaphragm   _____

5. A specific type of gum infection caused by a spirochete or
   bacillus   _____

6. Inflammation of the stomach lining   _____

7. Vomiting   _____

8. A structural problem that prevents food from exiting the stomach   _____

9. Inflammation of the liver, often caused by viruses   _____

10. Chronic liver condition associated with scarring and
    hepatocyte death   _____

EXERCISE 19-24

Write the appropriate term in each blank from the list below.

diverticulitis   obstipation   Crohn disease   ulcerative colitis

diverticulosis   irritable bowel syndrome   volvulus   intussusception

celiac disease   gastroenteritis

1. Slipping of part of the intestine into an adjacent part   _____

2. Twisting of the intestine   _____

3. A form of inflammatory bowel disease associated with inflammation of the large intestine and rectum   _____

4. An autoimmune disease involving inflammation of the small intestine   _____

5. Infection in the small pouches found in the intestinal wall   _____

6. Extreme constipation   _____

7. A condition characterized by large numbers of small pouches in the intestines   _____

8. Inflammation of the stomach and intestine   _____

9. Gluten intolerance   _____

10. A stress-related disorder associated with pain, constipation, and diarrhea   _____

16. USING THE CASE STUDY, DESCRIBE THE COLONOSCOPY PROCEDURE AND ITS ROLE IN DIAGNOSING CERTAIN COLON DISORDERS.

EXERCISE 19-25

Label each of the following statements as true (T) or false (F).

1. You should only get a colonoscopy if you have family members with colon cancer.   _____

2. Traditional colonoscopy uses an endoscope.   _____

3. Virtual colonoscopies use a small camera inside a pill that the patient swallows.   _____

4. All individuals over 50 years of age should consider getting a colonoscopy.   _____

17. SHOW HOW WORD PARTS ARE USED TO BUILD WORDS RELATED TO DIGESTION.

EXERCISE 19-26

Complete the following table by writing the correct word part or meaning in the space provided. Write a word that contains each word part in the "Example" column.

Word Part	Meaning	Example
1. _____	starch	_____
2. mes/o-	_____	_____
3. _____	intestine	_____
4. chole	_____	_____
5. bil/i	_____	_____
6. _____	bladder, sac	_____
7. _____	stomach	_____
8. _____	away from	_____
9. hepat/o	_____	_____
10. lingu/o	_____	_____

Making the Connections

The following concept map deals with the structure and regulation of the gastrointestinal system. Each pair of terms is linked together by a connecting phrase into a sentence. Complete the concept map by filling in the appropriate term or phrase. There is one right answer for each term (1, 2, 4, 7, 8, 12, 13). However, there are many correct answers for the connecting phrases. Write the connecting phrases along the arrows if possible. If your phrases are too long, you may want to write them in the margins or on a separate sheet of paper. Can you think of any other phrases to connect the terms? For instance, how could you connect "CCK" with "pancreas?"

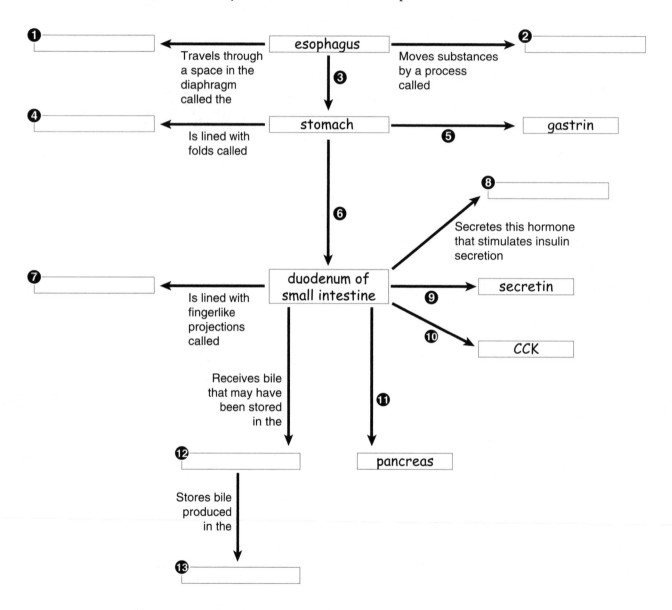

Optional Exercise: Make your own concept map based on the three processes of digestion and how they apply to proteins, sugars, and fats. Use the following terms and any others you would like to include: digestion, absorption, elimination, stomach, small intestine, large intestine, fats, carbohydrates, proteins, amylase, lipase, bile, hydrochloric acid, pepsin, trypsin, peptidase, and maltase.

Testing Your Knowledge

BUILDING UNDERSTANDING

I. MULTIPLE CHOICE

Select the best answer.

1. Which teeth would NOT be found in a 20-year-old male?
   a. bicuspid
   b. cuspid
   c. deciduous
   d. incisor

   1. _____

2. Which of the following is the correct order of tissue from the innermost to the outermost layer in the wall of the digestive tract?
   a. submucosa, serous membrane, smooth muscle, mucous membrane
   b. smooth muscle, serous membrane, mucous membrane, submucosa
   c. serous membrane, smooth muscle, submucosa, mucosa
   d. mucous membrane, submucosa, smooth muscle, serous membrane

   2. _____

3. Where is the parotid gland located?
   a. inferior and anterior to the ear
   b. under the tongue
   c. in the cheek
   d. in the oropharynx

   3. _____

4. Which term describes gallstone formation?
   a. cholecystitis
   b. enteritis
   c. hepatitis
   d. cholelithiasis

   4. _____

5. Which of the following is NOT a portion of the peritoneum?
   a. mesocolon
   b. mesentery
   c. hiatus
   d. greater omentum

   5. _____

6. What are the two active chemicals found in gastric juice?
   a. amylase and pepsin
   b. pepsin and hydrochloric acid
   c. maltase and secretin
   d. bile and trypsin

   6. _____

7. Which of these chemicals is released in response to gastric inhibitory peptide?
   a. pepsin
   b. insulin
   c. bicarbonate
   d. gastrin

   7. _____

8. Which of these instruments is used to examine the lower portion
   of the colon?   8. _____
   a. sigmoidoscope
   b. bronchoscope
   c. catheter
   d. electrocardiogram
9. Where does most fat digestion occur in adults?   9. _____
   a. mouth
   b. stomach
   c. small intestine
   d. transverse colon
10. Which of the following does NOT occur in the mouth?   10. _____
    a. mastication
    b. digestion of starch
    c. absorption of nutrients
    d. ingestion
11. Which of the following is an enzyme?   11. _____
    a. bile
    b. gastrin
    c. trypsin
    d. secretin
12. Which of the following is associated with the intestine?   12. _____
    a. rugae
    b. lacteals
    c. LES
    d. greater curvature
13. Which of these hormones stimulates appetite?   13. _____
    a. ghrelin
    b. insulin
    c. CCK
    d. GIP

II. COMPLETION EXERCISE

Write the word or phrase that correctly completes each sentence.

1. The process by which ingested nutrients are broken down into smaller components is called
   _____.

2. Any inflammation of the liver is called _____.

3. The lower part of the colon bends into an S shape, so this part is called the
   _____.

4. A temporary storage section for indigestible and unabsorbable waste products of digestion is
   the _____.

5. The wavelike movement that rapidly propels food down the esophagus is called
   _____.

6. A common cause of tooth loss is infection of the gums and bone around the teeth, a condition
   called _____.

7. The most common form of stomach cancer is _____.

8. Most digestive juices contain substances that cause the chemical breakdown of foods without entering into the reaction themselves. These catalytic agents are _____.

9. The portion of the peritoneum extending between the stomach and liver is called the _____.

10. The process of swallowing is called _____.

11. Teeth are largely composed of a calcified substance called _____.

12. The esophagus passes through the diaphragm at a point called the _____.

13. The hormone that stimulates gallbladder contraction is called _____.

14. Small projecting folds in the plasma membrane of intestinal epithelial cells are called _____.

15. The stomach enzyme involved in protein digestion is called _____.

16. The chemical reaction that breaks nutrients apart using water is called _____.

UNDERSTANDING CONCEPTS

I. TRUE/FALSE

For each statement, write T for true or F for false in the blank to the left of each number. If a statement is false, correct it by replacing the underlined term, and write the correct statement in the blanks below the statement.

_____ 1. Ms. J is suffering from inflammation of the serous membrane lining the abdomen as a complication of appendicitis. This inflammation is called periodontitis.

_____

_____ 2. There are 32 deciduous teeth.

_____

_____ 3. The layer of the peritoneum attached to the liver is part of the visceral peritoneum.

_____

_____ 4. Mr. D suffers from an ulcer because of a weakness in his LES. This ulcer is most likely in the duodenum.

_____

_____ 5. Ms. Q is deficient in lactase. She will be unable to digest some carbohydrates.

_____

_____ 6. Trypsin is secreted by the gastric glands.

_____

_____ 7. Increased acidity in the chyme could be neutralized by the actions of the hormone <u>gastrin</u>.

_____

_____ 8. Fats are absorbed into <u>blood capillaries</u> in the intestinal villi.

_____

_____ 9. The middle section of the small intestine is called the <u>jejunum</u>.

_____

_____ 10. Folds in the stomach wall are called <u>villi</u>.

_____

_____ 11. The <u>common bile</u> duct delivers bile from the liver and gallbladder into the duodenum.

_____

_____ 12. Amylase is involved in the digestion of <u>carbohydrates</u>.

_____

_____ 13. The <u>pancreas</u> is responsible for the synthesis of urea.

_____

_____ 14. <u>Spastic</u> constipation results from atonic intestinal muscles and can be improved by exercise and increased fiber intake.

_____

II. PRACTICAL APPLICATIONS

Study each discussion. Then write the appropriate word or phrase in the space provided.

1. Mr. P, aged 42, came to the clinic complaining of pain in the "pit of his stomach." He was a tense man who divided up his long working hours with coffee and cigarette breaks. When asked about his alcohol consumption, Mr. P mentioned that he drank three or four beers and a glass of wine each night and significantly more on the weekends. Endoscopy showed inflammation of the innermost layer of the stomach. A general term indicating inflammation of the stomach lining is _____.

2. The name of the layer that is inflamed in Mr. P's stomach is the _____.

3. A damaged area was also found in the most proximal part of the small intestine. This section of the small intestine is called the _____.

4. Mr. P was diagnosed with an ulcer and given a prescription for antibiotics. The antibiotics will treat one of the causative factors in ulcer formation, a bacterium called _____.

5. The physician had almost completed her physical exam of Mr. P when she felt the edge of his liver about 5 cm (2 in) below the ribs. She also noticed that the whites of his eyes were tinged with yellow. This coloration of the sclera (and/or the skin) is termed _____.

6. The yellowish coloration results from the buildup of a yellowish pigment in blood. This pigment results from the breakdown of red blood cells and is normally secreted into the duodenum along with a substance that emulsifies fat. This substance is named _____.

7. Based on her clinical findings and Mr. P's self-reported alcohol abuse, the physician suspected that Mr. P was suffering from a chronic liver disease in which active liver cells are replaced with connective tissue. This disease is called _____.

III. SHORT ESSAYS

1. Describe some features of the small intestine that increase the surface area for absorption of nutrients.

_____

_____

_____

_____

2. List four differences between the digestion and absorption of fats and proteins.

_____

_____

_____

_____

CONCEPTUAL THINKING

1. Ms. M is taking a drug that blocks the action of cholecystokinin. Discuss the impact of this drug on digestion.

_____

_____

_____

_____

2. Is bile an enzyme? Explain your answer.

_____

_____

_____

_____

Expanding Your Horizons

Both inflammatory bowel disease and Crohn disease are autoimmune disorders of the gastro-intestinal tract that frequently develop in young adults. The most common treatments attempt to reduce the inflammatory response, but they have significant side effects, including significant weight gain and fatigue. Dr. Joel Weinstock has developed an alternative treatment—drinking worm cocktails. The hypothesis is that the worms secrete substances that inhibit the immune response. Everyone wins because the immune system leaves both the worms and the intestinal lining alone. You can read more about this work on Dr. Weinstock's website or in the article listed below.

- The Joel Weinstock Lab. Available at: http://sackler.tufts.edu/Faculty-and-Research/Faculty-Research-Pages/Joel-Weinstock

- Weinstock JV, Elliott DE. Helminths and the IBD hygiene hypothesis. Inflamm Bowel Dis 2009;15:128–133. Available at: http://onlinelibrary.wiley.com/doi/10.1002/ibd.20633/full

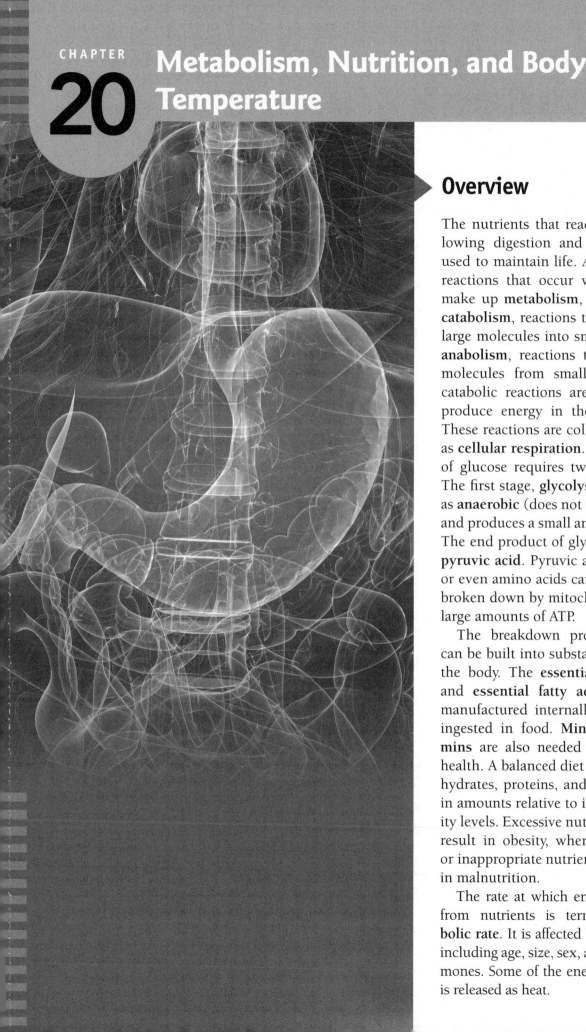

Metabolism, Nutrition, and Body Temperature

▶ Overview

The nutrients that reach the cells following digestion and absorption are used to maintain life. All the chemical reactions that occur within the cells make up **metabolism**, which includes **catabolism**, reactions that break down large molecules into smaller ones, and **anabolism**, reactions that build large molecules from smaller ones. Some catabolic reactions are specialized to produce energy in the form of ATP. These reactions are collectively known as **cellular respiration**. The catabolism of glucose requires two major stages. The first stage, **glycolysis**, is described as **anaerobic** (does not require oxygen) and produces a small amount of energy. The end product of glycolysis is called **pyruvic acid**. Pyruvic acid, fatty acids, or even amino acids can be completely broken down by mitochondria to yield large amounts of ATP.

The breakdown products of food can be built into substances needed by the body. The **essential amino acids** and **essential fatty acids** cannot be manufactured internally and must be ingested in food. **Minerals** and **vitamins** are also needed in the diet for health. A balanced diet includes carbohydrates, proteins, and fats consumed in amounts relative to individual activity levels. Excessive nutrient intake can result in obesity, whereas insufficient or inappropriate nutrient intake results in malnutrition.

The rate at which energy is released from nutrients is termed the **metabolic rate**. It is affected by many factors including age, size, sex, activity, and hormones. Some of the energy in nutrients is released as heat.

Heat production is greatly increased during periods of increased muscular or glandular activity. Most heat is lost through the skin, but heat is also dissipated through exhaled air and eliminated waste products (urine and feces). The **hypothalamus** maintains body temperature at approximately 37°C (98.6°F) by altering blood flow through the surface blood vessels and the activity of sweat glands and muscles. Abnormalities of body temperature are a valuable diagnostic tool. If thermo-regulatory mechanisms fail, **hyperthermia** (elevated body temperature) or **hypothermia** (decreased body temperature) can result. The presence of **fever**—an abnormally high body temperature result-ing from an elevated temperature set point—usually indicates infection but may also indicate a toxic reaction, a brain injury, or other disorders. The opposite of fever is **hypothermia**—an exceed-ingly low body temperature—which most often occurs when the body is exposed to very low out-side temperature. Hypothermia can cause serious damage to body tissues.

Addressing the Learning Objectives

1. DIFFERENTIATE BETWEEN CATABOLISM AND ANABOLISM.

EXERCISE 20-1

Fill in the blank after each statement—does it apply to catabolism (C) or to anabolism (A)?

1. The metabolic breakdown of complex compounds   _____

2. The metabolic building of simple compounds into substances
   needed by cells   _____

3. This process usually releases energy   _____

2. DIFFERENTIATE BETWEEN THE ANAEROBIC AND AEROBIC PHASES OF GLUCOSE CATABOLISM, AND GIVE THE END PRODUCTS AND THE RELATIVE AMOUNT OF ENERGY RELEASED BY EACH.

EXERCISE 20-2

Fill in the blank after each statement—does it apply to the anaerobic phase (AN) or to the aerobic phase (AE) of glucose catabolism?

1. This process generates two ATP per glucose molecule   _____

2. This process generates about 30 ATP per glucose molecule   _____

3. The end products are carbon dioxide and water   _____

4. The end product is pyruvic acid   _____

5. This process can occur in the absence of oxygen   _____

6. This process is also known as glycolysis   _____

7. This process requires oxygen   _____

8. This process occurs first   _____

3. DEFINE METABOLIC RATE, AND NAME SIX FACTORS THAT AFFECT IT.

EXERCISE 20-3

Write a definition for each term in the blanks provided. In the fourth line, list six determinants of the metabolic rate.

1. Metabolism _____

_____

2. Metabolic rate _____

_____

3. Basal metabolism _____

_____

4. Factors affecting metabolic rate _____

_____

4. EXPLAIN HOW CARBOHYDRATES, FATS, AND PROTEINS ARE METABOLIZED FOR ENERGY.

See Exercise 20-4.

5. COMPARE THE ENERGY CONTENTS OF CARBOHYDRATES, FATS, AND PROTEINS.

EXERCISE 20-4

Write the appropriate term in each blank from the list below. Not all terms will be used.

glycolysis   pyruvic acid   glycerol   lactic acid

glycogen   deamination   fat   protein

1. The storage form of glucose _____

2. A modification of amino acids that occurs before they can be oxidized for energy _____

3. An intermediate product of glucose catabolism that can be completely oxidized within the mitochondria _____

4. An organic substance produced from pyruvic acid during intense exercise _____

5. The nutrient type that generates the most energy per gram _____

6. A product of fat digestion that can be used for energy _____

7. The nutrient type that does not have a specialized storage form _____

6. LIST THE RECOMMENDED PERCENTAGES OF CARBOHYDRATE, FAT, AND PROTEIN IN THE DIET.

EXERCISE 20-5

Match each percentage to the corresponding nutrient type, by writing the appropriate letter in each blank.

1. Protein _____   a. 55% to 60%

2. Fat _____   b. less than 30%

3. Carbohydrate _____   c. 15% to 20%

7. DISTINGUISH BETWEEN SIMPLE AND COMPLEX CARBOHYDRATES, GIVING EXAMPLES OF EACH.

See Exercise 20-6.

8. COMPARE SATURATED AND UNSATURATED FATS.

EXERCISE 20-6

Write the appropriate term in each blank from the list below. Not all terms will be used.

trans-fatty acids   unsaturated fats   insulin   glycogen

monosaccharides   saturated fats   polysaccharides disaccharides

1. Fats that are usually of animal origin and are solid at room temperature   _____

2. Fats that are artificially saturated to prevent rancidity   _____

3. Carbohydrates with a low glycemic effect   _____

4. Glucose and fructose are examples of this type of nutrient   _____

5. Plant-derived fats that are usually liquid at room temperature   _____

6. The pancreatic hormone that promotes polysaccharide anabolism   _____

7. The pancreatic hormone that promotes polysaccharide catabolism   _____

9. DEFINE *ESSENTIAL AMINO ACID*.

See Exercise 20-7.

10. EXPLAIN THE ROLES OF MINERALS AND VITAMINS IN NUTRITION, AND GIVE EXAMPLES OF EACH.

EXERCISE 20-7

Write the appropriate term in each blank from the list below.

essential amino acids   essential fatty acids   antioxidants

nonessential amino acids   trace elements   vitamins   minerals

1. A class of substances that stabilizes free radicals   _____

2. Minerals required in very small amounts   _____

3. Complex organic molecules that are essential for metabolism   _____

4. Protein components that must be taken in as part of the diet   _____

5. Inorganic elements needed for body structure and many
   body functions, sometimes in large amounts   _____

6. Protein building blocks that can be manufactured by the body   _____

7. Linoleic acid is an example   _____

EXERCISE 20-8

Write the appropriate term in each blank from the list below. Not all terms will be used.

zinc   iodine   iron   potassium   calcium

folate   calciferol   riboflavin   vitamin A   vitamin K

1. The vitamin that prevents dry, scaly skin and night blindness   _____

2. The vitamin needed to prevent anemia, digestive disorders,
   and neural tube defects in the embryo   _____

3. Another name for vitamin D, the vitamin required
   for normal bone formation   _____

4. The mineral component of thyroid hormones   _____

5. A mineral important in blood clotting and muscle contraction   _____

6. The characteristic element in hemoglobin, the oxygen-carrying
   compound in the blood   _____

7. A mineral that promotes carbon dioxide transport and energy
   metabolism   _____

8. A vitamin involved in the synthesis of blood clotting factors
   that can be synthesized by colonic bacteria   _____

11. LIST SIX ADVERSE EFFECTS OF ALCOHOL CONSUMPTION.

EXERCISE 20-9

Write six adverse effects of excess alcohol consumption in the spaces below.

1. _____

2. _____

3. _____

4. _____

5. _____

6. _____

12. DESCRIBE FOUR NUTRITIONAL DISORDERS.

EXERCISE 20-10

Write the appropriate term in each blank from the list below. Not all terms will be used.

underweight	overweight	bulimia	food sensitivity	obese
anorexia nervosa	food allergy	marasmus	kwashiorkor	

1. A person with a body mass index of more than 30 _____

2. A person with a body mass index of less than 18.5 _____

3. Severe malnutrition in infancy _____

4. Protein deficiency, commonly observed in older children _____

5. A reaction to a particular food mediated by the immune system _____

6. A reaction to a particular food reflecting an enzyme deficiency or other nonimmune cause _____

7. A psychological disorder associated with abnormally low levels of food intake _____

8. A psychological disorder associated with overeating followed by vomiting or laxative use _____

13. EXPLAIN HOW HEAT IS PRODUCED AND LOST IN THE BODY.

EXERCISE 20-11

Write the appropriate term in each blank from the list below.

convection   evaporation   conduction   radiation

1. The direct transfer of heat from a warmer object to a cooler one _____

2. Heat loss resulting from the conversion of a liquid, such as perspiration, to a vapor _____

3. Heat loss resulting from moving air _____

4. Heat that travels from its source as heat waves _____

14. DESCRIBE THE ROLE OF THE HYPOTHALAMUS IN REGULATING BODY TEMPERATURE.

EXERCISE 20-12

Fill in the blank after each of the following body changes—which would the hypothalamus induce when the body is cold (C), and which would it induce when the body is excessively hot (H)?

1. Constriction of the skin's blood vessels _____

2. Increased sweat gland activity _____

3. Dilation of the skin's blood vessels _____

4. Increased skeletal muscle contraction (shivering) _____

15. EXPLAIN THE ROLE OF FEVER IN DISEASE.

EXERCISE 20-13

Write the appropriate term in each blank from the list below.

febrile   crisis   lysis

antipyretic   pyrogen

1. A substance that causes fever _____

2. The term that describes a person who has a fever _____

3. A sudden drop in temperature at the end of a period of fever _____

4. A gradual fall in temperature at the end of a period of fever _____

5. A class of drugs that treats fever _____

16. DESCRIBE RESPONSES TO EXCESSIVE HEAT AND COLD.

EXERCISE 20-14

Write the appropriate term in each blank from the list below.

fever	heat cramps	heat exhaustion
heat stroke	hypothermia	frostbite

1. A moderate, easily reversible disorder caused by body temperature rising above the set point   _____

2. An abnormally low body temperature, as may be caused by prolonged exposure to cold   _____

3. The final stage of excessive exposure to heat, characterized by central nervous system symptoms   _____

4. A change in body temperature that involves changing the hypothalamic set point   _____

5. A condition that may follow heat cramps if adequate treatment is not given   _____

6. A term that describes regional tissue freezing   _____

17. USING THE CASE STUDY AND THE TEXT, DEFINE ANOREXIA NERVOSA, AND LIST SOME OF ITS ADVERSE EFFECTS.

Also see Exercise 20-10.

EXERCISE 20-15

Claudia, the subject of the case study, suffers from anorexia nervosa and bulimia. List at least four problems she might encounter if she cannot control her disease. If possible, explain the link between the problem and inadequate nutrient intake.

1. _____

2. _____

3. _____

4. _____

18. SHOW HOW WORD PARTS ARE USED TO BUILD WORDS RELATED TO METABOLISM, NUTRITION, AND BODY TEMPERATURE.

EXERCISE 20-16

Complete the following table by writing the correct word part or meaning in the space provided. Write a word that contains each word part in the "Example" column.

Word Part	Meaning	Example
1. _____	heat	_____
2. -lysis	_____	_____
3. _____	sugar, sweet	_____
4. pyr/o	_____	_____

Making the Connections

The following concept map deals with nutrition and metabolism. Each pair of terms is linked together by a connecting phrase. Complete the concept map by filling in the appropriate term or phrase. There is one right answer for each term (1, 2, 5 to 7, 9, 11, 13, 14). However, there are many correct answers for the connecting phrases. Write the connecting phrases along the arrows if possible. If your phrases are too long, you may want to write them in the margins or on a separate sheet of paper.

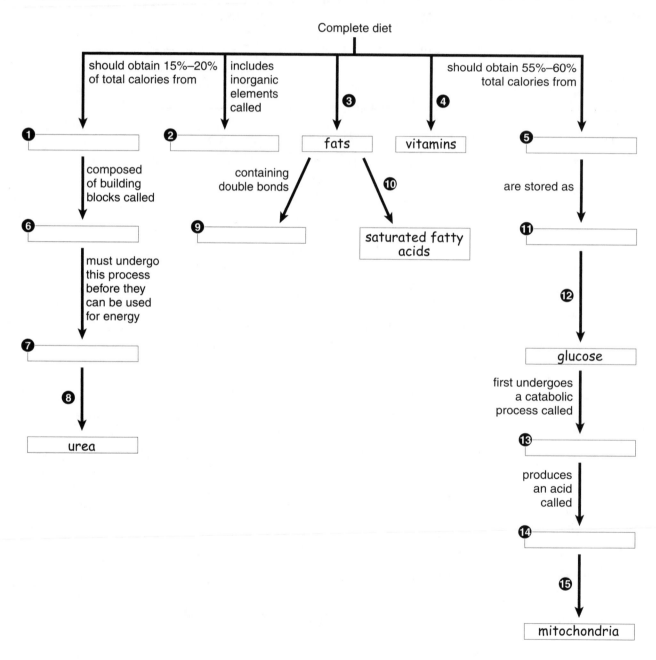

Optional Exercise: Make your own concept map based on the regulation of body temperature. Use the following terms and any others you would like to include: body temperature, fever, hypothermia, crisis, lysis, pyrogen, antipyretic, heat exhaustion, frostbite, hypothalamus, sweating, shivering, dilation, constriction, heat stroke, and heat cramps.

Testing Your Knowledge

BUILDING UNDERSTANDING

I. MULTIPLE CHOICE

Select the best answer.

1. Which region of the brain is involved in temperature regulation?　　　　1. _____
    a. hypothalamus
    b. cerebral cortex
    c. hippocampus
    d. thalamus
2. What does a complete protein contain?　　　　2. _____
    a. all the amino acids
    b. all the essential fatty acids
    c. a variety of minerals
    d. all the essential amino acids
3. What is deamination?　　　　3. _____
    a. an anabolic reaction
    b. the conversion of proteins into amino acids
    c. the conversion of glucose into glycogen
    d. the removal of a nitrogen group from an amino acid
4. If you have a fever, what might be your body temperature?　　　　4. _____
    a. 96°F
    b. 39°F
    c. 39°C
    d. 96°C
5. What is the end product of the anaerobic phase of glucose catabolism?　　　　5. _____
    a. glycogen
    b. pyruvic acid
    c. folic acid
    d. tocopherol
6. What are dietary trace elements?　　　　6. _____
    a. sugars with a high glycemic effect
    b. vitamins needed in very small amounts
    c. minerals needed in large quantity
    d. minerals needed in very small amounts
7. Which heat transfer process is increased by profuse sweating?　　　　7. _____
    a. radiation
    b. evaporation
    c. convection
    d. none of the above
8. What is a pyrogen?　　　　8. _____
    a. a substance that induces fever
    b. a cream used to treat frostbite
    c. a drug that reduces body temperature
    d. a substance that causes lysis

9. Which of these phrases describes unsaturated fats?

9. _____

   a. They are generally healthier than saturated fats.
   b. They can be converted into trans fats.
   c. They contain double bonds between the carbon atoms.
   d. All of the above.

10. Which of the following is an example of an anabolic reaction?

10. _____

   a. Glycerol and fatty acids are used to form a fat.
   b. Starches and glycogen are converted into glucose.
   c. A short peptide is converted into arginine and cysteine.
   d. Glucose is completely oxidized to carbon dioxide and water.

11. Which of these disorders would be very prevalent in a region encountering a severe food shortage?

11. _____

   a. obesity
   b. type 2 diabetes
   c. kwashiorkor
   d. none of the above

II. COMPLETION EXERCISE

Write the word or phrase that correctly completes each sentence.

1. Severe malnutrition beginning in infancy is called _____.

2. Tissue damage caused by exposure to cold is termed _____.

3. Shivering to generate additional body heat results from increased activity of the _____.

4. Organic substances needed in small amounts in the diet are the _____.

5. A gradual drop in temperature following a fever is known as _____.

6. Binge–purge syndrome is clinically known as _____.

7. An individual subjected to excessive heat who is dizzy and no longer sweating is probably suffering from _____.

8. The series of catabolic reactions that results in the complete breakdown of nutrients is called _____.

9. Heat that is moved away from the skin by the wind is lost by the process of _____.

10. Glycolysis occurs in the part of the cell called the _____.

11. Fatty acids that must be consumed in the diet are called _____.

12. The only nutrient that undergoes glycolysis is _____.

UNDERSTANDING CONCEPTS

I. TRUE/FALSE

For each statement, write T for true or F for false in the blank to the left of each number. If a statement is false, correct it by replacing the underlined term, and write the correct statement in the blanks below the statement.

_____ 1. A body temperature of 35°C would be considered <u>hypothermia</u>.

_____

_____ 2. Grape seed oil is liquid at room temperature. This oil is most likely a <u>saturated</u> fat.

_____

_____ 3. The conversion of glycogen into glucose is an example of a <u>catabolic</u> reaction.

_____

_____ 4. Most heat loss in the body occurs through the <u>skin</u>.

_____

_____ 5. The element nitrogen is found in all <u>sugars</u>.

_____

_____ 6. The end product of cellular respiration in the mitochondria is <u>pyruvic acid</u>.

_____

_____ 7. The rate at which energy is released from nutrients during <u>intense exercise</u> is called the basal metabolism.

_____

_____ 8. Pernicious anemia is caused by a deficiency in <u>vitamin B_{12}</u>.

_____

_____ 9. Mr. Q is suffering from a fever, but he has declined to take any medications. His blood contains large amounts of <u>antipyretics</u> that have induced the fever.

_____

_____ 10. The most serious disorder resulting from excessive heat is called <u>heat exhaustion</u>.

_____

_____ 11. A body mass index of 28 is indicative of <u>obesity</u>.

_____

II. PRACTICAL APPLICATIONS

Ms. S is researching penguin behavior at a remote location in Antarctica. She will be camping on the ice for two months. Study each discussion. Then write the appropriate word or phrase in the space provided.

1. Ms. S is spending her first night on the ice. She is careful to wear many layers of clothing in order to avoid a dangerous drop in body temperature, or _____.

2. She is out for a moonlight walk to greet the penguins when she surprises an elephant seal stalking a penguin. Frightened, she sprints back to her tent. Her muscles are generating ATP by an oxygen-independent pathway. Each glucose molecule is generating a small number of ATP molecules, or to be exact, _____.

3. The end product of this oxygen-independent pathway is a molecule called _____.

4. Ms. S is exercising so intensely that some of this end product is converted into a different acid. This acid, which can spill over into blood, is called _____.

5. Ms. S realizes that she lost her face mask. She notices a whitish patch on her nose. Her nose is suffering from _____.

6. After two weeks on the ice, Ms. S is out of fresh fruits and vegetables, and the penguins have stolen her multivitamin supplements. She has been reading accounts of early explorers suffering from scurvy and fears she will experience the same fate. Scurvy is due to a deficiency of _____.

7. Ms. S hikes to a distant penguin colony on her final day on the ice. She is dressed very warmly, and the sun is very bright. After several hours of hiking, Ms. S is sweating profusely. She is also experiencing a headache, tiredness, and nausea. She is probably suffering from _____.

8. She removes some clothing to cool off. Some excess heat will be lost through the evaporation of sweat. Heat will also be lost directly from her skin to the surrounding air by the process called _____.

III. SHORT ESSAYS

1. Is alcohol a nutrient? Defend your answer.

_____

_____

_____

_____

_____

2. A glucose molecule has been transported into a muscle cell. This cell has ample supplies of oxygen. Discuss the steps involved in using this glucose to produce energy. For each step, describe its location and oxygen requirements and name the substances produced.

_____

_____

_____

_____

_____

CONCEPTUAL THINKING

1. Your friend wants to lose some weight. She is following a diet that contains 20% carbohydrates, 40% fat, and 40% protein. Why is this diet designed to cause weight loss? You may have to review the actions of pancreatic hormones (Chapter 12) to answer this question.

_____

_____

_____

2. An invasion of influenza viruses has caused Mr. L to develop a fever. Using information from Chapters 17 and 20, describe the steps between the virus appearing in the bloodstream and Mr. L's increase in body temperature.

_____

_____

_____

Expanding Your Horizons

How does your diet measure up? Go to http://www.ChooseMyPlate.gov to find nutritional information and guidelines established by the U.S. Department of Agriculture. Some nongovernmental agencies, such as the Harvard School of Public Health, have criticized the government's efforts and proposed a graphic, the Healthy Eating Plate, that differs from the USDA guidelines. You can read about their critique at http://www.hsph.harvard.edu/nutritionsource/healthy-eating-plate/healthy-eating-plate-vs-usda-myplate/index.html (or do a website search for "Healthy Eating Plate").

Overview

The majority (50% to 70%) of a person's body weight is **water**. This water is a solvent, a transport medium, and a participant in metabolic reactions. A variety of substances are dissolved in this water, including electrolytes, nutrients, gases, enzymes, hormones, and waste products. Body fluids are distributed in two main compartments: (1) the **intracellular fluid** compartment within the cells and (2) the **extracellular fluid** compartment located outside the cells. The latter category includes blood plasma, interstitial fluid separating cells, lymph, and fluids in special compartments, such as the humors of the eye, cerebrospinal fluid, serous fluids, and synovial fluids.

Water balance is maintained by matching fluid intake with fluid output. Fluid intake is stimulated by the thirst center in the **hypothalamus**, but humans voluntarily control their fluid intake and may consume excess or insufficient fluids. Normally, the amount of fluid taken in with food and beverages equals the amount of fluid lost through the skin and the respiratory, digestive, and urinary tracts. When there is an imbalance between fluid intake and fluid output, serious disorders such as **edema**, **hyponatremia**, and **dehydration** may develop.

The composition of intracellular and extracellular fluids is an important factor in homeostasis. These fluids must have the proper levels of electrolytes and must be kept at a constant pH. The kidneys are the main regulators of body fluids. They alter the retention of specific electrolytes and water in response to hormones such as aldosterone, antidiuretic hormone (ADH), atrial natriuretic peptide (ANP), and angiotensin II (ATII). Buffers and the respiratory system aid the kidneys in maintaining constant blood pH. The normal pH of body fluids is a slightly alkaline 7.4. When regulating mechanisms fail to control shifts in pH, either **acidosis** or **alkalosis** results.

Fluid therapy is used to correct fluid and electrolyte imbalances and to give a patient nourishment.

Addressing the Learning Objectives

1. COMPARE INTRACELLULAR AND EXTRACELLULAR FLUIDS.

See Exercises 21-1 and 21-4.

2. LIST FOUR TYPES OF EXTRACELLULAR FLUIDS.

EXERCISE 21-1: Main Fluid Compartments (Text Fig. 21-1)

1. The three beakers represent the three major body fluid compartments. These three compartments constitute 40%, 15%, and 5% of the average person's body weight. Based on the size of the beaker, write in the correct number for each percentage.
2. Label the four different fluid compartments.

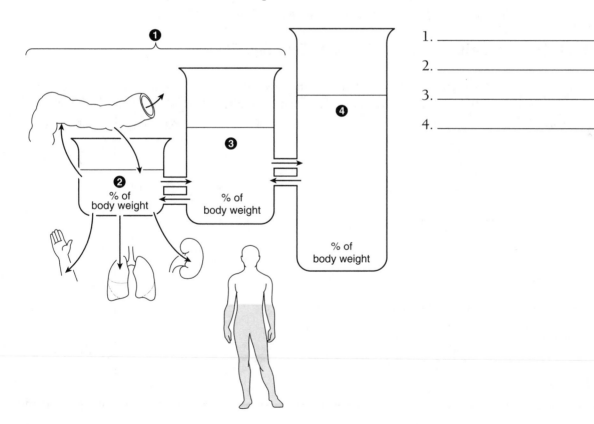

1. _____
2. _____
3. _____
4. _____

3. DEFINE *ELECTROLYTES*, AND DESCRIBE SOME OF THEIR FUNCTIONS.

EXERCISE 21-2

Write the appropriate term in each blank from the list below.

cation   anion   potassium   electrolyte

sodium   calcium   chloride

  1. A general term describing any positively charged ion   _____

  2. A general term describing any negatively charged ion   _____

  3. A component of stomach acid   _____

  4. The most abundant cation inside cells   _____

  5. A compound that forms ions in solution   _____

  6. A cation involved in bone formation   _____

  7. The most abundant cation in the fluid surrounding cells   _____

4. NAME FOUR SYSTEMS THAT ARE INVOLVED IN WATER BALANCE.

EXERCISE 21-3: Daily Gain and Loss of Water (Text Fig. 21-3)

Write the names of the different sources of water gain and loss on the appropriate numbered lines in different colors. Color the diagram with the corresponding colors.

1. _____

2. _____

3. _____

4. _____

5. _____

6. _____

7. _____

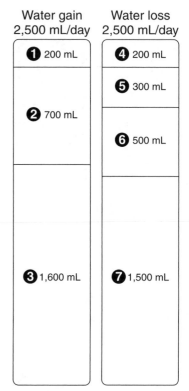

Water gain 2,500 mL/day

❶ 200 mL

❷ 700 mL

❸ 1,600 mL

Water loss 2,500 mL/day

❹ 200 mL

❺ 300 mL

❻ 500 mL

❼ 1,500 mL

5. EXPLAIN HOW THIRST IS REGULATED.

EXERCISE 21-4

Write the appropriate term in each blank from the list below. Not all terms will be used.

hypothalamus   brain stem   interstitial   intracellular   extracellular

water   blood plasma   body fluid concentration   body fluid volume

1. The substance that makes up about 4% of a person's body weight _____

2. The substance that makes up 50% to 70% of a person's body weight _____

3. The part of the brain that controls the sense of thirst _____

4. Term that specifically describes fluids in the microscopic spaces between cells _____

5. The thirst center is stimulated when this is increased _____

6. The thirst center is stimulated when this is decreased _____

7. Term that describes the fluid within the body cells _____

6. DESCRIBE THE ROLE OF HORMONES IN ELECTROLYTE BALANCE.

EXERCISE 21-5

Write the appropriate term in each blank from the list below. Each term will be used more than once. The number of blanks indicates the number of answers for that question.

angiotensin II (ATII)   **antidiuretic hormone (ADH)**

atrial natriuretic peptide (ANP)   **aldosterone**

1. Production increases when blood pressure decreases   _____

    _____

    _____

2. Production increases when blood pressure increases   _____

3. Produced by the adrenal cortex   _____

4. Secreted by the posterior pituitary gland   _____

5. Produced by the heart   _____

6. Stimulates aldosterone and ADH production   _____

7. Increases potassium loss in the urine   _____

8. Deficiency causes diabetes insipidus   _____

9. Directly or indirectly increases water retention   _____

    _____

    _____

10. Acts directly at the kidney to increase sodium retention   _____

7. DESCRIBE THREE METHODS FOR REGULATING THE pH OF BODY FLUIDS.

EXERCISE 21-6

Write the appropriate term in each blank below. Each term will be used twice.

buffer   kidney   lung

1. After hydrogen reacts with bicarbonate, this organ eliminates the product   _____

2. Synthesizes bicarbonate   _____

3. Hemoglobin is an example   _____

4. Undergoes a chemical reaction with a hydrogen ion   _____

5. Directly eliminates H+ ions from the body   _____

6. This organ uses up a buffer molecule every time it rids the body of a hydrogen ion   _____

8. COMPARE ACIDOSIS AND ALKALOSIS, INCLUDING POSSIBLE CAUSES.

See Exercise 21-7.

9. DESCRIBE FOUR DISORDERS INVOLVING BODY FLUIDS.

EXERCISE 21-7

Write the appropriate term in each blank from the list below.

polydipsia   pulmonary edema   ascites   Addison disease
hyponatremia   metabolic acidosis   respiratory alkalosis   respiratory acidosis
metabolic alkalosis   effusion

1. A collection of fluid within the abdominal cavity   _____

2. A condition that results from inadequate ventilation   _____

3. The accumulation of fluid in the lungs, such as may result from congestive heart failure   _____

4. The result of prolonged hyperventilation   _____

5. Excessive thirst   _____

6. A disorder resulting from aldosterone deficiency   _____

7. The escape of fluid into a cavity or space   _____

8. A condition in which body fluids are abnormally dilute   _____

9. An acid–base disorder that can result from untreated diabetes mellitus or starvation   _____

10. An acid–base disorder resulting from prolonged vomiting   _____

10. CITE INSTANCES IN WHICH FLUIDS ARE ADMINISTERED THERAPEUTICALLY, AND CITE FOUR FLUIDS THAT ARE USED.

EXERCISE 21-8

Write the appropriate term in each blank from the list below.

dextrose   normal saline   Ringer lactate   25% serum albumin

1. A sugar that is often administered intravenously   _____

2. An isotonic solution that is the first to be administered in emergencies   _____

3. A hypertonic solution used to treat edema by drawing fluid out of the interstitial spaces   _____

4. An isotonic solution that will help treat pH imbalances   _____

11. REFERRING TO THE OPENING CASE STUDY, EXPLAIN THE DANGERS OF OVERHYDRATING.

EXERCISE 21-9

Write the correct term in each blank using information from your textbook.

Drinking too much water causes the blood concentration of sodium to (1) _____ and the blood volume to (2) _____. Water moves freely between the blood and the fluid surrounding the cells, known as (3) _____. Excess fluid passing into body tissues causes swelling, known as (4) _____. Normally, these changes in blood volume and solute concentration should (5) _____ the production of (6) _____ from the hypothalamus/posterior pituitary, resulting in (7) _____ urine production and the restoration of homeostasis. However, these compensations were inadequate in Ethan's case.

12. SHOW HOW WORD PARTS ARE USED TO BUILD WORDS RELATED TO BODILY FLUIDS.

EXERCISE 21-10

Complete the following table by writing the correct word part or meaning in the space provided. Write a word that contains each word part in the "Example" column.

Word Part	Meaning	Example
1. _____	poison	_____
2. intra-	_____	_____
3. _____	water	_____
4. extra-	_____	_____
5. osmo-	_____	_____
6. _____	many	_____
7. _____	condition, process	_____
8. _____	partial, half	_____

Making the Connections

The following concept map deals with the types of body fluids and the sources of fluid input and output. Each pair of terms is linked together by a connecting phrase. Complete the concept map by filling in the appropriate term or phrase. There is one right answer for each term. However, there are many correct answers for the connecting phrases (4 to 7, 9, 11, 14, 15).

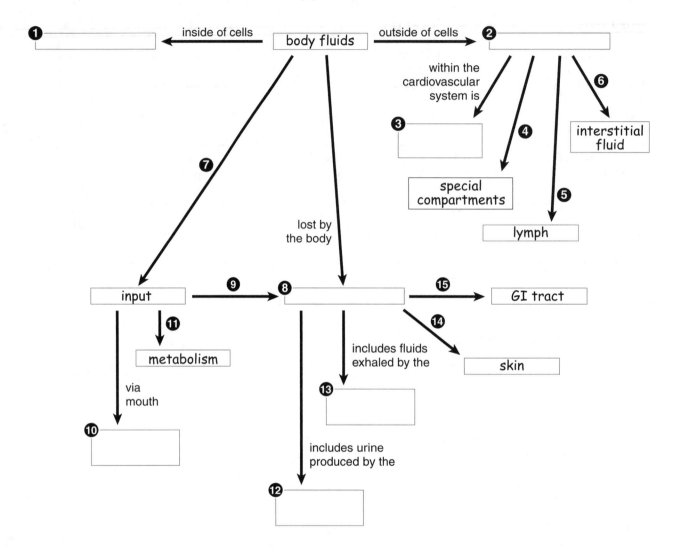

Optional Exercise: Make your own concept map based on the regulation of acid–base balance. Use the following terms and any others you would like to include: acidosis, alkalosis, metabolic, respiratory, carbon dioxide, kidney, bicarbonate, and hemoglobin.

Testing Your Knowledge

BUILDING UNDERSTANDING

I. MULTIPLE CHOICE

Select the best answer.

1. Which of these structures produces aldosterone?
   a. hypothalamus
   b. pituitary
   c. kidney
   d. adrenal cortex

   1. _____

2. Which of these is an action of antidiuretic hormone?
   a. increases salt retention by the kidney
   b. inhibits thirst
   c. decreases urine output
   d. increases potassium reabsorption by the kidney

   2. _____

3. Which minerals are required for normal bone formation?
   a. calcium
   b. phosphate
   c. neither calcium nor phosphate
   d. both calcium and phosphate

   3. _____

4. Which of the following is NOT a buffer?
   a. hemoglobin
   b. bicarbonate
   c. oxygen
   d. phosphate

   4. _____

5. Which of the following fluids is NOT in the extracellular compartment?
   a. cytoplasm
   b. cerebrospinal fluid
   c. lymph
   d. interstitial fluid

   5. _____

6. What can cause respiratory acidosis?
   a. obstructive pulmonary disease
   b. kidney failure
   c. excessive exercise
   d. hyperventilation

   6. _____

7. Which of these changes would cause edema?
   a. increased venous return to the heart
   b. increased capillary permeability
   c. increased plasma protein concentration
   d. hypernatremia

   7. _____

III. COMPLETION EXERCISE

Write the word or phrase that correctly completes each sentence.

1. A negatively charged ion is called a(n) _____.

2. The receptors that detect an increase in body fluid concentration are called _____.

3. The ion that plays the largest role in maintaining body fluid volume is _____.

4. Water balance is partly regulated by a thirst center located in a region of the brain called the _____.

5. Accumulation of excessive fluid in the abdominal cavity is _____.

6. A blood pH above 7.45 results in _____.

7. An abnormally low blood sodium concentration is known as _____.

8. The hormone that decreases blood volume is _____.

UNDERSTANDING CONCEPTS

I. TRUE/FALSE

For each statement, write T for true or F for false in the blank to the left of each number. If a statement is false, correct it by replacing the underlined term, and write the correct statement in the blanks below the statement.

_____ 1. Adherence to low-carbohydrate diets can result in underlined metabolic acidosis.

_____

_____ 2. The backup of fluid in the lungs that often accompanies congestive heart failure is called lymphedema.

_____

_____ 3. The exhalation of carbon dioxide makes the blood more acidic.

_____

_____ 4. Insufficient production of antidiuretic hormone results in Addison disease.

_____

_____ 5. The fluids found in joint capsules and in the eyeball are examples of extracellular fluids.

_____

_____ 6. The organ that excretes the largest amount of water per day is the skin.

_____

II. PRACTICAL APPLICATIONS

Study each discussion. Then write the appropriate word or phrase in the space provided.

1. Ms. S, aged 26, was teaching English in rural India. She developed severe diarrhea, probably as a result of a questionable pakora from a roadside stand. On admission to a hospital, she was found to be suffering from a severe fluid deficit, a condition called _____.

2. The physician's first concern was to increase Ms. S's plasma volume. Her fluid deficit was addressed by administering a 0.9% sodium chloride solution, commonly known as _____.

3. This solution will not change the volume of Ms. S's cells, because it is _____.

4. Next, the physician addressed Ms. S's acid–base balance. Prolonged diarrhea is often associated with a change in blood pH called _____.

5. The pH of a blood sample was tested. The laboratory technician was looking for a pH value outside the normal range. The normal pH range for blood is _____.

6. Ms. S's blood pH was found to be abnormal, and sodium bicarbonate was added to her IV. Bicarbonate is an example of a substance that helps maintain a constant pH. These substances are known as _____.

III. SHORT ESSAYS

1. Describe some circumstances under which fluid therapy might be given, and cite some fluids that are used.

_____

_____

_____

_____

_____

2. Compare respiratory acidosis and alkalosis, and cite some causes of each.

_____

_____

_____

_____

_____

CONCEPTUAL THINKING

1. Ms. J is stranded on a desert island with limited water supplies.
   a. How will the volume and concentration of her body fluids change?

   _____

   _____

   _____

   b. How will her hypothalamus and pituitary gland respond to these changes?

   _____

   _____

   _____

   c. What will be the net effect of these responses?

   _____

   _____

   _____

2. Your summer research project is to design new drugs to alter blood pressure by altering water retention and thus blood volume. You know that increasing blood volume increases blood pressure and vice versa. You test three different compounds, listed below. Predict the effect of each compound on blood pressure, and defend your answer.

   (Note: Agonists mimic the effect of the hormone, and antagonists block the effect of the hormone).
   a. an ADH agonist

   _____

   _____

   _____

   b. an aldosterone antagonist

   _____

   _____

   _____

c. an atrial natriuretic peptide (ANP) agonist

_____

_____

_____

d. an angiotensin II (ATII) antagonist

_____

_____

_____

Expanding Your Horizons

How much salt do you eat in a day? The average American consumes far more salt than the recommended 1,100 to 3,300 mg/day. One frozen entree, for instance, can contain 700 mg of salt or more! Track your salt intake over a few days, and see how you compare. Use the nutritional labels on prepared foods and this website below to gauge the sodium content of different foods.

- United States Department of Agriculture. USDA National Nutrient Database for Standard Reference, Release 26. Available at http://ndb.nal.usda.gov/ndb/nutrients/index

The Urinary System

▶ Overview

The urinary system comprises two **kidneys**, two **ureters**, one **urinary bladder**, and one **urethra**. This system is thought of as the body's main excretory mechanism; it is, in fact, often called the **excretory system**. The kidney, however, performs other essential functions; it maintains electrolyte and acid–base balance by controlling the amount and composition of urine, and it produces hormones. The kidneys also regulate blood pressure by producing **renin**, an enzyme that activates the protein **angiotensin**. The active form of this hormone (angiotensin II) increases blood pressure via actions on both the cardiovascular system and the renal system.

The functional unit of the kidney is the **nephron**, which consists of a small cluster of capillaries, the **glomerulus**, and the **renal tubule**. Blood enters the glomerulus from the **afferent arteriole** and leaves via the **efferent arteriole** and subsequently passes through the **peritubular capillaries** surrounding its nephron. Water and dissolved solutes filter from blood in the glomerulus into the proximal portion of the renal tubule, the **glomerular capsule**. The resulting fluid (the filtrate) passes through the rest of the renal tubule—the proximal tubule, nephron loop, and distal tubule—before entering the **collecting duct** and draining into the renal pelvis as **urine**. The filtrate is substantially modified as it passes through the renal tubule. Most of the water and solutes return to the bloodstream by the process of **reabsorption**; some solutes pass from peritubular blood into the tubule by the process of **secretion**. Finally, under the influence of antidiuretic hormone (ADH), the tubular fluid is concentrated just enough to maintain normal body fluid balance.

Prolonged or serious diseases of the kidney have devastating effects on overall body function and health. Renal dialysis and kidney transplantation are effective methods for saving the lives of people who otherwise would die of kidney failure and uremia.

Addressing the Learning Objectives

1. DESCRIBE THE ORGANS OF THE URINARY SYSTEM, AND GIVE THE FUNCTIONS OF EACH.

EXERCISE 22-1: Male Urinary System (Text Fig. 22-1)

1. Write the name of each structure in the line next to the bullet. (Hint: Structures 1 and 4 are veins.)
2. If you wish, color arteries red, veins blue, and structures encountering urine yellow.
3. Put arrows in the small boxes to indicate the direction of blood/urine movement.

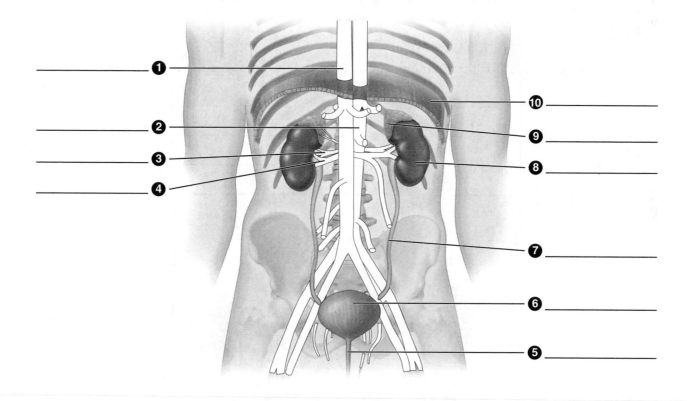

2. LIST FOUR SYSTEMS THAT ELIMINATE WASTE, AND NAME THE SUBSTANCES EACH ELIMINATES.

EXERCISE 22-2

Fill in the blank after each substance—is it eliminated by the urinary (U), digestive (D), respiratory (R), and/or integumentary (I) system(s)? There may be more than one answer for each substance.

1. Carbon dioxide   _____

2. Bile   _____

3. Water   _____

4. Salts   _____

5. Digestive residue   _____

6. Nitrogenous wastes   _____

3. LIST FOUR ACTIVITIES OF THE KIDNEYS IN MAINTAINING HOMEOSTASIS.

EXERCISE 22-3

In the spaces below, list four body parameters that the kidney helps maintain within tight limits.

1. _____

2. _____

3. _____

4. _____

4. DESCRIBE THE LOCATION AND INTERNAL ORGANIZATION OF THE KIDNEYS.

EXERCISE 22-4: Kidney Structure (Text Fig. 22-2, modified)

Write the name of each structure on the appropriate lines below.

1. _____
2. _____
3. _____
4. _____
5. _____
6. _____
7. _____
8. _____
9. _____
10. _____
11. _____

EXERCISE 22-5

Write the appropriate term in each blank from the list below. Not all terms will be used.

ureter	urinary bladder	urethra	renal capsule	renal cortex
retroperitoneal space	nephron	renal pelvis	adipose capsule	renal medulla

1. A funnel-shaped basin that drains into the ureter _____

2. The tube that carries urine from the bladder to the outside _____

3. The area behind the peritoneum that contains the ureters and the kidneys _____

4. A microscopic functional unit of the kidney _____

5. A tube connecting a kidney with the bladder _____

6. The crescent of fat that helps support the kidney _____

7. The inner region of the kidney _____

8. The outer region of the kidney _____

5. DESCRIBE A NEPHRON.

EXERCISE 22-6: A Cortical Nephron (Text Fig. 22-3A)

1. Label each structure by writing the appropriate term in each box. Use red for the only labeled structure that encounters blood.
2. Add arrows to the diagram indicating the direction of fluid flow.

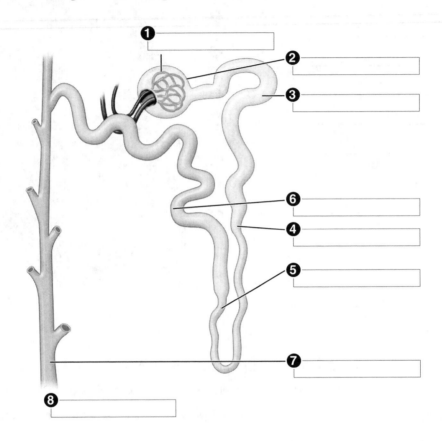

6. TRACE THE PATH OF A DROP OF BLOOD AS IT FLOWS THROUGH THE KIDNEY.

EXERCISE 22-7: Blood Supply to a Cortical Nephron (Text Fig. 22-3B)

1. Label each structure by writing the appropriate term in each box. Use red for structures that encounter arterial blood, purple for structures encountering mixed blood, and blue for veins.
2. Label the parts of the nephron.
3. Add arrows to the diagram to trace the path of blood as it flows through the kidney.

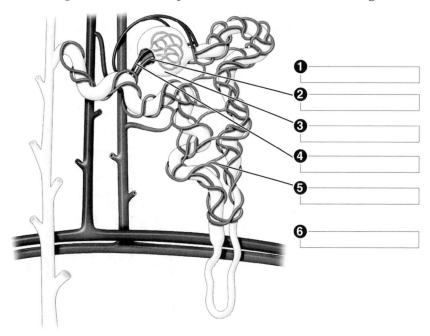

❶ _____
❷ _____
❸ _____
❹ _____
❺ _____

❻ _____

EXERCISE 22-8

Write the appropriate term in each blank from the list below. Not all terms will be used.

afferent arteriole	proximal tubule	peritubular capillaries	nephron loop
glomerulus	glomerular capsule	efferent arteriole	renal artery
renal vein	collecting duct		

1. A hollow bulb at the proximal end of the renal tubule _____

2. The blood vessels connecting the afferent and efferent arterioles _____

3. The portion of the nephron receiving filtrate from the glomerular capsule _____

4. The vessel that branches to form the glomerulus _____

5. The vessel that drains the kidney _____

6. The blood vessels that exchange substances with the nephron _____

7. A tube that receives urine from the distal tubule _____

8. The portion of the nephron that dips into the medulla _____

7. NAME THE FOUR PROCESSES INVOLVED IN URINE FORMATION, AND DESCRIBE WHAT HAPPENS DURING EACH.

EXERCISE 22-9: Glomerular Filtration (Text Fig. 22-5)

1. Write the names of the different structures on the appropriate numbered lines in different colors, and lightly shade the structures on the diagram. Use different shades of red for structures 1 to 3 and yellow for part 7. Do not color over the symbols. Some structures have appeared on earlier diagrams. You may want to use the same color scheme.
2. Color the symbols beside "soluble molecules," "proteins," and "blood cells" in different dark colors, and color the corresponding symbols on the diagram.

1. _____

2. _____

3. _____

4. _____

5. _____

6. _____

7. _____

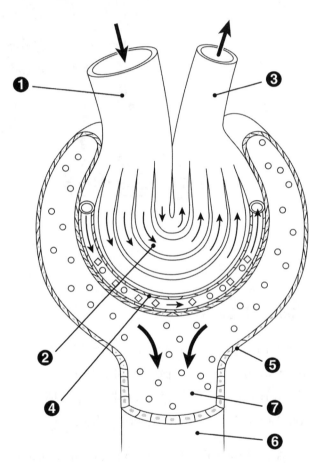

○ Soluble molecules
◇ Proteins
○ Blood cells

EXERCISE 22-10: Summary of Urine Formation (Text Fig. 22-7)

1. Label the structures and fluids by writing the appropriate terms on lines 5 to 11.
2. Write the name of the hormone that controls water reabsorption in the collecting duct on line 12.
3. Write and briefly describe the different kidney processes in lines 1 to 4.

1. _____

2. _____

3. _____

4. _____

5. _____

6. _____

7. _____

8. _____

9. _____

10. _____

11. _____

12. _____

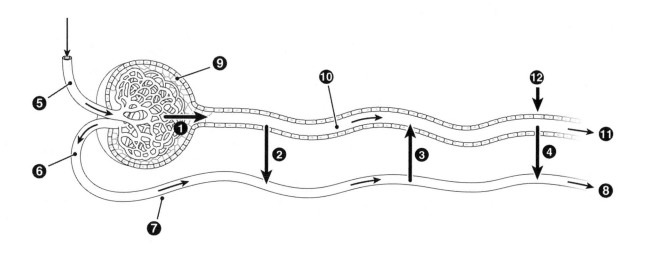

8. EXPLAIN THE ROLES OF JUXTAMEDULLARY NEPHRONS AND ANTIDIURETIC HORMONE (ADH) IN URINE FORMATION.

EXERCISE 22-11

Select the correct member of each term pair, and write it in the blank.

Specialized nephrons called (1) _____ (cortical/juxtamedullary) nephrons have especially long nephron loops. These nephrons establish a(n) (2) _____ (osmotic/pressure) gradient in the deeper portion of the kidney, known as the renal (3) _____ (cortex/medulla). This gradient is crucial for the final step of urine production, known as (4) _____ (reabsorption/concentration). As the filtrate travels to the renal pelvis via the (5) _____ (collecting duct/juxtamedullary apparatus), it encounters progressively more concentrated (6) _____ (interstitial fluid/cytosol). Water leaves the tubule and eventually enters the blood. However, water can only leave the tubule through specialized channels called (7) _____ (carriers/aquaporins). The hormone (8) _____ (aldosterone/ADH) stimulates the insertion of these channels into the tubule cells, so this hormone promotes the production of (9) _____ (concentrated/dilute) urine.

9. DESCRIBE THE COMPONENTS AND FUNCTIONS OF THE JUXTAGLOMERULAR (JG) APPARATUS.

EXERCISE 22-12: The Juxtaglomerular (JG) Apparatus (Text Fig. 22-8)

Write the names of the different structures on the appropriate numbered lines in different colors, and color the structures on the diagram. Use different shades of red for structures 1 to 3. (Hint: Blood flows from structure 1 toward structure 2.) Use a dark color for structure 8. Some structures have appeared on earlier diagrams. You may want to use the same color scheme.

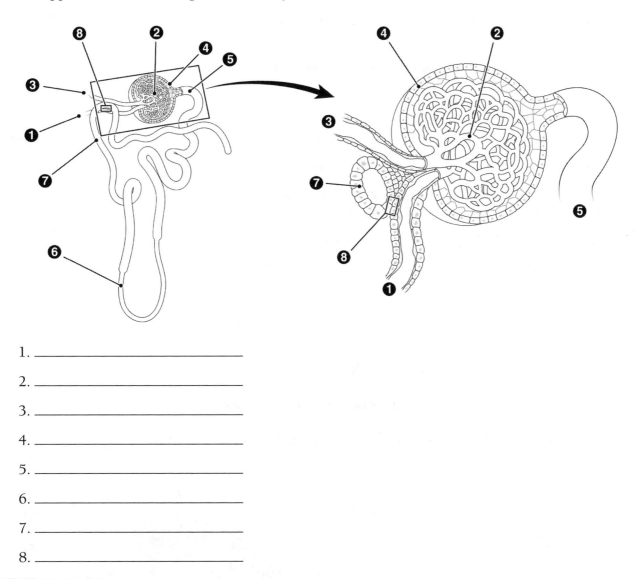

1. _____

2. _____

3. _____

4. _____

5. _____

6. _____

7. _____

8. _____

EXERCISE 22-13

Write the appropriate term in each blank from the list below. Not all terms will be used.

renin	juxtaglomerular apparatus	urea	EPO
filtration	tubular reabsorption	tubular secretion	antidiuretic hormone
angiotensin	aldosterone		

1. An enzyme produced by the kidney _____

2. The hormone that increases sodium reabsorption in the distal tubule _____

3. The process that returns useful substances in the filtrate to the bloodstream _____

4. The process by which substances leave the glomerulus and enter the glomerular capsule _____

5. The structure in the kidney that produces renin _____

6. The hormone produced in the kidney that stimulates erythrocyte production by the bone marrow _____

7. The process by which the renal tubule actively moves substances from the blood into the nephron to be excreted _____

8. The hormone that increases the permeability of the collecting duct to water _____

10. DESCRIBE THE PROCESS OF MICTURITION.

EXERCISE 22-14: The Male Urinary Bladder (Text Fig. 22-9)

Label the indicated parts. Hint: Bullet 3 labels "wrinkles" in the bladder wall.

1. _____
2. _____
3. _____
4. _____
5. _____
6. _____
7. _____
8. _____
9. _____
10. _____

Peritoneum

EXERCISE 22-15

Label each of the following statements as true (T) or false (F).

1. The internal urethral sphincter is formed of skeletal muscle.   _____

2. The external urethral sphincter is formed by the pelvic floor muscles.   _____

3. When the bladder fills with liquid, stretch receptors are activated.   _____

4. When the bladder fills with liquid, muscles in the bladder wall contract.   _____

5. Urination will occur when the external urethral sphincter is contracted.   _____

11. LIST NORMAL AND ABNORMAL CONSTITUENTS OF URINE.

EXERCISE 22-16

Fill in the blank after each term—is it a normal (N) or an abnormal (A) constituent of urine?

1. Glucose   _____

2. Blood   _____

3. Pigment   _____

4. Nitrogenous waste products   _____

5. Albumin   _____

6. Many casts   _____

7. White blood cells   _____

8. Electrolytes   _____

12. DISCUSS SIX TYPES OF URINARY SYSTEM DISORDERS.

EXERCISE 22-17

Write the appropriate term in each blank from the list below. Not all terms will be used.

specific gravity	glycosuria	albuminuria	hematuria
pyelonephritis	ureterocele	acute glomerulonephritis	hypospadias
hydronephrosis	white blood cells		

1. The presence of this material in the urine indicates pyuria   _____

2. An indication of the amount of dissolved substances in the urine   _____

3. Inflammation of the renal pelvis and the kidney tissue itself   _____

4. An inflammation of part of the nephron subsequent to a streptococcal infection   _____

5. The presence of an abundant blood protein in the urine   _____

6. The presence of blood in the urine   _____

7. Accumulation of fluid in the renal pelvis and calyces   _____

8. A condition in which the urethral opening is found on the underside of the penis or on the scrotum   _____

EXERCISE 22-18

Write the appropriate term in each blank from the list below. Not all terms will be used.

nocturia	acute renal failure	polycystic kidney disease	chronic renal failure
renal calculi	polyuria	cystitis	urethritis
cystoscopy	cystectomy	lithotripsy	

1. A gradual loss of nephrons   _____

2. Elimination of very large amounts of urine   _____

3. Kidney stones; solids formed when uric acid or calcium salts precipitate out of the urine   _____

4. A sudden decrease in kidney function resulting from a medical or surgical emergency   _____

5. Elimination of urine during the night   _____

6. Removal of the bladder   _____

7. The use of an endoscope to examine the bladder internally   _____

8. Inflammation of the urinary bladder   _____

9. Use of shock waves to shatter kidney stones   _____

10. Genetic disorder resulting in fluid-filled sacs in the kidney   _____

13. LIST SIX SIGNS OF CHRONIC RENAL FAILURE.

EXERCISE 22-19

List and define the six most important signs and symptoms of kidney failure in the spaces below.

1. _____
2. _____
3. _____
4. _____
5. _____
6. _____

14. EXPLAIN THE PRINCIPLE AND PURPOSE OF RENAL DIALYSIS.

EXERCISE 22-20

Label each of the following statements as true (T) or false (F).

1. Dialysis separates dissolved molecules based on their ability to pass through a semipermeable membrane. _____

2. During dialysis, substances move from the region of low concentration to the region of high concentration. _____

3. In hemodialysis, dialysis fluid is pumped into the peritoneal space. _____

4. In dialysis, accumulated waste products move from the blood into the dialysis solution. _____

15. REFERRING TO THE CASE STUDY, DESCRIBE HOW URETHRAL BLOCKAGE CAN AFFECT KIDNEY FUNCTION.

EXERCISE 22-21

Fill in the blanks in the following description using your textbook.

Adam was suffering from the enlargement of his (1) _____ gland, a doughnut-shaped structure inferior to his urinary bladder. When Adam attempted to urinate, a process described as (2) _____, he voluntarily relaxed his (3) _____ _____ sphincter. However, urine did not readily flow because his enlarged gland restricted flow through the (4) _____. He was unable to void his urinary bladder, and urine backed up and distended the (5) _____, which carry urine to the bladder. This distention is called (6) _____. Urine further backed up into his kidneys, causing (7) _____, or "water in the kidneys." Adam's urologist diagnosed his enlarged prostate using an instrument called a(n) (8) _____ and removed pieces of his prostate using an instrument called a(n) (9) _____.

16. SHOW HOW WORD PARTS ARE USED TO BUILD WORDS RELATED TO THE URINARY SYSTEM.

EXERCISE 22-22

Complete the following table by writing the correct word part or meaning in the space provided. Write a word that contains each word part in the "Example" column.

Word Part	Meaning	Example
1. _____	night	_____
2. dia-	_____	_____
3. _____	renal pelvis	_____
4. trans-	_____	_____
5. nephr/o	_____	_____
6. _____	sac, bladder	_____
7. _____	backward, behind	_____
8. _____	next to	_____
9. ren/o	_____	_____
10. -cele	_____	_____

Making the Connections

The following concept map deals with the organization of the kidney. Each pair of terms is linked together by a connecting phrase into a sentence. Complete the concept map by filling in the appropriate term or phrase. There is one right answer for each term. However, there are many correct answers for the connecting phrases (2, 3, 6). Bullets 12 through 15 refer to the four processes of urine formation. Figure out which bullet refers to which process.

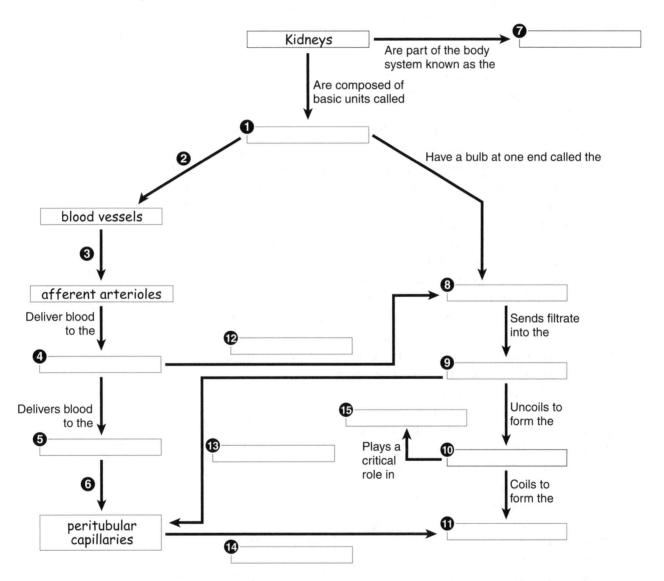

Optional Exercise: Make your own concept map based on the urinary system. Use the following terms and any others you would like to include: urinary system, kidneys, nephrons, ureters, urinary bladder, urethra, adipose capsule, renal capsule, renal pelvis, rugae, trigone, urinary meatus, penis, micturition, internal urethral sphincter, and external urethral sphincter.

Testing Your Knowledge

BUILDING UNDERSTANDING

I. MULTIPLE CHOICE

Select the best answer.

1. What is the correct order of urine flow from its source to the outside of the body?
   a. urethra, bladder, kidney, ureter
   b. bladder, kidney, urethra, ureter
   c. kidney, ureter, bladder, urethra
   d. kidney, urethra, bladder, ureter

   1. _____

2. What is the correct order of filtrate flow through the nephron?
   a. nephron loop, distal tubule, proximal tubule, collecting duct
   b. glomerular capsule, proximal tubule, nephron loop, distal tubule
   c. proximal tubule, distal tubule, nephron loop, glomerular capsule
   d. glomerular capsule, distal tubule, proximal tubule, collecting duct

   2. _____

3. How does ADH promote urine concentration?
   a. It stimulates sodium reabsorption in the proximal tubule, and water follows the sodium.
   b. It stimulates the insertion of water channels into the cells of the collecting duct.
   c. It stimulates the production of the medullary gradient.
   d. It stimulates renin production.

   3. _____

4. The juxtaglomerular apparatus consists of cells in what areas?
   a. proximal tubule and efferent arteriole
   b. renal artery and afferent arteriole
   c. collecting tubules and renal vein
   d. distal tubule and afferent arteriole

   4. _____

5. Which of the following is NOT a function of the kidneys?
   a. red blood cell destruction
   b. blood pressure regulation
   c. elimination of nitrogenous wastes
   d. modification of body fluid composition

   5. _____

6. Which of these processes moves substances from the distal tubule to the peritubular capillaries?
   a. secretion
   b. filtration
   c. reabsorption
   d. excretion

   6. _____

7. Which of these substances is an *abnormal* urine constituent?
   a. sodium
   b. hydrogen
   c. albumin
   d. urea

   7. _____

8. Which of these terms describes the process of expelling urine?　8. _____
    a. reabsorption
    b. micturition
    c. dehydration
    d. defecation
9. Which of these processes moves substances down a pressure gradient?　9. _____
    a. tubular secretion
    b. glomerular filtration
    c. tubular reabsorption
    d. urine concentration
10. Which of these cellular transport mechanisms allows substances to leave the patient's blood during dialysis?　10. _____
    a. active transport
    b. diffusion
    c. filtration
    d. all of the above

II. COMPLETION EXERCISE

Write the word or phrase that correctly completes each sentence.

1. The congenital anomaly in which the ureteral openings bulge into the bladder is known as _____.

2. One indication of nephritis is the presence in the urine of solid materials molded in the kidney tubules. They are called _____.

3. The form of bladder inflammation that is not infectious and is found exclusively in women is called _____.

4. When the bladder is empty, its lining is thrown into the folds known as _____.

5. The vessel that brings blood to the glomerulus is the _____.

6. Large kidney stones that fill the renal pelvis and extend into the calyces are called _____.

7. The distal tubule empties its fluid into the _____.

8. The device that employs shock waves to destroy kidney stones is called a(n) _____.

9. The kidney process that always requires active transport is _____.

10. The capillaries involved in filtration are called the _____.

11. The capillaries that surround the nephron are called the _____.

UNDERSTANDING CONCEPTS

I. TRUE/FALSE

For each statement, write T for true or F for false in the blank to the left of each number. If a statement is false, correct it by replacing the underlined term, and write the correct statement in the blanks below the statement.

_____ 1. Polyuria could be caused by a deficiency of the hypothalamic hormone underlined{erythropoietin}.

_____

_____ 2. The movement of hydrogen ions from the peritubular capillaries into the distal tubule is an example of underlined{tubular secretion}.

_____

_____ 3. The type of dialysis that does NOT use dialysis tubing is called underlined{hemodialysis}.

_____

_____ 4. Urine with a specific gravity of 0.05 is underlined{more} concentrated than urine with a specific gravity of 0.04.

_____

_____ 5. The portion of the nephron called the underlined{juxtaglomerular apparatus} is responsible for creating the medullary gradient.

_____

_____ 6. Voluntary control of urination involves the underlined{internal} urethral sphincter.

_____

_____ 7. The triangular-shaped region in the floor of the bladder is called the underlined{renal pelvis}.

_____

_____ 8. Most water loss occurs through the actions of the underlined{digestive} system.

_____

_____ 9. The presence of glucose in the urine, known as underlined{sugaruria}, is indicative of diabetes mellitus.

_____

II. PRACTICAL APPLICATIONS

Study each discussion. Then write the appropriate word or phrase in the space provided.

➤ Group A

1. K, aged 11, was brought to the emergency room by her mother, Ms. L, because her urine was red. She had not consumed beets recently, so the nurse suspected that the red coloration was due to the presence of blood in the urine, a condition called _____.

2. The presence of blood was confirmed by a laboratory study of her urine. This test is called a(n) _____.

3. Ms. L mentioned that K had recently recovered from "strep throat." Tests indicated that K was suffering from the most common disease of the kidneys, namely _____.

4. Ms. L was quite concerned by this diagnosis because she had had several similar episodes during her own childhood. She was assured that her daughter would recover with appropriate treatment but was advised to have her own urine tested as a precaution. The tests demonstrated protein in Ms. L's urine, consistent with damage to the kidney tubules. The presence of protein in the urine is called _____.

5. Urinary protein is indicative of a problem with the first process involved in urine formation. This process creates tubular fluid from blood and is called _____.

6. In particular, there has been damage to the walls of capillary bundle in the nephron. This capillary bundle is called the _____.

➤ Group B

1. Young S, age 1, was brought to the hospital with an enlarged abdomen. His blood pressure was shown to be extremely high, a disorder called _____.

2. High blood pressure can result from increased renin production. Renin is produced by a specialized region of the kidney called the _____.

3. Tests showed that S was suffering from a slow, progressive, irreversible loss of nephrons. This disease is called _____.

4. By the age of 5, S's kidneys were no longer functioning to any appreciable degree. S must visit a clinic three times a week, where excess waste products are removed from his blood by diffusion. This process is called _____.

5. Two years later, S received a kidney from his mother. The process by which an organ is transferred from one individual to another is called _____.

III. SHORT ESSAYS

1. Name five signs of chronic renal disease. Explain why each sign indicates the kidney is no longer functioning properly.

   _____

   _____

   _____

   _____

   _____

2. Mr. B has uncontrolled diabetes mellitus. Name two abnormal constituents you would expect to find in his urine, and explain why each would be present.

   _____

   _____

   _____

   _____

   _____

CONCEPTUAL THINKING

1. Ms. W has just eaten a large bag of salty popcorn. Her blood is now too salty and must be diluted. Is it possible for the kidney to increase water reabsorption without increasing salt absorption? Explain.

   _____

   _____

   _____

   _____

2. Mr. R is taking penicillin to cure a throat infection. He must take the drug frequently because the kidney clears penicillin very efficiently. That is, all of the penicillin that enters the renal artery leaves the kidney in the urine. However, only some penicillin molecules will be filtered into the glomerular capsule. How can the kidney excrete all of the penicillin it receives?

   _____

   _____

   _____

   _____

3. Mrs. J was recently diagnosed with cystitis, urethritis, and pyelonephritis. She bemoans her bad luck at getting three separate infections at once. What would you tell her?

_____

_____

_____

_____

_____

Expanding Your Horizons

Renal failure is a very common disorder because the kidneys are sensitive to toxins and physical trauma. Patients with chronic renal failure face a life of illness and frequent dialysis unless they can receive a kidney transplant. Kidney transplants are quite common (over 15,000 were performed in 2001 in the United States alone). In some cases, the patient receives a kidney from a live, related donor or an acceptable tissue match. You can watch videos, read patient stories, or learn about the risks of live related kidney donations on the Internet. Do a search for "live kidney transplant" or consult the websites listed below.

- National Kidney Foundation. Living Donation. Available at: http://www.kidney.org/Transplantation/LivingDonors/index.cfm.
- Sundaram C. Kidney Transplant Donor Surgery. Available at: http://www.youtube.com/watch?v=ou8CC4XN9wk.
- Sentara Health Care. Living Donor Kidney Transplant Surgery. Available at: http://www.orlive.com/sentara-profile/videos/living-donor-kidney-transplant-surgery1.

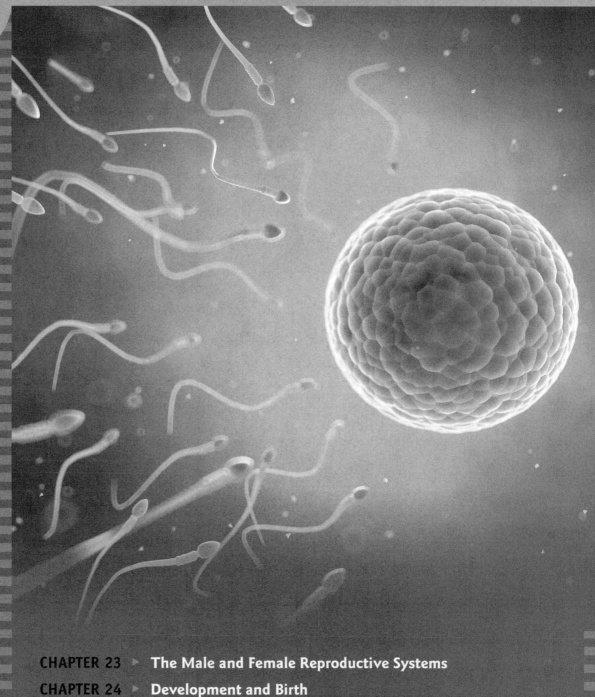

UNIT VII

Perpetuation of Life

The Male and Female Reproductive Systems

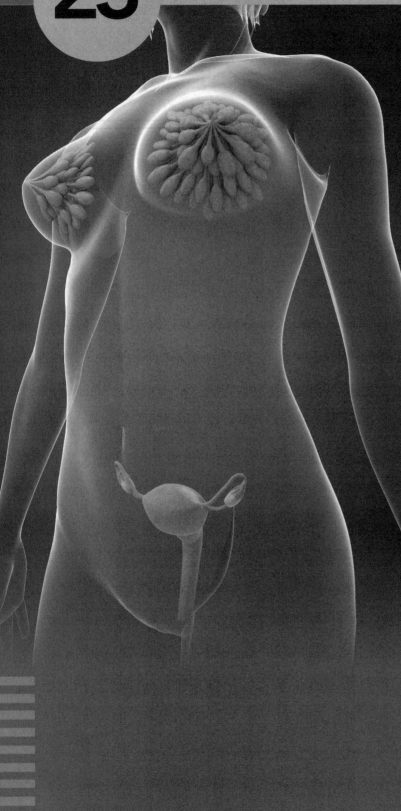

▶ Overview

Reproduction is the process by which life continues. Human reproduction is **sexual**; that is, it requires the union of two different *germ cells* or **gametes**. (Some simple forms of life can reproduce without a partner in the process of **asexual** reproduction.) These germ cells, the **spermatozoon** in males and the **ovum** in females, are formed by **meiosis**, a type of cell division in which the chromosome number is reduced by half. When fertilization occurs and the gametes combine, the original chromosome number is restored.

The **gonads** manufacture the gametes and also produce hormones. These activities are continuous in the male but cyclic in the female. The male gonad is the **testis**. The remainder of the male reproductive tract consists of passageways for storage and transport of spermatozoa; the male organ of copulation, the **penis**; and several glands that contribute to the production of **semen**.

The female gonad is the **ovary**. The ovum released each month at the time of **ovulation** travels through the **uterine tubes** to the **uterus**, where the ovum, if fertilized, develops. If no fertilization occurs, the ovum, along with the built-up lining of the uterus, is eliminated through the **vagina** as the **menstrual flow**.

Reproduction is under the control of hormones from the **anterior pituitary**, which in turn is controlled by the **hypothalamus** of the brain. These organs respond to **feedback** mechanisms, which maintain proper hormone levels.

Aging causes changes in both the male and female reproductive systems.

A gradual decrease in male hormone production begins as early as age 20 and continues throughout life. In the female, a more sudden decrease in activity occurs between ages 45 and 55 and ends in **menopause**, the cessation of menstruation and of the childbearing years.

The reproductive tract is subject to infection and inflammation, including **sexually transmitted infections (STIs)** and inflammation secondary to other infections such as mumps. Infections, congenital malformations, and hormonal imbalances can cause **infertility** or **sterility**.

Addressing the Learning Objectives

1. IDENTIFY THE MALE AND FEMALE GAMETES, AND STATE THE PURPOSE OF MEIOSIS.

EXERCISE 23-1

Label each of the following statements as true (T) or false (F). Exercise 23-4 discusses the male and female gametes.

1. Meiosis occurs only in germ cells.   _____

2. The cells formed by meiosis have the same number
   of chromosomes as does the parent cell.   _____

2. NAME THE ACCESSORY ORGANS AND GONADS OF THE MALE REPRODUCTIVE SYSTEM, AND CITE THE FUNCTION OF EACH.

Also see Exercise 23-13.

EXERCISE 23-2: Male Reproductive System (Text Fig. 23-1)

1. Write the name of each structure on the appropriate lines in different colors. Use the same color for structures 5 and 6. Use black for structures 7 and 12, because they will not be colored.
2. Write the names of the parts of the male reproductive system on the appropriate lines 7 to 19 in different colors. Use black for structure 7, because it will not be colored. Use black for structures 7 and 12, because they will not be colored.

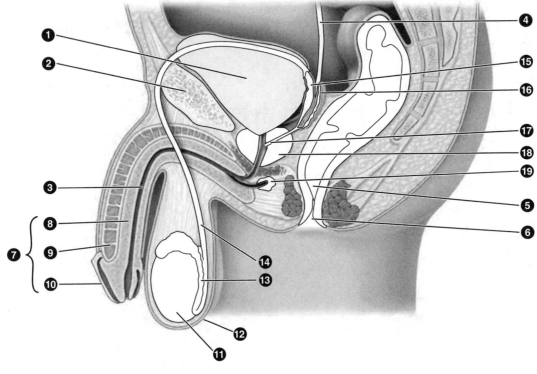

1. _____

2. _____

3. _____

4. _____

5. _____

6. _____

7. _____

8. _____

9. _____

10. _____

11. _____

12. _____

13. _____

14. _____

15. _____

16. _____

17. _____

18. _____

19. _____

EXERCISE 23-3: Cross-Section of the Penis (Text Fig. 23-2)

1. Write the names of the parts on the appropriate lines in different colors. When possible, use the same colors as in Exercise 23-2. (Hint: the nerve is solid, and the veins have larger lumens).
2. Color the structures on the diagram.

1. _____

2. _____

3. _____

4. _____

5. _____

6. _____

7. _____

8. _____

9. _____

10. _____

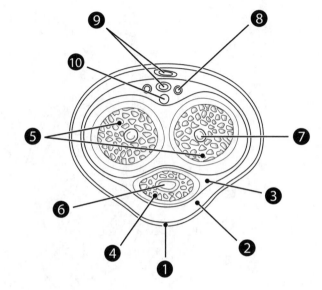

EXERCISE 23-4: Structure of the Testis (Text Fig. 23-3A and B)

1. Write the names of the parts on the appropriate lines in different colors. Use black for structures 1 and 4 because they will not be colored. Use the same color for structures 3 and 12 and related colors for structures 8 through 10.
2. Color the structures on the diagram. Label the indicated parts.

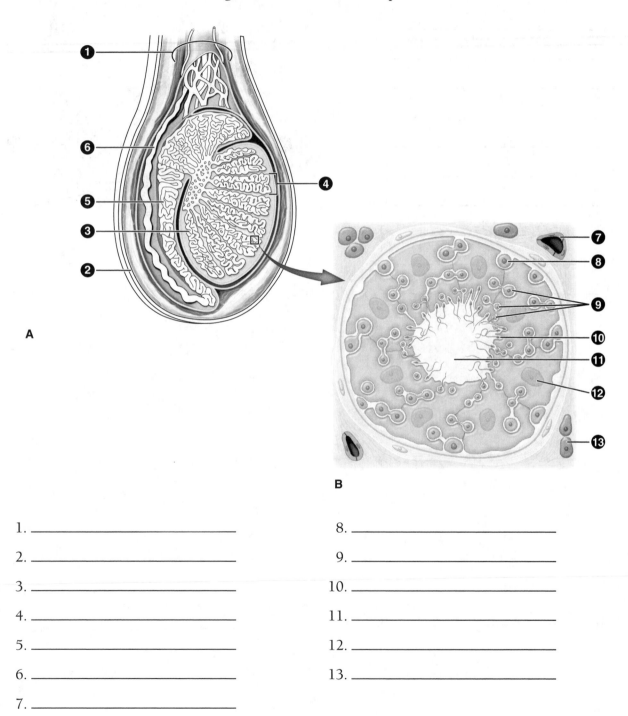

A

B

1. _____

2. _____

3. _____

4. _____

5. _____

6. _____

7. _____

8. _____

9. _____

10. _____

11. _____

12. _____

13. _____

EXERCISE 23-5

Write the appropriate term in each blank from the list below.

interstitial cell   sustentacular cell   spermatozoon

corpus spongiosum   corpus cavernosum   prepuce

1. A cell that nourishes and protects developing spermatozoa   _____

2. A cell that secretes testosterone   _____

3. The mature male gamete   _____

4. The portion of the penis containing the urethra   _____

5. This structure is removed by circumcision   _____

6. The penis contains two of this structure   _____

3. DESCRIBE THE COMPOSITION AND FUNCTION OF SEMEN.

EXERCISE 23-6

In the spaces below, list five functions of semen.

1. _____

2. _____

3. _____

4. _____

5. _____

4. DRAW AND LABEL A SPERMATOZOON.

EXERCISE 23-7: Diagram of a Human Spermatozoon (Text Fig. 23-4)

1. Write the names of the parts on the appropriate lines in different colors. Use black for structures 1 to 3 because they will not be colored.
2. Color the structures on the diagram.

1. _____

2. _____

3. _____

4. _____

5. _____

6. _____

7. _____

5. IDENTIFY THE TWO HORMONES THAT REGULATE THE PRODUCTION AND DEVELOPMENT OF THE MALE GAMETES.

EXERCISE 23-8

Write the name of the appropriate hormone in each blank. Each hormone may be used more than once. Note that questions 1 and 5 have *two* blanks because the statement applies to *two* hormones.

testosterone   **luteinizing hormone**   **follicle-stimulating hormone**

1. Produced by the anterior pituitary gland   _____

_____

2. Produced by the testicular interstitial cells   _____

3. Stimulates testicular interstitial cells   _____

4. Stimulates testosterone production   _____

5. Stimulates the formation and development of spermatozoa   _____

_____

6. Directly stimulates the development of secondary sex characteristics   _____

7. Stimulates sustentacular cells to produce growth factors   _____

6. DISCUSS THREE TYPES OF MALE REPRODUCTIVE SYSTEM DISORDERS, AND GIVE EXAMPLES OF EACH.

(Note that sexually transmitted infections, which apply equally to males and females, are discussed under Objective 11.)

EXERCISE 23-9

Write the appropriate term in each blank from the list below. Not all terms will be used.

cryptorchidism   phimosis   prostatitis   orchitis

epididymitis   torsion   BPH

1. The condition that results when the testes fail to descend into the scrotal sac during fetal life   _____

2. A noncancerous enlargement of the prostate gland   _____

3. A twisting of the spermatic cord   _____

4. Inflammation of the testis   _____

5. Tightness of the foreskin, which prevents it from being drawn back   _____

6. Condition resulting from infection of the prostate gland   _____

7. NAME THE ACCESSORY ORGANS AND GONADS OF THE FEMALE REPRODUCTIVE SYSTEM, AND CITE THE FUNCTION OF EACH.

EXERCISE 23-10: Female Reproductive System (Sagittal View) (Text Fig. 23-8A)

Label the indicated parts.

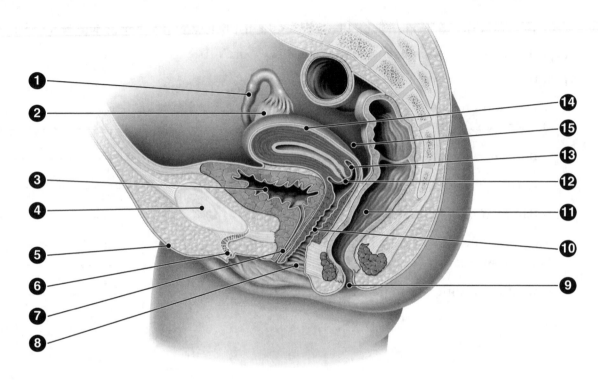

Sagittal view

1. _____   10. _____

2. _____   11. _____

3. _____   12. _____

4. _____   13. _____

5. _____   14. _____

6. _____   15. _____

7. _____

8. _____

9. _____

EXERCISE 23-11: Female Reproductive System (Frontal View) (Text Fig. 23-8B)

Label the indicated parts.

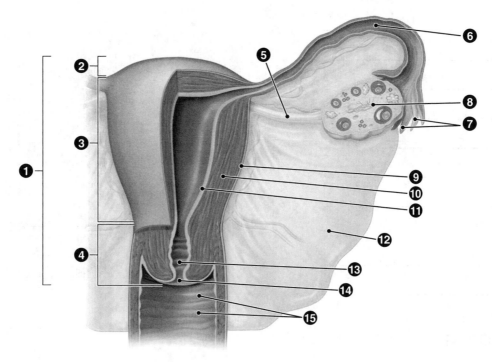

Frontal view

1. _____

2. _____

3. _____

4. _____

5. _____

6. _____

7. _____

8. _____

9. _____

10. _____

11. _____

12. _____

13. _____

14. _____

15. _____

EXERCISE 23-12: External Parts of the Female Reproductive System (Text Fig. 23-9)

Label the indicated parts.

1. _____

2. _____

3. _____

4. _____

5. _____

6. _____

7. _____

8. _____

9. _____

10. _____

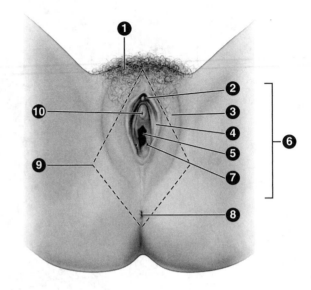

EXERCISE 23-13

Write the appropriate term in each blank from the list below. Not all terms will be used. Note that this exercise discusses both male and female structures.

epididymis   vas deferens   seminal vesicle   prostate gland   ejaculatory duct

fimbriae   uterine tube   bulbourethral gland   greater vestibular gland   seminiferous tubule

1. One of the two glands that secrete a thick, yellow, alkaline secretion _____

2. The gland that secretes a thin alkaline secretion and that can contract to aid in ejaculation _____

3. The coiled tube in which spermatozoa are stored as they mature and become motile _____

4. The tube that transports the female germ cells _____

5. Fringelike extensions that sweep the ovum into the uterine tube _____

6. A gland in the female reproductive tract that secretes into the vagina _____

7. A duct in the male that empties into the urethra _____

8. A gland located inferior to the prostate that secretes mucus during sexual stimulation _____

EXERCISE 23-14

Write the appropriate term in each blank from the list below

cervix   myometrium   fundus   rectouterine pouch

perineum   endometrium   posterior fornix hymen

1. The most inferior portion of the peritoneal cavity   _____

2. The vaginal region around the cervix   _____

3. The specialized tissue that lines the uterus   _____

4. The small, rounded part of the uterus located above the openings of the uterine tubes   _____

5. A fold of membrane found at the opening of the vagina   _____

6. The necklike part of the uterus that dips into the upper vagina   _____

7. The muscular wall of the uterus   _____

8. The pelvic floor, especially the region between the vaginal opening and the anus   _____

8. IN THE CORRECT ORDER, LIST THE HORMONES PRODUCED DURING THE MENSTRUAL CYCLE, CITING THE SOURCE AND FUNCTION OF EACH.

EXERCISE 23-15: The Menstrual Cycle (Text Fig. 23-11)

1. Name the three phases of the female reproductive cycle (bullets 1 to 3) and the uterine cycle (bullets 11 to 13).
2. Identify the hormone responsible for each of the numbered lines (bullets 4, 5, 9, 10), and write the hormone names on the appropriate lines in different colors. Trace over the lines in the graph with the corresponding colors.
3. Identify structures 6 through 8 by writing the appropriate term on each line.

1. _____

2. _____

3. _____

4. _____

5. _____

6. _____

7. _____

8. _____

9. _____

10. _____

11. _____

12. _____

13. _____

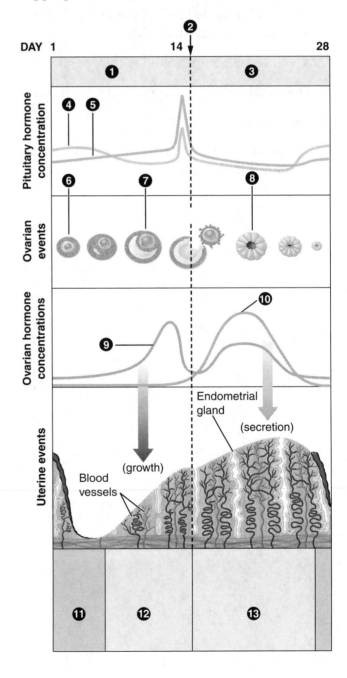

EXERCISE 23-16

Write the appropriate term in each blank from the list below. Not all terms will be used.

luteinizing hormone follicle-stimulating hormone estrogen progesterone

corpus luteum ovulation menstruation

1. A hormone that is produced only during the postovulatory phase of the menstrual cycle _____

2. An ovarian hormone that is produced during the preovulatory and postovulatory phases of the menstrual cycle _____

3. Discharge of an ovum from the surface of the ovary _____

4. The hormone that stimulates the development of the cells surrounding the ovum _____

5. The structure formed by the ruptured follicle after ovulation _____

6. The phase of the menstrual cycle when the endometrium is degenerating _____

9. DESCRIBE THE CHANGES THAT OCCUR DURING AND AFTER MENOPAUSE.

EXERCISE 23-17

In the spaces below, write five different changes that may be associated with menopause.

1. _____

2. _____

3. _____

4. _____

5. _____

10. CITE THE MAIN METHODS OF BIRTH CONTROL IN USE.

EXERCISE 23-18

Write the appropriate term in each blank from the list below. Not all terms will be used.

IUD   birth control patch   birth control ring   vasectomy

tubal ligation   condom   diaphragm

1. A method used to administer estrogen and progesterone through the skin   _____

2. A device implanted into the uterus that prevents fertilization and implantation   _____

3. A rubber cap fitted over the cervix   _____

4. The birth control method that is also highly effective against sexually transmitted infections   _____

5. A method used to administer birth control hormones internally   _____

6. A surgical method of birth control in females   _____

11. DISCUSS THREE TYPES OF FEMALE REPRODUCTIVE SYSTEM DISORDERS, AND GIVE EXAMPLES OF EACH.

EXERCISE 23-19

Write the appropriate term in each blank from the list below.

salpingitis   dysmenorrhea   fibroid

hysterectomy   premenstrual syndrome   amenorrhea

1. Depression, nervousness, and irritability preceding menses   _____

2. This term, describing a complete lack of menstrual flow, is usually associated with infertility   _____

3. An alternate name for myoma   _____

4. Surgical removal of the uterus   _____

5. Painful or difficult menstruation   _____

6. Inflammatory infection of a uterine tube   _____

12. USING THE TEXT AND INFORMATION IN THE CASE STUDY, DISCUSS POSSIBLE CAUSES OF INFERTILITY IN MEN AND WOMEN.

EXERCISE 23-20

Write the appropriate term in each blank from the list below. Not all terms will be used. Each term describes a possible cause of infertility.

endometriosis   PID   oligospermia   syphilis

gonorrhea   estrogen   testosterone

1. A deficiency in the number of sperm cells in the semen   _____

2. A sexually transmitted disease that can cause blockages in the male and female ductal systems   _____

3. Inflammation of the pelvic cavity resulting from untreated infections of the reproductive tract   _____

4. A deficiency in this hormone can cause male infertility   _____

5. A deficiency in this hormone can cause female infertility   _____

6. A condition in which the tissue normally lining the uterus grows in other sites, such as the external surface of the ovary, resulting in inflammation and sometimes infertility   _____

13. SHOW HOW WORD PARTS ARE USED TO BUILD WORDS RELATED TO THE REPRODUCTIVE SYSTEMS.

EXERCISE 23-21

Complete the following table by writing the correct word part or meaning in the space provided. Write a word that contains each word part in the "Example" column.

Word Part	Meaning	Example
1. _____	extreme end	_____
2. hyster/o	_____	_____
3. _____	egg	_____
4. metr/o	_____	_____
5. fer	_____	_____
6. test/o	_____	_____
7. _____	few, deficiency	_____
8. orchid/o, orchi/o	_____	_____
9. crypt/o-	_____	_____
10. salping/o	_____	_____

Making the Connections

The following concept map deals with the female reproductive system. Each pair of terms is linked together by a connecting phrase. Complete the concept map by filling in the appropriate term or phrase. There is one right answer for each term (1, 3, 5, 10). However, there are many correct answers for the connecting phrases.

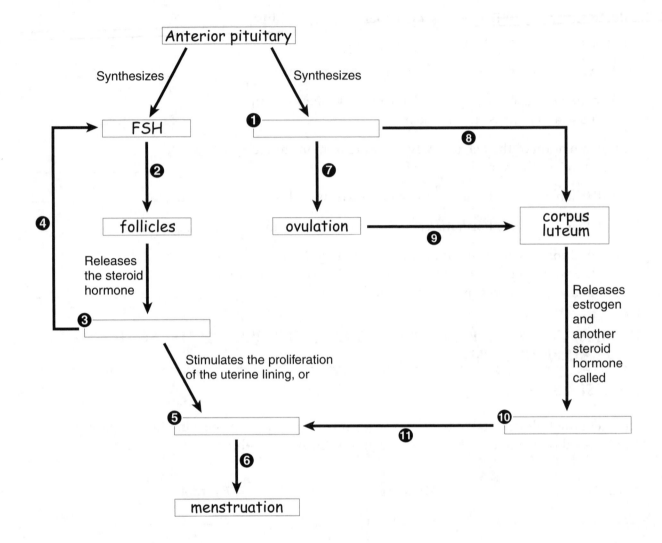

Optional Exercise: Make your own concept map based on male reproductive anatomy and physiology. Use the following terms and any others you would like to include: spermatozoon, testosterone, testis, FSH, LH, interstitial cell, epididymis, ductus deferens, seminiferous tubule, sustentacular cell, prostate gland, bulbourethral gland, and penis.

Testing Your Knowledge

BUILDING UNDERSTANDING

I. MULTIPLE CHOICE

Select the best answer.

1. Which of these hormones is secreted by the corpus luteum
   in large amounts?   1. _____
   a. testosterone
   b. progesterone
   c. FSH
   d. LH
2. Which of the following is NOT part of the uterus?   2. _____
   a. fimbriae
   b. cervix
   c. endometrium
   d. corpus
3. In the walls of which structure do spermatozoa develop?   3. _____
   a. prostate gland
   b. penis
   c. seminal vesicles
   d. seminiferous tubules
4. Which of these organs develops more malignancies than does any
   other male organ?   4. _____
   a. testis
   b. prostate
   c. urethra
   d. seminal vesicle
5. Which of these structures forms the glans penis?   5. _____
   a. corpus spongiosum
   b. corpus cavernosum
   c. pubic symphysis
   d. vas deferens
6. Which of these infectious agents causes genital warts?   6. _____
   a. *Rickettsia*
   b. human papillomavirus
   c. rubella
   d. chlamydia
7. Which of these structures houses the developing ovum?   7. _____
   a. corpus luteum
   b. sustentacular cells
   c. interstitial cells
   d. ovarian follicle

8. Which of these conditions can be diagnosed by a Pap smear?   8. _____
   a. pregnancy
   b. breast cancer
   c. cervical cancer
   d. infertility

9. Which of the following glands is found in the female?   9. _____
   a. greater vestibular glands
   b. seminal vesicles
   c. Cowper glands
   d. prostate

10. Which of these substances is normally absent from semen?   10. _____
    a. seminal fluid
    b. sugar
    c. a high concentration of hydrogen ions
    d. sperm

II. COMPLETION EXERCISE

Write the word or phrase that correctly completes each sentence.

1. The enzyme-containing cap on the head of a spermatozoon is called the
   _____.

2. The hormone that promotes development of spermatozoa in the male and development of ova in the female is _____.

3. The process of cell division that reduces the chromosome number by half is
   _____.

4. The testes are contained in an external sac called the _____.

5. The main male sex hormone is _____.

6. The pelvic floor in both males and females is called the _____.

7. The hormone that stimulates ovulation is _____.

8. In the male, the tube that carries urine away from the bladder also carries sperm cells. This tube is the _____.

9. The virus linked to cervical cancer is called _____.

10. The surgical removal of the foreskin of the penis is called _____.

UNDERSTANDING CONCEPTS

I. TRUE/FALSE

For each statement, write T for true or F for false in the blank to the left of each number. If a statement is false, correct it by replacing the underlined term, and write the correct statement in the blanks below the statement.

_____ 1. A low ability to reproduce is called <u>sterility</u>.

_____

_____ 2. The urethra is contained in the <u>corpus spongiosum</u> of the penis.

_____

_____ 3. The most superior region of the uterus is the <u>fundus</u>.

_____

_____ 4. Gametes are produced by the process of <u>mitosis</u>.

_____

_____ 5. The peritoneal folds that separate the female pelvis into anterior and posterior portions are called the <u>broad ligaments</u>.

_____

_____ 6. Progesterone levels are highest during the <u>preovulatory phase</u> of the menstrual cycle.

_____

_____ 7. The uterine phase following menstruation is the <u>secretory phase</u>.

_____

_____ 8. A deficiency in <u>luteinizing hormone</u> would result in testosterone deficiency.

_____

II. PRACTICAL APPLICATIONS

Study each discussion. Then write the appropriate word or phrase in the space provided.

➤ **Group A**

1. Ms. J, age 22, complained of painful menstrual cramps and excessive menstrual flow. The medical term for painful menstruation is _____.

2. Menstruation is the shedding of the uterine layer called the _____.

3. The physician suggested an antiinflammatory medication, but Ms. J had previously tried medication to no avail. An alternate approach was tried in which the passageway between the vagina and the uterus was artificially dilated. The part of the uterus containing this passageway is the

_____.

4. The excessive bleeding was also a disturbing symptom. Ultrasound revealed the presence of numerous small, benign uterine tumors. These tumors are called myomas or _____.

5. The tumors were removed by a special instrument that passes through the vagina into the uterine cavity. This instrument is called a(n) _____.

6. Unfortunately, the myomas recurred. Eventually, Ms. J chose to have her uterus removed, a procedure described as a(n) _____.

➤ Group B

1. Mr. and Ms. S, both age 35, had been trying to conceive for two years with no success. Analysis of Mr. S's semen revealed a low sperm count. This condition is called _____.

2. The fertility specialist asked Mr. S about any illnesses or infections that might be responsible for his low sperm count. Mr. S mentioned that his testes were very swollen when he had a mumps infection at the age of 17. Inflammation of the testis is called _____.

3. Since Mr. S's testes were of normal size, the mumps infection was probably not the cause of his infertility. Blood tests were performed, revealing a deficiency in the pituitary hormone that acts on sustentacular cells. This hormone is called _____.

4. Conversely, testosterone production was normal. The cells that produce testosterone are called _____.

5. Mr. S's hormone deficiency was successfully treated, and Mr. and Ms. S became the proud parents of triplets. Mr. S went to the clinic shortly after the birth, and his ductus deferens was cut bilaterally. This method of birth control is called a(n) _____.

III. SHORT ESSAYS

1. Discuss changes occurring in the ovary and in the uterus during the preovulatory phase of the menstrual cycle.

_____

_____

_____

2. Name the microorganisms that commonly cause infections of the reproductive tract in males and females.

_____

_____

_____

CONCEPTUAL THINKING

1. Mr. S is taking synthetic testosterone in order to improve his wrestling performance. To his alarm, he notices that his testicles are shrinking. Explain why this is happening.

_____

_____

_____

2. Compare and contrast the role of FSH in the regulation of the male and female gonads.

_____

_____

_____

Expanding Your Horizons

Have you heard of kamikaze sperm? Based on studies in animals, some animal behavior researchers have hypothesized that semen contains a population of spermatozoa that is capable of destroying spermatozoa from a competing male. A "sperm team" contains blockers, which are misshapen spermatozoa with multiple tails and/or heads. Other sperm, the kamikaze spermatozoa, seek out and destroy spermatozoa from other males. A very small population of egg-getting spermatozoa attempt to fertilize the ovum. Although some aspects of this hypothesis have been challenged, the presence of sperm subpopulations is now well established. You can read more about the kamikaze sperm hypothesis in the scientific articles listed below. The full text of each article is freely available on the Internet; do a website search for the author and a few words of the title.

- Baker RR, Bellis MA. Elaboration of the Kamikaze sperm hypothesis: a reply to Harcourt. *Anim Behav* 1989;37:865–867.
- Martin R. Kamikaze sperms or flawed products? *Psychol Today* 2013. Available at: http://www.psychologytoday.com/blog/how-we-do-it/201310/kamikaze-sperms-or-flawed-products
- Moore HD, Martin M, Birkhead TR. No evidence for killer sperm or other selective interactions between human spermatozoa in ejaculates of different males in vitro. *Proc R Soc Lond B Biol Sci* 1999;266:2343–2350.

Development and Birth

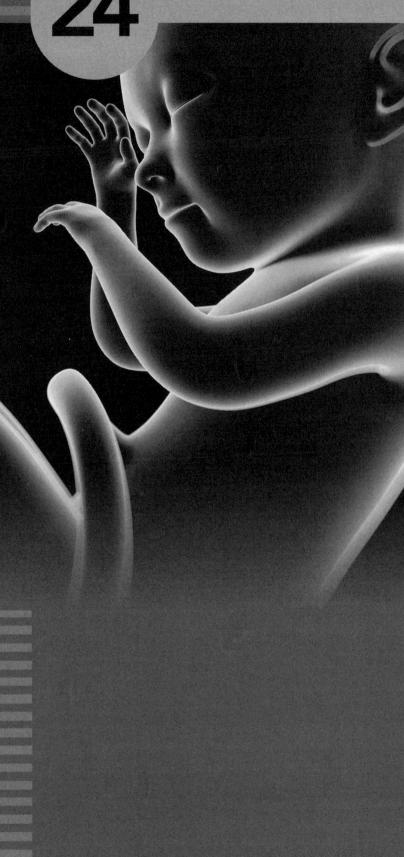

Overview

Pregnancy begins with fertilization of an ovum by a spermatozoon to form a **zygote**. This single cell divides to produce a **blastocyst**, which implants in the uterus. Over the next 38 weeks of **gestation**, the offspring develops first as an **embryo** and then as a **fetus**. During this period, it is nourished and maintained by the **placenta**, formed from tissues of both the mother and the embryo. The placenta secretes a number of hormones, including progesterone, estrogen, human chorionic gonadotropin, human placental lactogen, and relaxin. These hormones induce changes in the uterus and breasts to support the pregnancy and prepare for childbirth and milk production.

Childbirth or **parturition** occurs in four stages, beginning with contractions of the uterus and dilation of the cervix. Subsequent stages include expulsion of the infant, expulsion of the afterbirth, and control of bleeding. Several hormones, including prolactin and oxytocin, stimulate milk production, or **lactation**. Removal of milk from the breasts is the stimulus for continued production.

Although pregnancy and birth are much safer than they were in previous centuries, they are not without risk. Infectious diseases, endocrine abnormalities, tumors, and structural problems can harm the fetus and/or the mother.

Addressing the Learning Objectives

1. DESCRIBE FERTILIZATION AND THE EARLY DEVELOPMENT OF THE FERTILIZED EGG.

EXERCISE 24-1

Write the appropriate term in each blank from the list below. Not all terms will be used.

gestation   zygote   embryo

blastocyst   morula   trophoblast

fetus   ovum   inner cell mass

1. The fertilized egg _____

2. The developing offspring from the third month until birth _____

3. The entire period of development in the uterus _____

4. The structure that implants in the uterus _____

5. The developing offspring from implantation until the third month _____

6. The small ball of cells that develops from the zygote _____

7. The cells of the blastocyst that divide to form the baby _____

8. One of the blastocyst cells that divides to form the placenta and other supporting tissues _____

2. DESCRIBE THE STRUCTURE AND FUNCTION OF THE PLACENTA.

Also see Exercise 24-3.

EXERCISE 24-2

Write the appropriate term in each blank from the list below.

chorion   venous sinuses   chorionic villi

decidua   umbilical artery umbilical vein

1. The maternal contribution to the placenta _____

2. A vessel that carries oxygen-poor blood in the umbilical cord _____

3. A vessel that carries oxygen-rich blood in the umbilical cord _____

4. Projections of the fetal portion of the placenta containing fetal capillaries _____

5. Channels in the placenta containing maternal blood _____

6. The fetal tissue contributing to the placenta _____

3. DESCRIBE HOW FETAL CIRCULATION DIFFERS FROM ADULT CIRCULATION.

Also see Exercise 24-5.

EXERCISE 24-3: Fetal Circulation (Text Fig. 24-2)

1. Write the name of each labeled part of the fetal circulation and placenta on the numbered lines on the next page. The structures identified by empty bullets are labeled elsewhere in the diagram, so you need to write in the correct number.
2. Color the legend boxes red (high oxygenation), purple (medium oxygenation), and blue (low oxygenation).
3. Color the blood vessels on the diagram based on the oxygen content of the contained blood.

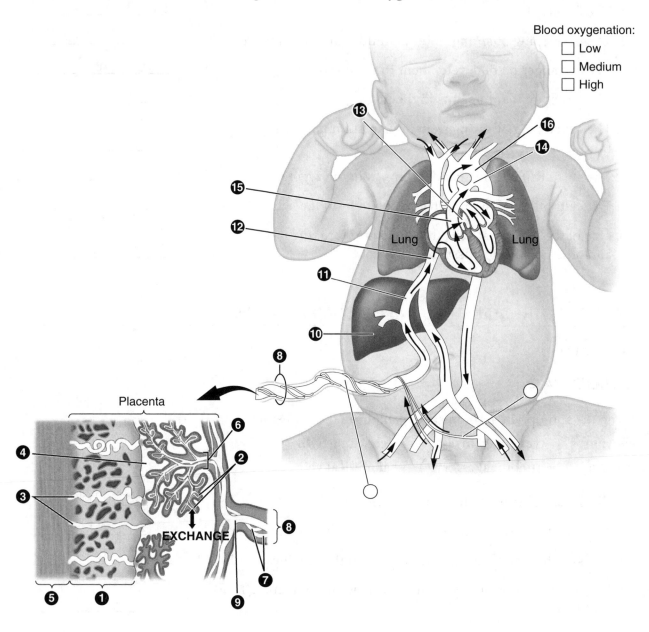

1. _____   9. _____

2. _____   10. _____

3. _____   11. _____

4. _____   12. _____

5. _____   13. _____

6. _____   14. _____

7. _____   15. _____

8. _____   16. _____

4. NAME FIVE HORMONES ACTIVE DURING PREGNANCY, AND DESCRIBE THE FUNCTION OF EACH.

EXERCISE 24-4

Write the appropriate term in each blank from the list below.

progesterone   relaxin   estrogen

human placental lactogen   human chorionic gonadotropin

1. The hormone that loosens the pubic symphysis and softens the cervix for parturition   _____

2. A hormone that prepares the breasts for lactation and increases glucose availability for the fetus   _____

3. A placental hormone that stimulates progesterone synthesis and is detected by pregnancy tests   _____

4. The hormone produced by the corpus luteum and the placenta that inhibits uterine contractions and prepares the breasts for lactation   _____

5. The hormone produced by the corpus luteum and the placenta that stimulates growth of the breasts and uterus   _____

5. BRIEFLY DESCRIBE THE CHANGES THAT OCCUR IN THE EMBRYO, FETUS, AND MOTHER DURING PREGNANCY.

EXERCISE 24-5: Midsagittal Section of a Pregnant Uterus (Text Fig. 24-5)

Write the name of each labeled part on the numbered lines.

1. _____   8. _____

2. _____   9. _____

3. _____   10. _____

4. _____   11. _____

5. _____   12. _____

6. _____   13. _____

7. _____   14. _____

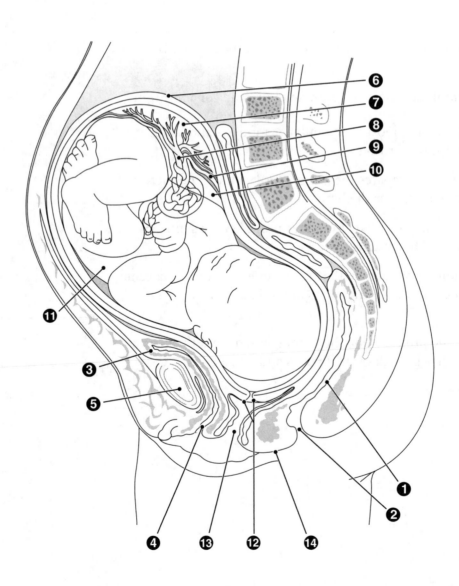

EXERCISE 24-6

Write the appropriate term in each blank from the list below. Not all terms will be used.

vernix caseosa   ultrasonography   ductus venosus   electrocardiography

ductus arteriosus   foramen ovale   amniotic sac

1. The structure that surrounds the developing offspring and serves as a protective cushion   _____

2. A small vessel joining the pulmonary artery to the descending aorta   _____

3. A small hole in the atrial septum of the fetus   _____

4. The cheeselike material that protects the skin of the fetus   _____

5. A technique commonly used to monitor fetal development   _____

6. The small vessel that enables blood to bypass the liver   _____

6. BRIEFLY DESCRIBE THE FOUR STAGES OF LABOR.

EXERCISE 24-7

Write the appropriate term in each blank from the list below. Not all terms will be used.

oxytocin   episiotomy   prostaglandin   cortisol

negative feedback   positive feedback   fraternal   identical

first stage   second stage   third stage   fourth stage

1. A fetal hormone that inhibits maternal progesterone secretion and stimulates uterine contractions   _____

2. The hormone that can initiate labor and stimulate milk ejection   _____

3. A substance produced by the myometrium that stimulates uterine contractions   _____

4. The stage of labor that begins when the cervix is completely dilated   _____

5. The stage of labor that ends with the expulsion of the afterbirth   _____

6. The stage of labor in which both the baby and the afterbirth have been expelled   _____

7. The stage of labor during which the cervix dilates   _____

8. The form of feedback regulating oxytocin secretion   _____

9. The term describing twins that share a placenta   _____

10. The term describing twins developing from different fertilized ova   _____

7. NAME FIVE HORMONES ACTIVE IN LACTATION, AND DESCRIBE THE ACTION OF EACH.

Also see Exercise 24-4.

EXERCISE 24-8

Fill in the blank after each statement—does it apply to prolactin (P) or oxytocin (O)?

1. Produced by the anterior pituitary gland   _____

2. Produced by the posterior pituitary gland   _____

3. Stimulates milk duct contraction   _____

4. Stimulates milk production   _____

5. Also stimulates the uterine contractions of childbirth   _____

EXERCISE 24-9: Section of the Breast (Text Fig. 24-10)

Write the name of each labeled part on the numbered lines.

1. _____

2. _____

3. _____

4. _____

5. _____

6. _____

7. _____

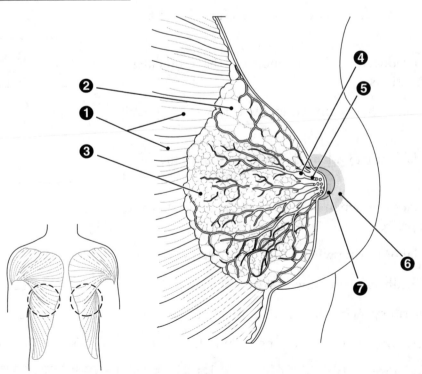

8. CITE THE ADVANTAGES OF BREAST-FEEDING.

EXERCISE 24-10

Briefly summarize four advantages of breast-feeding in the spaces below.

1. _____

2. _____

3. _____

4. _____

9. DESCRIBE FOUR DISORDERS ASSOCIATED WITH THE PLACENTA.

EXERCISE 24-11

Write the appropriate term in each blank from the list below. Not all terms will be used.

pregnancy-induced hypertension	choriocarcinoma	fetal death	ectopic
abortion	placenta previa	hydatidiform mole	
abruptio placentae	gestational diabetes mellitus	puerperal infection	

1. A disorder associated with high blood pressure and proteinuria _____

2. Infection related to childbirth _____

3. The term describing the loss of a 16-week fetus _____

4. A benign placental tumor _____

5. A malignant placental tumor _____

6. Term describing a pregnancy that develops outside the uterine cavity _____

7. The attachment of the placenta near or over the cervical opening _____

8. A metabolic disorder of pregnancy associated with high blood glucose levels _____

9. The premature detachment of the placenta from the uterine wall _____

10. EXPLAIN HOW BREAST CANCER IS DIAGNOSED AND TREATED.

EXERCISE 24-12

Fill in the blanks of the discussion below using your textbook.

Breast cancer is often detected by a radiographic study of the breast, known as a(n) (1) _____. Common risk factors for this disorder include age, family history, the number of menstrual cycles, and the presence of a mutation in two genes known as (2) _____ and _____. Treatment for breast cancer can involve surgical removal of the lump, known as a(n) (3) _____, or removal of the entire breast, known as a(n) (4) _____. If the nearby lymph nodes are also removed, this procedure is known as a(n) (5) _____ _____.

11. CITE FOUR POSSIBLE CAUSES OF LACTATION DISTURBANCES.

EXERCISE 24-13

List four possible causes of lactation disturbances in the lines below.

1. _____

2. _____

3. _____

4. _____

12. REFERRING TO THE CASE STUDY AND THE TEXT, DISCUSS POSSIBLE CAUSES OF HIGH-RISK PREGNANCIES.

Also see Exercise 24-11.

EXERCISE 24-14

In the lines below, list the four important signs of pregnancy-induced hypertension.

1. _____

2. _____

3. _____

4. _____

13. SHOW HOW WORD PARTS ARE USED TO BUILD WORDS RELATED TO DEVELOPMENT AND BIRTH.

EXERCISE 24-15

Complete the following table by writing the correct word part or meaning in the space provided. Write a word that contains each word part in the "Example" column.

Word Part	Meaning	Example
1. _____	outside, external	_____
2. zyg/o	_____	_____
3. _____	labor	_____
4. ox/y	_____	_____
5. chori/o	_____	_____
6. _____	body	_____
7. mamm/o	_____	_____
8. _____	breast	_____

Making the Connections

The following concept map deals with different aspects of pregnancy, parturition, and lactation. Each pair of terms is linked together by a connecting phrase. Complete the concept map by filling in the appropriate term or phrase. There is one right answer for each term. However, there are many correct answers for the connecting phrases (1, 2, 4, 5, 7, 8, 9).

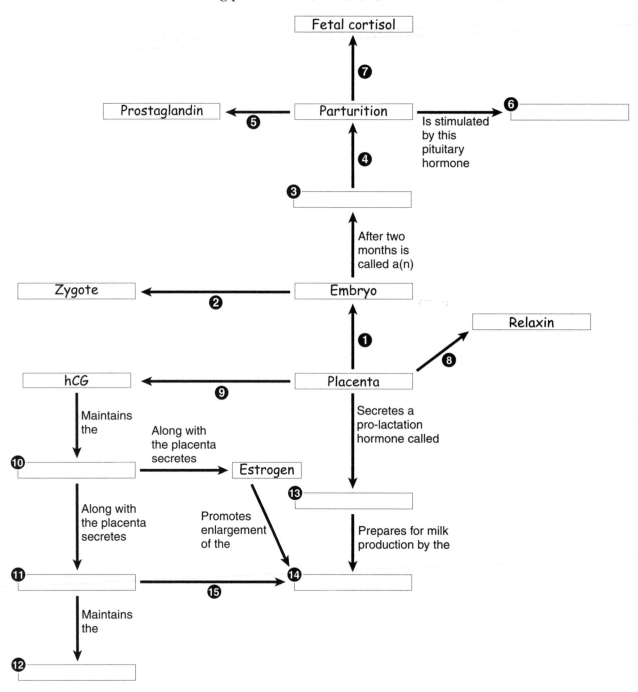

Testing Your Knowledge

BUILDING UNDERSTANDING

I. MULTIPLE CHOICE

Select the best answer.

1. Which of these substances stimulates contraction of the myometrium?   1. _____
   a. progesterone
   b. human placental lactogen
   c. prolactin
   d. prostaglandin
2. Which two organs are largely bypassed in the fetal circulation?   2. _____
   a. lungs and liver
   b. liver and kidney
   c. lungs and intestine
   d. kidney and intestine
3. What event occurs during the second stage of labor?   3. _____
   a. the onset of contractions
   b. expulsion of the afterbirth
   c. passage of the fetus through the vagina
   d. expulsion of the placenta
4. Which of these disorders develops from pregnancy-induced hypertension?   4. _____
   a. ectopic pregnancy
   b. eclampsia
   c. placenta previa
   d. abruptio placentae
5. Which of these phrases describes a viable fetus?   5. _____
   a. spontaneously aborted
   b. developing outside the uterus
   c. capable of living outside the uterus
   d. stillborn
6. Where would you find maternal blood?   6. _____
   a. umbilical cord
   b. chorionic villi
   c. venous sinuses of the placenta
   d. all of the above
7. Which of these is an action of human placental lactogen?   7. _____
   a. maintenance of the corpus luteum
   b. breast development
   c. decreased maternal blood glucose levels
   d. growth of the placenta
8. What is a morula?   8. _____
   a. a small ball of cells that develops from a zygote
   b. the single cell formed by the sperm and ovum
   c. a membrane surrounding the fetus that secretes fluid
   d. a small blood vessel that bypasses the fetal liver

II. COMPLETION EXERCISE

Write the word or phrase that correctly completes each sentence.

1. A surgical cut and repair of the perineum to prevent tearing is called a(n) _____.

2. By the end of the first month of embryonic life, the beginnings of the extremities may be seen. These are four small swellings called _____.

3. The mammary glands of the female provide nourishment for the newborn through the secretion of milk; this is a process called _____.

4. The stage of labor during which the afterbirth is expelled from the uterus is the

   _____.

5. Twins that develop from the same fertilized egg are called _____.

6. The normal site of fertilization is the _____.

7. The clear liquid that flows from the uterus when the mother's "water breaks" is technically called _____.

8. A hydatidiform mole can develop into a malignant tumor called a(n) _____.

UNDERSTANDING CONCEPTS

I. TRUE/FALSE

For each statement, write T for true or F for false in the blank to the left of each number. If a statement is false, correct it by replacing the underlined term, and write the correct statement in the blanks below the statement.

_____ 1. Ms. J is seven weeks' pregnant. Her uterus contains an underline{embryo}.

_____

_____ 2. The baby is expelled during the underline{fourth} stage of labor.

_____

_____ 3. underline{Fraternal} twins are genetically distinct.

_____

_____ 4. Blood is carried from the placenta to the fetus in the underline{umbilical vein}.

_____

_____ 5. underline{Oxytocin} secreted from the fetal adrenal gland may help induce labor.

_____

_____ 6. The underline{decidua} forms the fetal portion of the placenta.

_____

_____ 7. The underline{morula} implants in the uterus.

_____

II. PRACTICAL APPLICATIONS

Study each discussion. Then write the appropriate word or phrase in the space provided.

1. Mr. and Ms. L had been trying to conceive for two years. Finally, Ms. L realized her period was late and purchased a pregnancy detection kit. These kits test for the presence of a hormone produced exclusively by embryonic tissues that helps maintain the corpus luteum. This hormone is called _____.

2. The pregnancy test was positive. Six weeks had elapsed since Ms. L's last menstrual period. The gestational age of her new offspring at this time would be _____.

3. During a prenatal appointment 14 weeks later, Ms. L was able to see her future offspring using a technique for visualizing soft tissues without the use of x-rays. This technique is called _____.

4. The radiologist noted that the placenta was located near the cervix. Ms. L was told that her placenta would probably migrate upward as her pregnancy progressed. However, if the placenta remained at its current position, she was at risk for a syndrome called _____.

5. Fortunately, later exams showed that the placenta had migrated upward toward the fundus of the uterus. However, urinalysis revealed the presence of protein in Ms. L's urine, and her blood pressure was higher than normal. Ms. L may be suffering from a serious disorder called _____.

6. Ms. L's condition did not change with time, and her physician thus decided to deliver her baby through an incision made in the abdominal wall and the wall of the uterus. This operation is called a(n) _____.

7. The operation went smoothly, and baby L was born. Ms. L immediately began to breast-feed her infant. The first secretion from her breasts was not milk, but rather _____.

III. SHORT ESSAYS

1. Compare and contrast human placental lactogen (hPL) and human chorionic gonadotropin (hCG). Discuss the synthesis and action of each hormone.

   _____

   _____

   _____

   _____

2. Describe the site of synthesis and actions of the hormones involved in lactation and preparing the breasts for lactation.

   _____

   _____

   _____

   _____

CONCEPTUAL THINKING

1. Trace the path of a blood cell from the placenta to the fetal heart and back to the placenta. Assume that this blood cell passes through the foramen ovale.

   _____

   _____

   _____

   _____

2. Ms. J has suffered from repeated miscarriages. Blood tests reveal a deficiency in progesterone. Could this deficiency be implicated in Ms. J's miscarriages? Why or why not?

   _____

   _____

   _____

3. Describe how each maternal organ changes during pregnancy, and (if relevant) how the change facilitates fetal growth and survival.
   a. heart

   _____

   _____

   b. lungs

   _____

   _____

   c. kidney

   _____

   _____

   d. bladder

   _____

   _____

   e. digestive system

   _____

   _____

Expanding Your Horizons

Have you ever wondered why animals can give birth alone, but humans require all sorts of technological equipment? Women rarely give birth alone, even in societies with few technological advances. The difficulties of human birth reflect some of our evolutionary adaptations. For instance, the orientation of the human pelvis facilitates walking upright but requires the fetus to turn and bend during parturition. You can read more about "The Evolution of Birth" in *Scientific American*.

- Rosenberg KR, Trevathan WR. The evolution of human birth. Sci Am 2001;285:72–77.

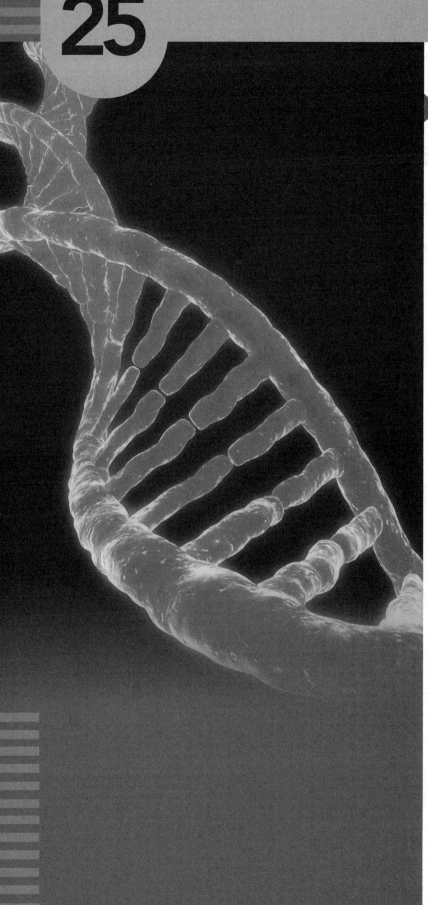

Heredity and Hereditary Diseases

▶ Overview

The scientific study of heredity has advanced with amazing speed in the past 50 years. Nevertheless, many mysteries remain. Gregor Mendel was the first person known to have carried out formal experiments in genetics. He identified independent units of heredity, which he called *factors* and which we now call **genes**.

The chromosomes in the nucleus of each cell are composed of a complex molecule, **DNA**. This material makes up the many thousands of genes that determine a person's traits and are passed on to offspring at the time of fertilization. We saw in Chapter 3 that DNA directs the formation of proteins. Many proteins are **enzymes**, which in turn make possible all the chemical reactions of metabolism. Other proteins serve structural roles. Defective genes, resulting from **mutation**, may disrupt normal enzyme activity or cell structure and result in hereditary disorders such as sickle cell anemia, albinism, and phenylketonuria. Some human traits are determined by a single pair of genes (one gene from each parent), but most are controlled by multiple pairs of genes acting together.

An **allele** is a particular version of a gene. Alleles may be classified as **dominant** or **recessive**. If one parent contributes a dominant allele, then any offspring who receives that allele will show the trait (e.g., Huntington disease). Traits carried by recessive alleles may remain hidden for generations and be revealed only if they are contributed by both parents (e.g., albinism, cystic fibrosis, PKU, and sickle cell anemia).

In some cases, treatment begun early may help prevent problems associated with a genetic disorder. Genetic counseling should be sought by all potential parents whose relatives are known to have an inheritable disorder.

Addressing the Learning Objectives

1. DEFINE A GENE, AND BRIEFLY DESCRIBE HOW GENES FUNCTION.

EXERCISE 25-1

Label each of the following statements as true (T) or false (F).

1. Genes are segments of proteins.   _____

2. Genes contain the blueprints to make proteins.   _____

3. Genes are composed of DNA.   _____

4. Genes code for specific traits.   _____

5. Humans have 46 autosomes.   _____

6. Humans have 46 chromosomes.   _____

2. EXPLAIN THE RELATIONSHIP BETWEEN GENOTYPE AND PHENOTYPE FOR DOMINANT AND RECESSIVE ALLELES.

EXERCISE 25-2

Write the appropriate term in each blank from the list below. Not all terms will be used.

heterozygous   homozygous   autosome   sex chromosome

carrier   recessive   dominant   allele

1. One version of a specific gene   _____

2. Term describing an allele that expresses its effect only if homozygous   _____

3. An allele pair consisting of two dominant or two recessive alleles   _____

4. Any chromosome except the X and Y chromosomes   _____

5. Term describing any allele pair composed of two different alleles   _____

6. Term describing an allele that expresses its effect if homozygous or heterozygous   _____

7. Term describing an individual heterozygous for a recessive trait   _____

EXERCISE 25-3

Fill in the blank after each statement—does it apply to a phenotype (P) or a genotype (G)?

1. The genetic makeup of an individual   _____

2. The characteristics that can be observed and/or measured   _____

3. Eye color   _____

4. Homozygous dominant   _____

5. Blood type   _____

6. Heterozygous   _____

3. EXPLAIN HOW CHROMOSOMES ARE DISTRIBUTED IN MEIOSIS.

EXERCISE 25-4

Label each of the following statements as true (T) or false (F).

1. Following meiosis, each reproductive cell contains 46 chromosomes.   _____

2. Following meiosis, each reproductive cell contains 23 chromosomes.   _____

3. Some reproductive cells contain all of the maternal chromosomes, while other reproductive cells contain all of the paternal chromosomes.   _____

4. Each reproductive cell contains a mix of maternal and paternal chromosomes.   _____

5. DNA replication always precedes meiosis.   _____

6. Meiosis involves two separate cell divisions.   _____

7. The paternal and maternal chromosomes separate into separate cells after the second meiotic division.   _____

4. PERFORM A GENETIC CROSS USING A PUNNETT SQUARE.

EXERCISE 25-5

Progressive retinal atrophy (PRA) is a common cause of blindness in dogs. PRA is an autosomal recessive disorder.

1. Complete this Punnett square based on the mating between a heterozygous female and an homozygous recessive male. Use Figure 25-3 in your textbook to guide you. Possible options for the genotype are PP, Pp, or pp. Possible options for the phenotype are Normal, Carrier, or Disease. The possible parental alleles and some of the answers have already been added.

		Mother	
		P	p
Father	p	Genotype:	Genotype:
		Phenotype: normal	Phenotype:
	p	Genotype:	Genotype: pp
		Phenotype:	Phenotype:

2. What is the probability (in a percentage) that a pup will have PRA? _____

3. What is the probability (in a percentage) that a pup will be homozygous dominant? _____

5. EXPLAIN HOW SEX IS DETERMINED IN HUMANS AND HOW SEX-LINKED TRAITS ARE INHERITED.

EXERCISE 25-6

Fill in the blank after each example—would the genotype result in a female phenotype (F) or a male phenotype (M), assuming that development proceeds normally?

1. Union between an X sperm and an X ovum   _____

2. Union between a Y sperm and an X ovum   _____

EXERCISE 25-7

Fill in the blank after each statement—does it refer to sex-linked traits (S) or autosomal (non–sex-linked) traits (A)?

1. A trait carried on the Y chromosome

   _____

2. A trait carried on the X chromosome

   _____

3. A trait carried on chromosomes other than the X or Y chromosomes

   _____

4. A trait for which males or females can be carriers

   _____

5. A trait for which only females can be carriers

   _____

6. LIST THREE FACTORS THAT MAY INFLUENCE THE EXPRESSION OF A GENE.

EXERCISE 25-8

In the spaces below, list three factors that influence gene expression.

1. _____

2. _____

3. _____

7. DEFINE MUTATION.

EXERCISE 25-9

In the spaces below, define *mutation* and *mutagen*. In your definition of mutation, describe some different types of mutations.

1. Mutation: _____

   _____

2. Mutagen: _____

   _____

8. EXPLAIN THE PATTERN OF MITOCHONDRIAL INHERITANCE.

See Exercise 25-10.

9. DIFFERENTIATE AMONG CONGENITAL, GENETIC, AND HEREDITARY DISORDERS, AND GIVE SEVERAL EXAMPLES OF EACH.

See Exercises 25-10 and 25-11.

10. GIVE SEVERAL EXAMPLES OF TERATOGENS, AND DESCRIBE THEIR EFFECTS.

EXERCISE 25-10

Write the appropriate term in each blank from the list below.

multifactorial   mitochondrial   mutagen   genetic

hereditary   teratogen   congenital

1. A gene that is only transmitted from mothers to their children   _____

2. Any substance that interferes with prenatal development   _____

3. A specific term describing a trait determined by multiple gene pairs acting together   _____

4. Term describing any disease caused by a change in the DNA sequence, whether or not it is inherited from a parent   _____

5. Term describing any trait present at birth   _____

6. Term that indicates specifically that a trait is inherited from a parent   _____

7. Any substance that induces a change in DNA   _____

11. DESCRIBE THE SYMPTOMS AND INHERITANCE PATTERNS OF SOME COMMON GENETIC DISEASES.

Also see Exercise 25-13.

EXERCISE 25-11

Write the appropriate term in each blank from the list below. Not all terms will be used.

Marfan syndrome   Huntington disease   osteogenesis imperfecta

spina bifida   cystic fibrosis   Tay-Sachs disease

neurofibromatosis   trisomy 21

1. A neurodegenerative disease that is genetic and hereditary but is not evident until midlife   _____

2. A genetic disorder that is not always hereditary   _____

3. A genetic disorder characterized by abnormal, disfiguring growths developing from nervous tissue   _____

4. A congenital disorder that is not genetic   _____

5. A hereditary disease affecting all connective tissue, resulting in problems in many body systems   _____

6. A genetic disease in which fat is abnormally deposited in CNS tissue   _____

7. A genetic disease that results in blockages of the bronchi, intestine, and pancreatic ducts   _____

12. DESCRIBE FOUR METHODS FOR DIAGNOSING FETAL DISORDERS.

EXERCISE 25-12

Write the appropriate term in each blank from the list below.

pedigree chart   amniocentesis   chorionic villus sampling

karyotype   nuchal transparency

1. A study of the chromosomes, as done on fetal cells, used to determine fetal sex and screen for significant chromosomal abnormalities   _____

2. Removal of fluid from the sac surrounding the fetus   _____

3. Removal of projections from a placental membrane   _____

4. A test that analyzes the amount of fluid accumulated behind the cervical vertebrae   _____

5. A complete, detailed family tree that can detect carriers _____

13. GIVE EXAMPLES OF METHODS CURRENTLY USED TO TREAT CERTAIN GENETIC DISORDERS.

EXERCISE 25-13

Write the appropriate term in each blank from the list below. Not all terms will be used.

PKU	Wilson disease	maple syrup urine disease
albinism	fragile X syndrome	Klinefelter syndrome
Turner syndrome		

1. The most common cause of mental retardation in males _____

2. A genetic disease treated with large doses of thiamine _____

3. Children with this disease must not consume a specific amino acid _____

4. A disease associated with abnormal copper accumulation that can be treated with drugs and dietary management _____

5. An individual with this disorder would have 45 chromosomes _____

6. An individual with this disorder would have at least 47 chromosomes _____

14. USING THE CASE STUDY AND INFORMATION IN THE TEXT, DESCRIBE THE ROLE OF A GENETIC COUNSELOR.

EXERCISE 25-14

Genetic counselors sometimes use pedigree charts or Punnett squares to advise their clients. Based on the information in the case study, draw a pedigree chart for the inheritance of the sickle cell trait that includes Cole and Cole's parents. Use Figure 25-11 as your guide, and label each generation. Use S for the dominant allele and s for the sickle cell allele. Squares indicate males, and circles indicate females. Use unfilled symbols for individuals without the disease and filled symbols for individuals with the disease.

15. SHOW HOW WORD PARTS ARE USED TO BUILD WORDS RELATED TO HEREDITY.

EXERCISE 25-15

Complete the following table by writing the correct word part or meaning in the space provided. Write a word that contains each word part in the "Example" column.

Word Part	Meaning	Example
1. _____	nucleus	_____
2. con-	_____	_____
3. _____	color	_____
4. -cele	_____	_____
5. phen/o	_____	_____
6. _____	self	_____
7. _____	tapping, perforation	_____
8. _____	other, different	_____
9. homo-	_____	_____
10. dactyl/o	_____	_____

Making the Connections

The following concept map deals with some aspects of heredity. Each pair of terms is linked together by a connecting phrase. Complete the concept map by filling in the appropriate term or phrase. There is one right answer for each term (4 to 6, 11). However, there are many correct answers for the connecting phrases.

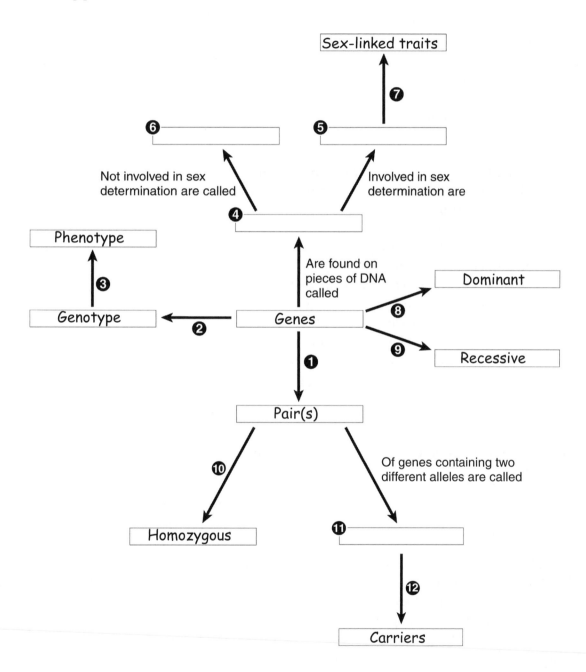

Optional Exercise: Make your own concept map based on aspects of genetic diseases. Use the following terms and any others you would like to include: karyotype, amniocentesis, chromosomes, chorionic villus sampling, hereditary, congenital, genetic, genetic engineering, pedigree, genetic screening, Huntington disease, Down syndrome, cystic fibrosis, and phenylketonuria.

Testing Your Knowledge

BUILDING UNDERSTANDING

I. MULTIPLE CHOICE

Select the best answer.

1. Two parents with normal pigmentation can give birth to a child with albinism. Which of the following terms does NOT describe albinism?
   a. recessive trait
   b. dominant trait
   c. congenital disorder
   d. genetic disorder

   1. _____

2. What is Klinefelter syndrome?
   a. a disorder in which masses grow along the nerves
   b. a disorder that involves the sex chromosomes
   c. a condition of having extra fingers
   d. a form of dwarfism

   2. _____

3. What determines gender in humans?
   a. the number of X chromosomes
   b. the sex chromosome carried by the ovum
   c. the number of autosomes
   d. the sex chromosome carried by the spermatozoon

   3. _____

4. If an imaginary animal has 28 chromosomes per cell, how many chromosomes will be found in each spermatozoon?
   a. 7
   b. 14
   c. 28
   d. 56

   4. _____

5. Which of these terms describes traits that are determined by more than one gene pair?
   a. sex-linked
   b. recessive
   c. multifactorial
   d. dominant

   5. _____

6. Spina bifida is present at birth but is not passed on from parents to offspring. How is spina bifida classified?
   a. congenital but not hereditary
   b. hereditary but not congenital
   c. hereditary and congenital
   d. neither hereditary nor congenital

   6. _____

II. COMPLETION EXERCISE

Write the word or phrase that correctly completes each sentence.

1. The larger sex chromosome is called the _____.

2. A change in a gene or chromosome is called a(n) _____.

3. The number of autosomes in the human genome is _____.

4. The process of cell division that halves the chromosome number is called _____.

5. An agent that induces a change in chromosome structure is called a(n) _____.

6. If the genotype of both parents is Cc, the chance that an offspring will have the genotype CC is _____.

UNDERSTANDING CONCEPTS

I. TRUE/FALSE

For each statement, write T for true or F for false in the blank to the left of each number. If a statement is false, correct it by replacing the underlined term, and write the correct statement in the blanks below the statement.

_____ 1. An individual with Down syndrome received <u>23</u> chromosomes from one parent and 23 chromosomes from the other parent.

_____

_____ 2. All <u>congenital</u> diseases run in families.

_____

_____ 3. Sex-linked traits appear almost exclusively in <u>males</u>.

_____

_____ 4. The sex of the offspring is determined by the sex chromosome contributed by the <u>father</u>.

_____

_____ 5. Carriers for a particular trait are always <u>homozygous</u> for the dominant allele.

_____

_____ 6. A recessive trait is expressed in individuals <u>heterozygous</u> for the recessive allele.

_____

II. PRACTICAL APPLICATIONS

Study each discussion. Then write the appropriate word or phrase in the space provided.

➤ **Group A**

1. Mr. and Ms. J have consulted a genetic counselor. Ms. J is eight weeks' pregnant, and cystic fibrosis runs in both of their families. Cystic fibrosis is a disease in which an individual may carry the disease allele but not have cystic fibrosis. A term to describe this type of trait is _____.

2. First, the genetic counselor prepared a diagram summarizing the incidence of cystic fibrosis in the two families and the relationships between family members. This type of diagram is called a(n) _____.

3. Mr. and Ms. J were then screened for the presence of the cystic fibrosis allele. It was determined that both Mr. and Ms. J have the allele even though they do not have cystic fibrosis. For the cystic fibrosis trait, they are both considered to be _____.

4. Ms. J wanted to know if her baby would have cystic fibrosis. Cells were taken from the hairlike projections of the membrane that surrounds the embryo. This technique is called _____.

5. The fetus was shown to carry one normal allele and one cystic fibrosis allele. The fact that the alleles are different means that they can be described as _____.

6. As part of the prenatal screening, the chromosomes were obtained from cells in metaphase and photographed for analysis. This form of evaluation is called a(n) _____.

7. The analysis revealed the presence of two XX chromosomes. The gender of the baby is therefore _____.

➤ **Group B**

1. Mr. and Ms. B have a child with three copies of chromosome 21. The child has a chromosomal abnormality known as _____.

2. This abnormality, like any change in DNA, is known as a(n) _____.

3. Chromosome 21 is one of 22 pairs of chromosomes known as _____.

4. The remainder of the child's chromosomes are normal. The total number of chromosomes in the child's cells is _____.

5. The chromosome abnormality arose during the production of gametes. The form of cell division that occurs exclusively in germ cell precursors is called _____.

6. The family knew in advance about their baby's chromosome abnormality because of prenatal screening and testing. The first indication was an abnormal accumulation of fluid at the back of the fetus's neck. This test, which uses ultrasound, is known as the _____.

III. SHORT ESSAYS

1. Name two methods used to treat genetic disorders. If possible, provide specific examples of diseases treated by each method.

_____

_____

_____

2. Some traits in a population show a range instead of two clearly alternate forms. List some of these traits, and explain what causes this variety.

_____

_____

_____

CONCEPTUAL THINKING

1. Are identical twins identical individuals? Defend your answer, using the terms "genotype" and "phenotype."

_____

_____

_____

2. Young S, female, is afflicted with Edwards syndrome, or trisomy 18. She was born with a small jaw, misshapen ears, malfunctioning kidneys, and a heart defect. None of her relatives suffer from this disorder, which resulted from the presence of an extra chromosome 18 in the spermatozoon. Is trisomy 18 (a) congenital, (b) hereditary, and/or (c) genetic? Defend your answer.

_____

_____

_____

Expanding Your Horizons

The movie *Gattaca* (1997) describes a futuristic world where genetic screening and engineering are the norm. Everyone has a genetic description (including life span) available to potential mates, and virtually everyone (except the hero) is genetically engineered. Genetic screening determines one's future on Gattaca, and genetic engineering has essentially abolished independent thought and creativity. Science fiction aside, genetic screening and engineering are becoming a technological reality in our society.

For a modest fee, numerous companies will screen your genome and provide information regarding ancestry. In the past, they also provided information about genetic risk factors for various diseases (some of which were quite speculative), but in November 2013, they ceased providing health information pending a USDA regulatory review. Concerns include the following: Will genetic screening soon be required to obtain health insurance? Will some parents use prenatal screening to select for a super baby? How can we keep genetic information secure from computer hackers? Learn more in the resources below.

- Wong K. Finding my inner:Neandertal. Sci Am 2013. Available at: http://www.scientificamerican.com/blog/post/dna-finding-my-inner-neandertal
- 23andme. Available at: https://www.23andme.com
- Hall SS. Revolution postponed. Sci Am 2010;303:60–67.
- Aldhous P, Reilly M. How my Genome was hacked. New Scientist 2009;201:6–9.